Fruits Quality and Sensory Analysis

Fruits Quality and Sensory Analysis

Guest Editors

Paola Sánchez-Bravo
Luis Noguera-Artiaga

MDPI

Basel • Beijing • Wuhan • Barcelona • Belgrade • Novi Sad • Cluj • Manchester

Guest Editors

Paola Sánchez-Bravo
Universidad Miguel
Hernández de Elche (UMH)
Orihuela
Spain

Luis Noguera-Artiaga
Universidad Miguel
Hernández de Elche (UMH)
Orihuela
Spain

Editorial Office
MDPI AG
Grosspeteranlage 5
4052 Basel, Switzerland

This is a reprint of the Special Issue, published open access by the journal *Horticulturae* (ISSN 2311-7524), freely accessible at: https://www.mdpi.com/journal/horticulturae/special_issues/Fruits_Quality_Sensory_Analysis.

For citation purposes, cite each article independently as indicated on the article page online and as indicated below:

Lastname, A.A.; Lastname, B.B. Article Title. *Journal Name* **Year**, *Volume Number*, Page Range.

ISBN 978-3-7258-2993-4 (Hbk)
ISBN 978-3-7258-2994-1 (PDF)
https://doi.org/10.3390/books978-3-7258-2994-1

Contents

About the Editors

Paola Sánchez-Bravo

Paola Sánchez-Bravo obtained her degree in Food Science and Technology in 2016 and her Master's in Advanced Research Techniques in Fruit Growing in 2017. Then, she obtained her PhD in 2024 from the Miguel Hernandez University of Elche obtaining the following accolades: a distinction "Cum Laude", the Extraordynary PhD Award, and International Mention. Paola is a postdoctoral researcher (Margarita Salas) with the Miguel Hernández University of Elche, working with the Spanish National Research Council, Centro de Edafología y Biología Aplicada del Segura (CEBAS-CSIC), where she focuses her scientific research on the development of new functional beverages. Her involvement and interest is mostly related to the food science field and sensory analysis, which is reflected in her publications on dairy products, plant-based foods, and beverages—focused on the study of their sensory and volatile profiles—as well as bioactive compounds and consumer studies. Most of them are related to consumer behavior, water sustainability, and the determination of the volatile profile of fruits. She is also prone to collaborating with interdisciplinary research teams at both a national and international level, as can also be proven by the authorship of her papers. Paola has participated in the training of sensory analysis panels for olive oil, table olives, chocolate, wine, paprika, and tomato. She also has more than 2000 hours of experience in sensory analysis as a panelist and she has carried out more than 100 national and international consumer studies.

Luis Noguera-Artiaga

Luis Noguera Artiaga is a graduate in Food Science and Technology (2014) and has a Master's in Advanced Research Techniques in Fruit Growing (2015) from Miguel Hernández University of Elche. He obtained his PhD from the University of Sonora (Mexico) in 2019, and obtained another PhD from the Miguel Hernández University of Elche in 2020. Luis is an Assistant Professor at the Institute of Agro-Food and Agro-Environmental Research and Innovation of the Miguel Hernández University of Elche. His research is based on the study of the sensory analysis of food and the impact that agricultural practices have on its quality. Among other topics, he studied the influence that different types of irrigation have on the quality of agri-food products, studying their physical, chemical, functional, bio-functional, and sensory properties. In addition, he studied the impact that aroma has on the flavor and quality of food by analyzing its volatile organic compounds. Luis has participated in the training of sensory analysis panels in wine, vegetable, paprika, beer, ice cream, and chocolate companies. He has conducted more than 100 consumer studies (national and international) and he has accumulated more than 5,000 hours of experience in sensory analysis as a panelist in descriptive and discriminative tests on all types of foods.

horticulturae

MDPI

Editorial

Fruits Quality and Sensory Analysis

Paola Sánchez-Bravo [1,2,*] and Luis Noguera-Artiaga [1,*]

1 Department of Agro-Food Technology, Escuela Politécnica Superior de Orihuela (EPSO), Centro de Investigación e Innovación Agroalimentaria y Agroambiental (CIAGRO-UMH), Universidad Miguel Hernández de Elche (UMH), Carretera de Beniel, km 3.2, 03312 Orihuela, Spain
2 Phytochemistry and Healthy Food Lab., Research Group on Quality, Safety, and Bioactivity of Plant Foods, Department of Food Science and Technology, CEBAS-CSIC, Campus Espinardo, 25, 30100 Murcia, Spain
* Correspondence: paola.sanchezb@umh.es (P.S.-B.); lnoguera@umh.es (L.N.-A.)

Citation: Sánchez-Bravo, P.; Noguera-Artiaga, L. Fruits Quality and Sensory Analysis. *Horticulturae* **2024**, *10*, 1279. https://doi.org/10.3390/horticulturae10121279

Received: 15 October 2024
Revised: 8 November 2024
Accepted: 27 November 2024
Published: 1 December 2024

1. Introduction

With the objective of using sensory analysis tools to dually consider the changes in quality that fruit undergoes (functional, nutritional, and sensory) throughout its growth, processing, and shelf life and consumer acceptance, given consumers are increasingly demanding sustainable and higher-quality food, we edited this Special Issue on "Fruits Quality and Sensory Analysis" (URL: https://www.mdpi.com/journal/horticulturae/special_issues/Fruits_Quality_Sensory_Analysis (accessed on 10 October 2024)). The participating papers studied tomato, apricot, coffee, sweet cherry, pistachio, kiwi, mushroom, pear, and orange, among other things.

2. An Overview of the Published Articles

The correlation between the sensory profile of a food and its phytochemical profile, as well as its influence on consumer preferences, highlights the importance of expanding the research on this topic. An example of this was a study carried out on tomato fruits from three *Physalis* species (*P. ixocarpa*, *P. angulata*, and *P. philadelphica*) (Contribution 1), in which it was found that certain metabolite expressions are involved in taste variations that directly impact consumer preferences. Hence, preserving biodiversity is important not only in terms of fruit production but also in terms of contributing to meeting consumer demands in terms of fruit's aroma and flavor (Contribution 1). In line with this, studies have been carried out on genetically improving the metabolic profile and volatile compound contents in plum and apricot hybrids, demonstrating that hybrid seeds inherit the metabolites present in parental seeds, as well as the volatile compound profile from one of the parents (Contribution 2).

On the other hand, in coffee cultivation (the main beverage worldwide after water), a positive correlation was observed between average growing temperature (22–26 °C) and phenol and flavonoid concentrations, as well as key sensory attributes, highlighting the importance of good cultivation practices to optimizing coffee's functional and organoleptic quality (Contribution 3). Similar results were found for sweet cherries planted in China (Contribution 4) in that the type of soil and fertilization, as well as climatic parameters, had an effect on their functional (β-carotene and flavonoids), nutritional (minerals, TSS, and titratable acidity), and sensory (maturity, size, and edibility rate) quality (Contribution 4). Furthermore, Caranqui-Aldaz et al. (Contribution 5) evaluated the quality of the "mortiño" fruit (*Vaccinium floribundum* Kunth) grown in the moorland of Chimborazo volcano in Ecuador, revealing that altitude also affects its contents of bioactive compounds (flavonols, hydroxycinnamic acids, and anthocyanins), sugars, organic acids, and minerals.

In evaluating correlations with the expression of certain metabolites in fruits, it was found that in kiwi, the contents of certain minerals and metabolites (fructose, maltose, mannobiose, tagatose, and citrate) are key to its sensory quality and provide a strong flavor, while others (such as serine) have a negative effect (Contribution 6). Furthermore, the relationship between the color of fruit's skin and its physicochemical and sensory qualities was also evaluated in this Special Issue. Alonso-Salinas et al. evaluated the pigmentation intensity of the blood orange "Sanguinelli", showing that the more reddish the color of its skin was, the higher its sugar and anthocyanin contents were, which correlated with a greater consumer perception of sweeter fruit, while the other parameters studied did not depend on its external pigmentation (acidity, firmness, size, or juice yield); in this way, they provided a simpler and more intuitive way for the industry to classify fruit batches (Contribution 7).

The use of rootstocks and how they affect the yield, growth, and sensory quality of fruit have been widely established. In this Special Issue, a further step was taken to advance this scientific knowledge in demonstrating that the cultivar–rootstock combination influences the tree vigor, yield, yield efficiency, weight, kernel oil content, and fatty acid profile in pistachios (Contribution 8).

The current trend of increasing fruit conservation through postharvest treatments was also covered by this Special Issue. Alonso-Salinas et al. (Contribution 9) used potassium permanganate filters in devices with ultraviolet radiation and a constant air flow to promote contact between ethylene and oxidizing agents (at 1 and 8 °C), optimizing the maintenance of the physicochemical quality (weight, size, firmness, antioxidant capacity, etc.) of "Ercolini" pears in postharvest conservation and even improving their sensory quality attributes (flavor and odor), associated with green fruit (Contribution 9). This Special Issue also features two bibliographic reviews that provide a comprehensive overview of the strategies for delaying ethylene-mediated ripening in climacteric fruits and their impact on their shelf life, sensory attributes, and volatile compound contents (Contribution 10), as well as the use of new emerging technologies (active packaging, natural antioxidant and antimicrobial compounds, high hydrostatic pressure, UV-C radiation, and ozone) to effectively prolong the quality of fresh fruit (physicochemical, microbiological, nutritional, and sensory) during storage (Contribution 11).

Computer vision has also been used to classify the quality of one of the most popular fungus species, *Oudemansiella raphaniepes*. Six convolution neural network (CNN) models were used, and based on their detection time and accuracy in classifying images, it was shown that they can be a great tool when it comes to identifying and classifying these fungi, thus clearing the path for the use of this technique with other products to minimize production costs (Contribution 12).

3. Conclusions

In summary, reducing losses in food quality and maintaining health-related compound contents, along with consumer consumption, are indispensable, and research is required to improve how fruit quality and its parameters are affected during cultivation or processing, since in order to obtain high-quality food products, these need to be controlled from the moment it is planted and harvested. Following these developments and global trends towards more nutritious, functional, and sustainable foods, as demonstrated by most of the authors that contributed to this Special Issue, further research into each and every one of the key aspects mentioned is required to help preserve the integrity of bioactive ingredients throughout fruit's growth, processing, and shelf life, as well as its sensory quality to ensure consumer acceptance of food products.

Acknowledgments: Paola Sánchez-Bravo was funded by the grant for the recall of the Spanish university system for the training of young doctors (Margarita Salas, 04912/2021), funded by the European Union's NextGenerationEU and the Ministry of Universities of Spain.

Conflicts of Interest: The authors declare no conflicts of interest.

List of Contributions:

1. Arias-Martínez, S.; Oyoque-Salcedo, G.; Gutiérrez-Cárdenas, O.G.; Oregel-Zamudio, E.; Torres-García, J.R. Comparative metabolomic fingerprinting analysis of tomato fruits from physalis species in mexico's balsas basin. *Horticulturae* **2024**, *10*, 600.

2. Mihaylova, D.; Popova, A.; Dincheva, I.; Pandova, S. Hs-spme-gc–ms profiling of volatile organic compounds and polar and lipid metabolites of the "stendesto" plum–apricot kernel with reference to its parents. *Horticulturae* **2024**, *10*, 257.
3. Mihai, R.A.; Ortiz-Pillajo, D.C.; Iturralde-Proaño, K.M.; Vinueza-Pullotasig, M.Y.; Sisa-Tolagasí, L.A.; Villares-Ledesma, M.L.; Melo-Heras, E.J.; Cubi-Insuaste, N.S.; Catana, R.D. Comprehensive assessment of coffee varieties (coffea arabica l.; coffea canephora l.) from coastal, andean, and amazonian regions of ecuador; a holistic evaluation of metabolism, antioxidant capacity and sensory attributes. *Horticulturae* **2024**, *10*, 200.
4. Nie, Y.; Huang, J.; Liu, R.; Wang, P.; Liu, P.; Lu, M.; Sun, J. Physicochemical properties of geographical indication (gi) sweet cherries in china and their influencing factors of cultivar, climate type, and soil condition. *Horticulturae* **2023**, *9*, 1118.
5. Caranqui-Aldaz, J.M.; Muelas-Domingo, R.; Hernández, F.; Martínez, R. Chemical composition and polyphenol compounds of vaccinium floribundum kunth (ericaceae) from the volcano chimborazo paramo (ecuador). *Horticulturae* **2022**, *8*, 956.
6. Titeli, V.S.; Michailidis, M.; Tanou, G.; Molassiotis, A. Physiological and metabolic traits linked to kiwifruit quality. *Horticulturae* **2023**, *9*, 915.
7. Tarancón, P.; Cebrián, B.; Fernández-Serrano, P.; Besada, C. Relation between rind pigmentation and internal quality of blood orange 'sanguinelli': Physicochemical and sensory studies. *Horticulturae* **2022**, *8*, 448.
8. Ouni, S.; Noguera-Artiaga, L.; Carbonell-Barrachina, A.; Ouerghui, I.; Jendoubi, F.; Rhouma, A.; Chelli-Chaabouni, A. Cultivar and rootstock effects on growth, yield and nut quality of pistachio under semi-arid conditions of south mediterranean. *Horticulturae* **2022**, *8*, 606.
9. Alonso-Salinas, R.; Acosta-Motos, J.R.; Pérez-López, A.J.; Noguera-Artiaga, L.; Núñez-Delicado, E.; Burló, F.; López-Miranda, S. Effect of combination of kmno4 oxidation and uv-c radiation on postharvest quality of refrigerated pears cv. 'Ercolini'. *Horticulturae* **2022**, *8*, 1078.
10. Alonso-Salinas, R.; López-Miranda, S.; Pérez-López, A.J.; Acosta-Motos, J.R. Strategies to delay ethylene-mediated ripening in climacteric fruits: Implications for shelf life extension and postharvest quality. *Horticulturae* **2024**, *10*, 840.
11. Iturralde-García, R.D.; Cinco-Moroyoqui, F.J.; Martínez-Cruz, O.; Ruiz-Cruz, S.; Wong-Corral, F.J.; Borboa-Flores, J.; Cornejo-Ramírez, Y.I.; Bernal-Mercado, A.T.; Del-Toro-Sánchez, C.L. Emerging technologies for prolonging fresh-cut fruits' quality and safety during storage. *Horticulturae* **2022**, *8*, 731.
12. Li, T.; Huang, H.; Peng, Y.; Zhou, H.; Hu, H.; Liu, M. Quality grading algorithm of oudemansiella raphanipes based on transfer learning and mobilenetv2. *Horticulturae* **2022**, *8*, 1119.

horticulturae

MDPI

Article

Comparative Metabolomic Fingerprinting Analysis of Tomato Fruits from *Physalis* Species in Mexico's Balsas Basin

Sergio Arias-Martínez [1,†], Guadalupe Oyoque-Salcedo [1,2,†], Oscar Giovanni Gutiérrez-Cárdenas [3], Ernesto Oregel-Zamudio [1,*] and Jesús Rubén Torres-García [1,4,*]

1 Instituto Politécnico Nacional, Centro Interdisciplinario de Investigación para el Desarrollo Integral Regional (CIIDIR), Unidad Michoacán, Justo Sierra 28, Col. Centro, Jiquilpan 59510, Michoacán, Mexico; sariasm@ipn.mx (S.A.-M.); goyoque@ipn.mx (G.O.-S.)
2 Tecnológico Nacional de México, Instituto Tecnológico de Roque, Carretera Celaya, Juventino Rosas Km. 8, Celaya 38110, Guanajuato, Mexico
3 Universidad de La Ciénega del Estado de Michoacán de Ocampo, Genómica Alimentaria, Sahuayo 59103, Michoacán, Mexico; oggutierrez25@gmail.com
4 Investigadores por México, Consejo Nacional de Humanidades, Ciencias y Tecnología (CONAHCYT), Ciudad de México 03940, Estado de México, Mexico
* Correspondence: eoregel@ipn.mx (E.O.-Z.); jrtorresg@ipn.mx (J.R.T.-G.); Tel.: +52-353-533-0218 (E.O.-Z.); +52-222-548-4626 (J.R.T.-G.)
† These authors contributed equally to this work.

Abstract: This study investigated the chemical and sensory distinctions in tomato fruits from three *Physalis* species (*P. ixocarpa, P. angulata, and P. philadelphica*) found in Michoacán, Mexico, using metabolomic fingerprinting through GC-MS analysis. The objective was to identify organoleptic differences that could influence consumer preferences, highlighting the significance of these species' unique traits. These species represented a valuable genetic reservoir for potential hybridization or selection aimed at enhancing commercial varieties by focusing on organoleptic properties rather than traditional selection criteria like fruit size or yield. This research emphasizes the importance of preserving Mexican biodiversity and providing insights into domestication processes that prioritize flavor and sensory qualities. By analyzing metabolite profiles and their correlation with taste preferences, this study contributes to understanding how these differences could be leveraged in breeding programs to develop new tomato varieties with preferred flavors. It was suggested that variations in taste among the species are mainly due to differences in metabolite expression. This knowledge underscores the importance of organoleptic properties in the selection and domestication of edible fruits, offering a pathway toward the conservation and enhancement of tomato varieties through the exploitation of genetic diversity for organoleptic improvement.

Keywords: *Physalis* species; metabolomic fingerprinting; SPME-GC/MS analysis; organoleptic properties; consumer preferences; genetic diversity

Citation: Arias-Martínez, S.; Oyoque-Salcedo, G.; Gutiérrez-Cárdenas, O.G.; Oregel-Zamudio, E.; Torres-García, J.R. Comparative Metabolomic Fingerprinting Analysis of Tomato Fruits from *Physalis* Species in Mexico's Balsas Basin. *Horticulturae* **2024**, *10*, 600. https://doi.org/10.3390/horticulturae10060600

Academic Editors: Paola Sánchez-Bravo and Luis Noguera-Artiaga

Received: 11 May 2024
Revised: 29 May 2024
Accepted: 31 May 2024
Published: 7 June 2024

1. Introduction

The genus *Physalis* is renowned for its rich diversity and its significant roles in the agriculture and food industries. It is native to and has diversified extensively in Mexico, comprising approximately 65 species [1,2]. This genus exhibits remarkable biological diversity, including numerous edible species that have been cultivated and domesticated over time [3,4]. The phenotypic variability among these species is considerable, influenced by inherent biological characteristics and selective pressures from agricultural practices aimed at optimizing agronomic yields and environmental resilience [5].

Domestication efforts have typically focused on improving fruit yield and size, as well as modifying plant and fruit morphology to enhance adaptability and productivity. Furthermore, organoleptic qualities such as acidity, sweetness, and bitterness are critically assessed, as these traits significantly impact consumer preferences and guide the selection of particular cultivars [6]. Within this genus, *Physalis philadelphica* stands out as the most commercially significant species, characterized by larger fruits and a mildly acidic taste profile, making it an essential component of many traditional Mexican dishes [5,7,8].

The most well-known *Physalis* species is *P. philadelphica*, which is distributed worldwide and from which commercial varieties have been developed [7]. Unlike other less domesticated species, it is characterized by a larger fruit (up to six times bigger) and a slightly acidic taste [5,8]. Despite being the only commercially cultivated *Physalis* species, it is a fundamental ingredient in many dishes of traditional Mexican cuisine [9].

In Mexico, it is common to find various locally consumed *Physalis* varieties [10]. These showcase distinct flavors and chemical properties. In the Ciénega region of Michoacán, which belongs to the Balsas Basin, it is still possible to find species, such as *P. angulata* and *P. ixocarpa*, cultivated and marketed under the name "tomate milpero" [11,12].

Previous research suggests that the domestication of these species is still ongoing, though not necessarily with a focus on increasing fruit size or yield [13]. However, both species are valued for their organoleptic properties in contrast to commercial tomatoes.

This study aims to analyze the chemical differences in the fruits of the three main *Physalis* species marketed in the Cienega region of Michoacán. Employing metabolomic fingerprinting techniques through Solid Phase Microextraction, Gas Chromatography–Mass Spectrometry (SPME-GC/MS), we seek to uncover variations in the organoleptic properties of the fruits. This methodology not only enables the discovery of new traits that could be integrated into commercial varieties but also supports the conservation of a vital segment of Mexican biodiversity, thereby safeguarding a valuable genetic reservoir for future hybridization and selection efforts.

2. Materials and Methods

2.1. Sample Collection

Biological samples were procured from local markets in the Cienega region of Michoacán. Fruits were sourced directly from local producers and collectors, ensuring their provenance from municipalities adjacent to the region. Selection criteria for the fruit specimens included an absence of disease or insect damage and a uniform stage of maturity. Maturity was verified through consistent indicators of fruit quality, such as turgency and uniformity of color. These stringent criteria ensured the collection of high-quality specimens suitable for subsequent analysis.

2.2. Metabolomic Fingerprinting Analisis using SPME-GC/MS

2.2.1. Sample Preparation and Extraction

Fruits were meticulously cleaned by rinsing them three times with distilled water before they were air-dried in a shaded area to prevent chemical degradation from sunlight exposure. Once dried, the fruits were rapidly frozen using liquid nitrogen to halt all metabolic activity and were stored at −80 °C to preserve their biochemical integrity. Subsequently, the fruits were lyophilized in a vacuum chamber to remove moisture without degrading sensitive compounds. The dried samples were then finely ground using a Retch mill to produce a uniform powder, facilitating consistent sample handling and extraction. A total of 100 milligrams of the lyophilized and ground tissue from each fruit sample was precisely weighed and transferred into amber vials to protect the sensitive compounds from light degradation. For statistical robustness, five replicates were prepared for each population studied. The vials were then incubated at 40 °C for one hour to equilibrate.

2.2.2. Solid-Phase Microextraction (SPME)

A 50/30 μm DVB/CAR/PDMS fiber (Supelco, Bellefonte, PA, USA) was used for extraction. The fiber was preconditioned in the GC/MS at 230 ± 1 °C for 15 min before each run. For analysis, the fiber was exposed for 30 min in the headspace of the sample at a controlled temperature of 30 ± 1 °C. The fiber was then stored in a holder (57330-u,

Supelco, PA, USA) to prevent contamination and subsequently introduced into the GC/MS injector at 230 ± 1 °C to desorb the volatile organic compounds (VOCs). All analyses were performed in triplicate to ensure reproducibility.

2.2.3. Gas Chromatography–Mass Spectrometry (GC/MS)

A Clarus 680 gas chromatograph (Perkin-Elmer Inc., Massa, MA, USA) equipped with an Elite-5 MS capillary column (30 m length, 0.32 mm ID, 0.25 µm film thickness, operational temperature range from −60 to 320/350 °C) was used. Helium was employed as the carrier gas at a constant flow rate of 1 mL/min with an initial hold time of 0.05 min. The column temperature started at 30 ± 1 °C for 2 min and was then ramped up to 140 °C at 9 °C/min and held for 5 min. The injector temperature was maintained at 230 ± 1 °C. The mass spectrometer (Clarus SQ8T, Perkin-Elmer Inc., Massa, MA, USA) operated with an electron impact ionization source at 70 eV in the scan mode, analyzing a range from 30 to 400 m/z. The temperatures of the transfer line and the ionization source were set at 230 and 250 °C, respectively, optimizing the transfer and ionization of analytes. This comprehensive setup allowed for detailed profiling of the metabolomic fingerprint of each fruit sample, providing insights into their unique chemical compositions.

2.3. *Data Analysis*

To discern differences in the metabolomic fingerprints of the samples, we adopted a Non-Targeted Metabolomic approach. Original (.raw) files were converted to the CDF format for examination on the XCMS Online platform (https://xcmsonline.scripps.edu) [14]. This platform facilitated feature detection, retention time adjustment, and the peak alignment of original chromatograms [15]. To avoid false positives in metabolite identification, only those with q values ≤ 0.05 were considered. Post-detection by XCMS Online, data normalization, and an ANOVA test were executed to pinpoint pivotal metabolites in relation to treatments. We manually annotated the chosen metabolites using the NIST library with a threshold of 0.8. Annotated results underwent analysis in RStudio ver. 2023.03.0 utilizing the MetaboanalystR ver. 3.3.0 library [16]. Detected metabolites were categorized by their chemical groupings, followed by a comparative analysis of their relative and absolute expression proportions. A subsequent ANOVA discerned significant metabolites in each trial.

We computed a correlation matrix to detect relationships between diverse metabolites. Results were charted using the corrplot library in R, exclusively highlighting correlations with statistically significant differences. To ascertain inter-species distinctions, we constructed a heatmap utilizing the ion matrix of the most differentiated metabolites. The heatmap's construction used automatically normalized and scaled data. Dendrograms employed the Minkowski correlation as a distance metric, paired with the Ward clustering algorithm, and set branch significance at $p \leq 0.05$.

3. Results

3.1. *Metabolite Profiling across Physalis Species*

Across the three studied *Physalis* species, a total of 50 metabolites were identified, which belong to 15 chemical groups. The aldehydes group contained the highest number of metabolites, numbering 11, followed by alcohol with 10. Notably, eight chemical groups comprise only a single metabolite (Figure 1). All species exhibited identical metabolites, with no evidence of any compound's absence or exclusivity to a particular species.

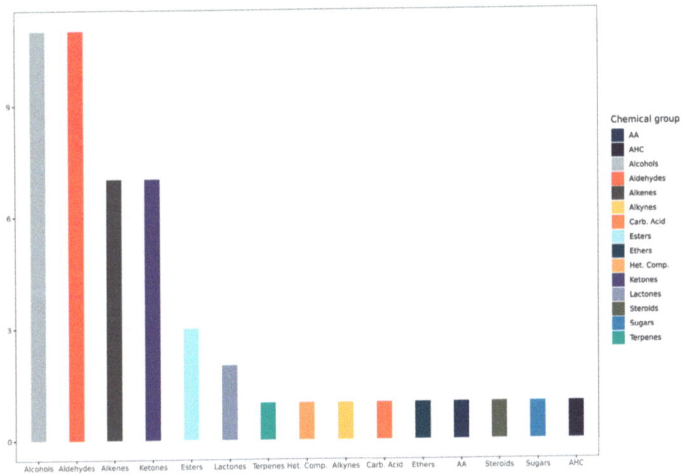

Figure 1. Metabolomic profile detected by GC-MS in the fruits of three *Physalis* species (*angulata*, *ixocarpa*, and *philadelphica*).

3.2. Metabolite Distribution by Chemical Group

Significant differences ($p < 0.05$) in metabolite expression were observed across almost all chemical groups, excluding esters and terpenes (Figure 2). The groups expressing the highest levels, in descending order, were alcohols, ketones, aldehydes, alkenes, and lactones. All other chemical groups displayed reduced expressions in comparison. In all chemical groups, *P. angulata* manifested the most pronounced expression compared to the other two *Physalis species*.

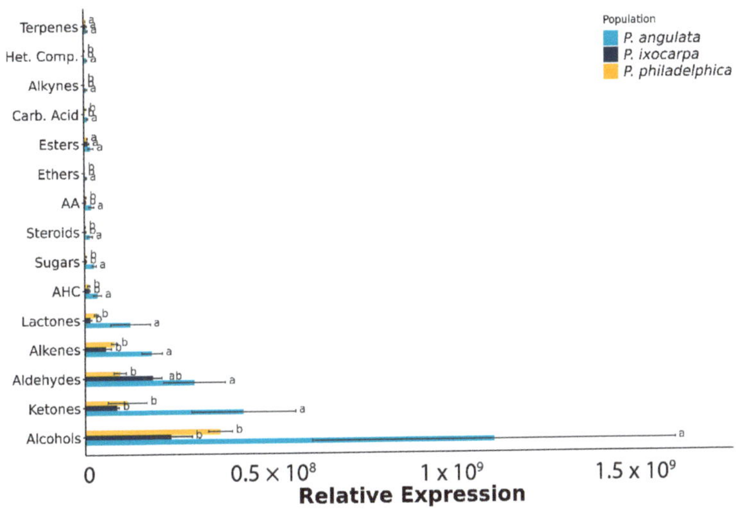

Figure 2. Metabolite expression by chemical group in fruits of three *Physalis* species (*angulata*, *ixocarpa*, and *philadelphica*). Error bars indicate standard errors of the mean. Statistically significant differences are denoted by distinct letters, as determined by Tukey's post hoc test ($p < 0.05$).

Concerning relative compound abundance in the fruits, a similar trend to absolute expression was noticed, with alcohols, aldehydes, and ketones being the most predominant groups, together comprising 75% of the total compound abundance. The remaining chemical groups contributed a lower proportion (Figure 3).

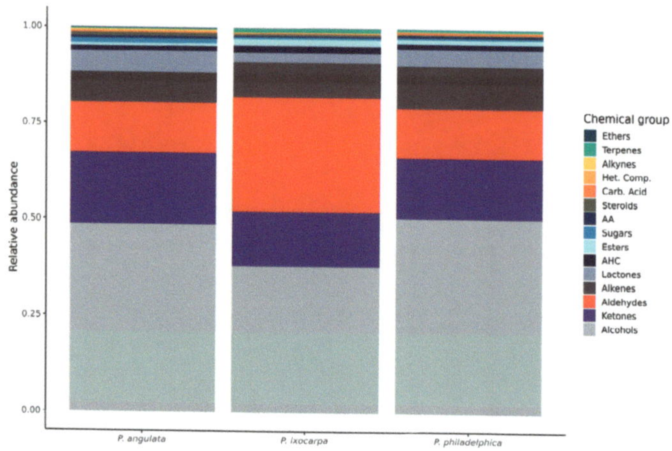

Figure 3. Relative compound abundance by chemical group in fruits of three *Physalis* species (*angulata*, *ixocarpa*, and *philadelphica*).

Within the alcohol group, the species *P. angulata* and *P. philadelphica* exhibited a higher relative abundance compared to *P. ixocarpa*. As for aldehydes, *P. ixocarpa* showed a marked abundance of these compounds compared to the other two species. Specifically, for *P. angulata* and *P. philadelphica*, ketones emerged as the second most abundant group, followed by aldehydes. Conversely, in *P. ixocarpa*, aldehydes were second in abundance, trailed by ketones. Alkenes, lactones, and sugars displayed some differences in proportion, but their overall abundance was limited.

3.3. Alcohol and Aldehyde Groups

Alcohol represents the most abundant group in the fruits of all three species. Within this group, four specific alcohols accounted for nearly 75% of the total alcohol abundance: 1-Penten-3-ol, 1-phenyl-; 2,5-Dimethylcyclohexanol; Benzenemethanol, 4-ethyl-; and 2-Isopropyl-5-methyl-1-heptanol (Figure 4). Notable differences among species were observed in less abundant alcohols, such as 2-Cyclohexen-1-ol and 1-methyl-, with *P. ixocarpa* displaying the highest abundance.

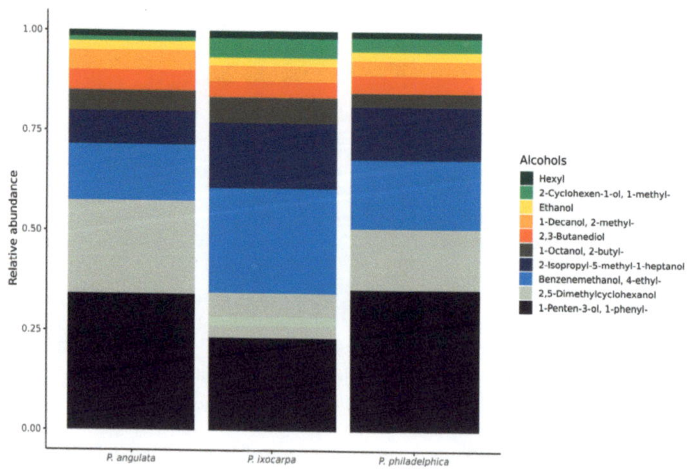

Figure 4. Distribution of specific alcohols in fruits of three *Physalis* species (*angulata*, *ixocarpa*, and *philadelphica*).

Aldehydes emerged as the second most abundant group, holding significant importance due to their contribution to fruit flavors (Figure 5). A greater variability was observed among the aldehydes than the alcohols, stemming from divergent proportions across species. *P. angulata* and *P. philadelphica* shared similar aldehyde distributions. A notable difference was evident in the benzaldehyde levels between these two species. Conversely, *P. ixocarpa* presented a distinct aldehyde profile, with furfural as the predominant aldehyde, followed by 2,5-dihydroxybenzaldehyde. This species exhibited a significant decrease in levels of octanal, benzaldehyde 4-methyl, 2-heptenal, and 2-hexenal relative to the other species while showing an increase in 2,5-dihydroxybenzaldehyde.

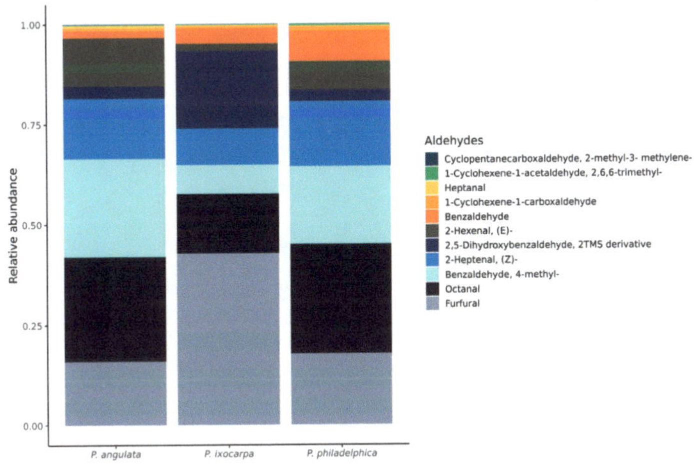

Figure 5. Comparative analysis of aldehyde distribution in fruits of three *Physalis* species (*angulata*, *ixocarpa*, and *philadelphica*).

3.4. Metabolite Correlations

Chemical group correlations across each of the three populations were presented as a matrix, only including statistically significant correlations ($p < 0.05$). In *P. ixocarpa*, four positive correlations between chemical groups were found. Noteworthy interactions were observed between sugar concentrations with amino acids and between alcohols and ketones (Figure 6).

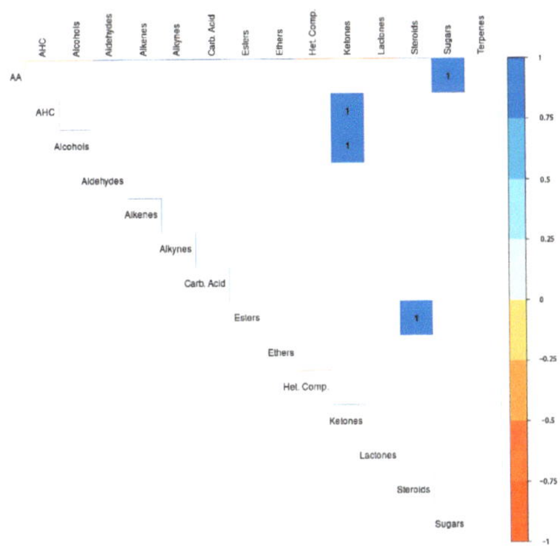

Figure 6. Matrix representation of significant chemical group correlations in fruits of *P. ixocarpa*.

P. angulata exhibited both positive and negative significant correlations. Perhaps the most crucial correlation, especially regarding flavor context, was the inverse relationship between aldehydes and sugars, which showed a correlation coefficient of −1 (Figure 7).

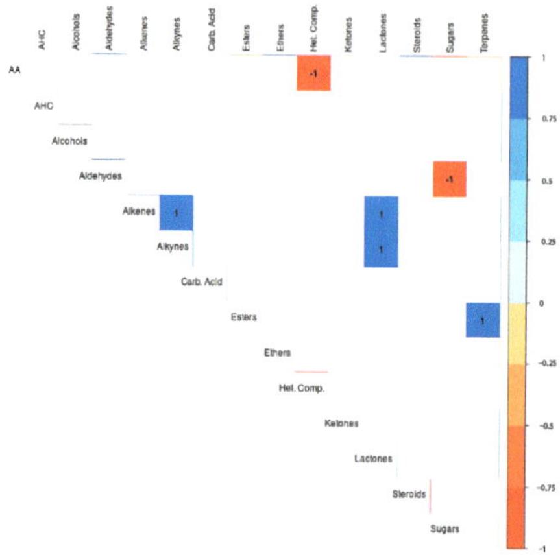

Figure 7. Matrix representation of significant chemical group correlations in fruits of *P. angulata*.

In *P. philadelphica*, eight significant correlations were identified, with the interactions between sugars with amino acids and aldehydes being particularly important (Figure 8).

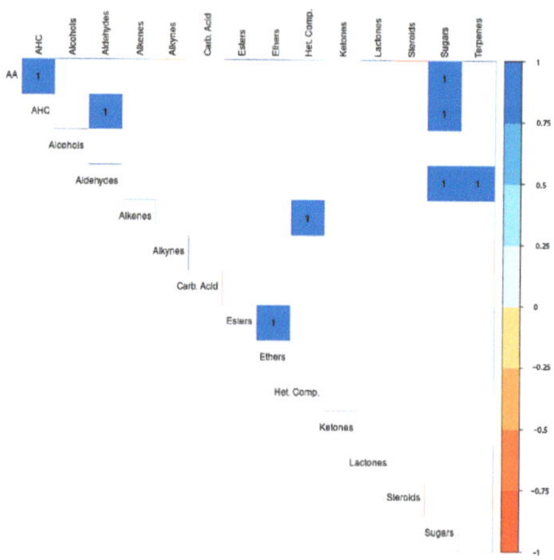

Figure 8. Matrix representation of significant chemical group correlations in fruits of *P. philadelphica*.

3.5. Heatmap Analysis

Of the 50 detected metabolites in *Physalis* species' fruits, 20 displayed significant differences in relative expression ($p \leq 0.05$). The heatmap was constructed using the normalized expression of metabolites, which showed significant variance. The expression levels of metabolites are color-coded, ranging from blue (indicating low expression) to red (indicating high expression) (Figure 9). In the figure, each row signifies an individual metabolite, and each column corresponds to a specific *Physalis species*. The top dendrogram indicates a global classification (metabolomic fingerprinting), suggesting that *P. ixocarpa* and *P. philadelphica* are more closely related than *P. angulata*, which exhibits a contrasting expression pattern. The side dendrogram delineates the metabolites according to their expression levels. Here, metabolites from *P. ixocarpa* and *P. philadelphica* express higher levels relative to *P. angulata*. Specific metabolites with heightened expression in *P. ixocarpa* include 2-Cyclohexen-1-ol, 1-methyl-; Terpinen-4-ol; 2-hydroxy-2,4-dimethyl-hept-6-en-3-one; (+)-4-Carene; methyl salicylate; furfural; and Benzaldehyde. In contrast, *P. philadelphica* showed elevated expression in 1-Undecene, 7-methyl-. In this section, *P. angulata* predominantly underexpressed these metabolites.

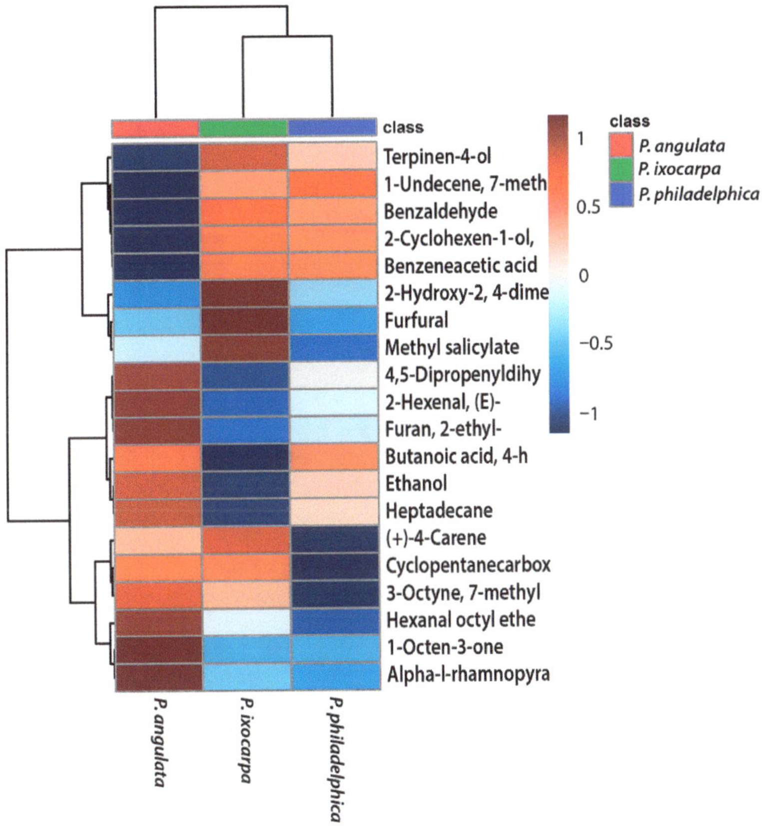

Figure 9. Comparative metabolomic heatmap of fruits of three *Physalis* species based on the relative expression of 20 significant metabolites. The heatmap is generated from normalized data, with expression levels color-coded from blue (low expression) to red (high expression). Each row corresponds to a specific metabolite, and each column represents an individual *Physalis* species. The top dendrogram showcases global metabolomic fingerprinting.

The bottom portion of the heatmap depicts an inverse trend where *P. angulata* shows a pronounced overexpression of metabolites like Ethanol; 4,5-Dipropenyldihydro-furan-2-one; 2-hexenal, (E)-; Heptadecane; 3-Octyne, 7-methyl-; Hexanal octyl ether; Furan, 2-ethyl-; and 1-Octen-3-one. Conversely, *P. ixocarpa* and *P. philadelphica* largely exhibit reduced or even negative expression levels.

4. Discussion

In this study, the chemical composition of the fruits from three species of the genus *Physalis* found in the Balsas basin region of Mexico was analyzed. Significant differences were identified in the absolute expression and relative abundance of various metabolites. However, there were no differences in the presence or absence of metabolites across the three studied species. As such, variations in taste can be attributed to the proportions of these metabolites. These observed discrepancies might primarily arise from the organoleptic characteristics selected during domestication, given that the three species undergo varying levels of selection.

The Balsas basin has a rich history of domesticating plant species with agricultural value. *P. philadelphica* is the most renowned species due to its global production and consumption [17]. Conversely, the species *P. angulata* and *P. ixocarpa* are fruits primarily produced and consumed in specific regions of Mexico [3,4,8], and they exhibit morphological and organoleptic characteristics distinct from *P. philadelphica*, which are valued by locals [13].

The lack of selection based on size suggests that the primary focus in the selection process revolves around traits associated with fruit flavors [5,13]. Both *Physalis* species are reported to have more pleasant tastes than their commercial counterparts. These attributes could be harnessed in breeding programs to establish commercial varieties [18].

The absence of exclusive metabolites among the three species suggests a low differentiation level, indicating no evolution of new biochemical pathways [19]. Differences are likely due to metabolic regulation, where certain metabolites are produced preferentially due to selection (domestication) processes. This phenomenon has been documented in species with extensive selection histories, where the search for novel organoleptic features remains a primary goal [20,21]. Such is the case with chili peppers and other fruit-bearing plants [22]; defining taste is complex as it involves a vast array of compounds, and their proportion is crucial.

Significant differences were observed in the relative expression of nearly all detected chemical groups. *P. angulata* exhibited the highest abundance in most chemical groups. However, fruits from this species are not the preferred choice for residents of the Cienega de Michoacán region. Local inhabitants report that *P. angulata* fruits have a more acidic and bitter taste compared to the *P. ixocarpa* fruits (J. R. Torres-García, pers. observ.). This flavor variation might be due to the dominant presence of alcohols, overshadowing the tastes from other pleasant-tasting compounds that are present in smaller proportions [6].

In addition to noticeable concentration differences among this species, perhaps the most critical aspect is the balance of various metabolites within the fruit. For instance, sugars and terpenes, known for their contribution to food flavor, were found in relatively smaller concentrations compared to other chemical groups [23].

In all three species, the chemical groups of aldehydes and alcohols had the highest number of metabolites, making them the most abundant in both absolute and relative metabolite expressions. As previously mentioned, all three species had the same compounds, but their proportions varied. In the case of alcohol, the most prevalent were 1-Penten-3-ol and 1-phenyl- in *P. angulata* and *P. philadelphica*, both of which impart a bitter taste. In contrast, in *P. ixocarpa*, higher concentrations of alcohol like Benzenemethanol 4-ethyl-, and 2-isopropyl-5-methyl-1-heptanol were detected, which contribute pleasant aromas and fruity flavors to foods [21].

Correlations indicate that even if the same metabolites are shared, their expression level and interrelation differ between species. Various studies have suggested that domestication can influence the flavors or organoleptic characteristics of plants [20,21]. In most of these, it is argued that this is due to differences in gene regulation related to flavor compound synthesis.

In the case of *P. ixocarpa*, the positive correlation between sugars and amino acids is relevant both nutritionally and organoleptically [24]. An increase in the fruit's total sugars is associated with an increase in the number of amino acids, specifically lysine, which is an essential amino acid. Thus, a better taste also implies a higher nutritional value. Furthermore, the correlation between alcohols and ketones is closely linked to flavor, as both chemical groups significantly influence fruit flavor. These correlations are crucial since they affect both taste and nutrition, which are two essential characteristics for genetic improvement [18]. For *P. angulata*, the negative correlation found between aldehydes and sugars could be a reason why its flavor is not as popular in the Cienega de Michoacán region. Both chemical groups contribute to organoleptic characteristics, providing a pleasant taste and aroma [6]. The negative correlation suggests that it is challenging to maximize one's concentration without compromising the other, limiting the selection of improved taste varieties. For *P. philadelphica*, relevant correlations were observed in both nutritional (amino acid–sugars) and organoleptic characteristics (aldehyde–sugars, aldehyde–terpenes). It is the species of *Physalis* genus that has the highest production and global consumption, which might explain the observed correlations, possibly due to improvement schemes [18].

Classification based solely on metabolites with significant statistical differences ($p \leq 0.05$) allows for a more objective evaluation between species. It was observed that *P. ixocarpa* and *P. philadelphica* have more similar metabolic profiles compared to *P. angulata*.

Specifically, certain compounds were identified to be more significant than others in terms of classification and flavor attributes. In the case of *P. ixocarpa*, higher concentrations of certain compounds pivotal to flavor contribution were observed, such as 2-Cyclohexen-1-ol, 1-methyl-, which possesses a sweet and woody aroma [21] and is frequently found in plant essential oils. Terpinen-4-ol exudes a herbaceous and spicy aroma and has been recognized for its antimicrobial potential, potentially influencing food preservation [24]. Benzaldehyde, with its bitter almond scent, is extensively employed in the food industry and can be found in other aromatic fruits, such as the Huangjiu [25]. Another notable metabolite, (+)-4-Carene, carries a sweet and resinous scent. Among the compounds most expressed in this fruit are furfural, which emits a sweet and almondy aroma, and methyl salicylate, with its sweet minty scent, which is commonly utilized in the food and cosmetic industries. However, two compounds were also identified that do not contribute pleasant flavors. These include 2-hydroxy-2,4-dimethyl, which has a burnt scent, and Cyclopentanecarboxaldehyde, 2-methyl-3-methylene-, which emits a strong and unpleasant odor. In the case of the latter, its expression level was found to be equal to that in *P. angulata*.

In *P. angulata*, there is a higher expression of compounds such as Ethanol; 4,5-Dipropeny ldihydro-furan-2-one; 2-hexenal, (E)-; Heptadecane; 3-Octyne, 7-methyl-; Hexanal octyl ether; Furan, 2-ethyl-; and 1-Octen-3-one. These compounds typically impart mild and pleasant flavors. However, at high concentrations, they can produce unpleasant tastes. On the other hand, compounds known to deliver unpalatable flavors include 3-Octyne, 7-methyl-; Heptadecane; Furan, 2-ethyl-; 1-Octen-3-one; Butanoic acid, 4-hydroxy-; and Cyclopentanecarboxaldehyde, 2-methyl-3-methylene-. The latter displays an expression level similar to that in *P. ixocarpa*, suggesting that domestication processes focusing on flavor have not encompassed all metabolites, and characteristics of wild relatives can still be found. Conversely, the compound Cyclopentanecarboxaldehyde, 2-methyl-3-methylene- showed the lowest expression in the *P. philadelphica* population. This compound is an aldehyde that can also influence the aroma and taste of tomato fruits. Nonetheless, due to its low expression, its contribution to the flavor of fruits from this population may be less significant.

In light of the observations from this comprehensive study, it becomes evident that the process of domestication and genetic selection within the *Physalis* genus has led to significant variations in metabolite profiles among species, even when the same metabolites are universally present. These variations are intricately connected to perceived tastes and organoleptic characteristics and are further influenced by the relative concentrations of different metabolites. It is crucial to recognize that the quest for optimizing flavor does not solely rely on the presence or absence of specific compounds but depends heavily on their relative proportions and interrelations. As the world continues to globalize and the demand for diverse and unique flavors increases, understanding these intricate relationships in crops like *Physalis* becomes paramount. Embracing this knowledge can pave the way for creating optimized crop varieties that cater not just to local preferences but can find acceptance on a global palate. Furthermore, this study underscores the value of revisiting and appreciating the biochemical complexity inherent in our traditional crops, as there remains a vast untapped potential to harness their genetic diversity for the betterment of global agriculture and food systems.

5. Conclusions

In conclusion, this study on the metabolomic profiling and sensory characteristics of tomato fruits from three distinct *Physalis* species in the Cienega region of Michoacán, Mexico, provides clear evidence of the significant chemical diversity that exists among the species studied. Utilizing advanced metabolomic fingerprinting techniques via SPME-GC/MS, we were able to identify specific variations in the organoleptic properties of the fruits, which directly impact consumer preferences. These findings highlight the importance of these species as valuable genetic reservoirs that could be exploited to enhance commercial tomato varieties, focusing on improving organoleptic qualities over traditional selection criteria such as fruit size or yield. Moreover, this study underscores the relevance of preserving Mexican biodiversity and leveraging its potential to strengthen the sustainability and acceptance of native varieties in local and global markets. Integrating these insights into breeding programs could lead to the development of tomato varieties that not only meet market demands in terms of flavor and aroma but also contribute to the conservation of native species, enriching global agricultural diversity.

Author Contributions: Conceptualization, O.G.G.-C., E.O.-Z. and J.R.T.-G.; methodology, E.O.-Z. and J.R.T.-G.; software, E.O.-Z. and J.R.T.-G.; validation, S.A.-M. and G.O.-S.; formal analysis, S.A.-M. and G.O.-S.; investigation, O.G.G.-C.; resources, O.G.G.-C., S.A.-M. and G.O.-S.; data curation, S.A.-M. and G.O.-S.; writing—original draft preparation, E.O.-Z. and J.R.T.-G.; writing—review and editing, E.O.-Z. and J.R.T.-G.; visualization, O.G.G.-C.; supervision, E.O.-Z. and J.R.T.-G.; project administration, E.O.-Z. and J.R.T.-G.; funding acquisition, E.O.-Z. and J.R.T.-G. All authors have read and agreed to the published version of the manuscript.

Funding: This research received financial support from the Secretaría de Investigación y Posgrado of the Instituto Politécnico Nacional (IPN), which provided the necessary funds to carry out this study.

Data Availability Statement: The data presented in this study are available in this article.

Acknowledgments: We appreciate the support provided by the staff of the Instituto Politécnico Nacional and the laboratory technicians at CIIDIR-IPN Unidad Michoacán and extend our gratitude to the producers from the Cienega region of Michoacán for supplying the essential biological samples for this research. Their collaboration has been invaluable.

Conflicts of Interest: The authors declare no conflicts of interest.

References

1. Martínez, M.; Vargas-Ponce, O.; Rodríguez, A.; Chiang, F.; Ocegueda, S. Solanaceae family in Mexico. *Bot. Sci.* **2017**, *95*, 131–145. [CrossRef]
2. Martínez, M.; Vargas-Ponce, O.; Zamora-Tavares, P. Taxonomic revision of Physalis in Mexico. *Front. Genet.* **2023**, *14*, 1080176. [CrossRef]
3. Zamora-Tavares, P.; Vargas-Ponce, O.; Sánchez-Martínez, J.; Cabrera-Toledo, D. Diversity and genetic structure of the husk tomato (*Physalis philadelphica* Lam.) in Western Mexico. *Genet. Resour. Crop Evol.* **2015**, *62*, 141–153. [CrossRef]
4. del Pilar Zamora-Tavares, M.; Martínez, M.; Magallón, S.; Guzmán-Dávalos, L.; Vargas-Ponce, O. Physalis and physaloids: A recent and complex evolutionary history. *Mol. Phylogenetics Evol.* **2016**, *100*, 41–50. [CrossRef]
5. Solís-Montero, L.; Aceves-Chong, L.; Vega-Polanco, M.; Vargas-Ponce, O. Changes in reproductive traits in *Physalis philadelphica;* An unexpected shift toward self-Incompatibility in a domesticated annual fruit crop. *Front. Plant Sci.* **2021**, *12*, 658406. [CrossRef]
6. Ulrich, D.; Olbricht, K. Fruit organoleptic properties and potential for their genetic improvement. In *Breeding for Fruit Quality;* John Wiley & Sons, Inc.: Hoboken, NJ, USA, 2011; pp. 39–59.
7. González-Pérez, J.E.; Guerrero-Beltrán, J.Á. Tomatillo or husk tomato (*Physalis philadelphica* and *Physalis ixocarpa*): A review. *Sci. Hortic.* **2021**, *288*, 110306. [CrossRef]
8. Sandoval-Padilla, I.; Pérez-Alquicira, J.; Rodríguez, A.; del Pilar Zamora-Tavares, M.; Vargas-Ponce, O. The plastome of the husk tomato (*Physalis philadelphica* Lam., Solanaceae): A comparative analysis between wild and cultivated pools. *Genet. Resour. Crop Evol.* **2022**, *69*, 1391–1405. [CrossRef]
9. Hernández, J.F.S.; Yáñez, S.B. Aprovechamiento tradicional de las especies de Physalis en México. *Rev. Geogr. Agríc.* **2009**, *43*, 81–86.
10. Valdivia-Mares, L.E.; Zaragoza, F.A.R.; González, J.J.S.; Vargas-Ponce, O. Phenology, agronomic and nutritional potential of three wild husk tomato species (Physalis, Solanaceae) from Mexico. *Sci. Hortic.* **2016**, *200*, 83–94. [CrossRef]

11. Vargas-Ponce, O.; Sánchez Martínez, J.; Zamora Tavares, M.d.P.; Valdivia Mares, L.E. Traditional management of a small-scale crop of *Physalis angulata* in Western Mexico. *Genet. Resour. Crop Evol.* **2016**, *63*, 1383–1395. [CrossRef]
12. Saavedra, J.d.C.M.; Zaragoza, F.A.R.; Toledo, D.C.; Hernández, C.V.S.; Vargas-Ponce, O. Agromorphological characterization of wild and weedy populations of *Physalis angulata* in Mexico. *Sci. Hortic.* **2019**, *246*, 86–94. [CrossRef]
13. Martínez-Vega, A.; Oregel-Zamudio, E.; García-Ruíz, I.; Villapando-Arteaga, E.; Torres-García, J. Genetic and metabolomic differentiation of *Physalis ixocarpa* Brot. ex Hornem. populations in Michoacan State, Mexico. *Genet. Resour. Crop Evol.* **2022**, *69*, 1867–1877. [CrossRef]
14. Forsberg, E.M.; Huan, T.; Rinehart, D.; Benton, H.P.; Warth, B.; Hilmers, B.; Siuzdak, G. Data processing, multi-omic pathway mapping, and metabolite activity analysis using XCMS Online. *Nat. Protoc.* **2018**, *13*, 633–651. [CrossRef]
15. Tautenhahn, R.; Patti, G.J.; Rinehart, D.; Siuzdak, G. XCMS Online: A web-based platform to process untargeted metabolomic data. *Anal. Chem.* **2012**, *84*, 5035–5039. [CrossRef]
16. Pang, Z.; Chong, J.; Li, S.; Xia, J. MetaboAnalystR 3.0: Toward an optimized workflow for global metabolomics. *Metabolites* **2020**, *10*, 186. [CrossRef]
17. Ariza-Mejía, D.; Oyoque-Salcedo, G.; Angóa-Pérez, V.; Mena-Violante, H.G.; Álvarez-Bernal, D.; Torres-García, J.R. Diversity and Potential Function of the Bacterial Rhizobiome Associated to *Physalis ixocarpa* Broth. in a Milpa System, in Michoacan, Mexico. *Agronomy* **2022**, *12*, 1780. [CrossRef]
18. Zhao, J.; Sauvage, C.; Zhao, J.; Bitton, F.; Bauchet, G.; Liu, D.; Huang, S.; Tieman, D.M.; Klee, H.J.; Causse, M. Meta-analysis of genome-wide association studies provides insights into genetic control of tomato flavor. *Nat. Commun.* **2019**, *10*, 1534. [CrossRef]
19. Colantonio, V.; Ferrão, L.F.V.; Tieman, D.M.; Bliznyuk, N.; Sims, C.; Klee, H.J.; Munoz, P.; Resende, M.F., Jr. Metabolomic selection for enhanced fruit flavor. *Proc. Natl. Acad. Sci. USA* **2022**, *119*, e2115865119. [CrossRef]
20. Bennett, A.B. Taste: Unraveling tomato flavor. *Curr. Biol.* **2012**, *22*, R443–R444. [CrossRef]
21. Klee, H.J.; Tieman, D.M. The genetics of fruit flavour preferences. *Nat. Rev. Genet.* **2018**, *19*, 347–356. [CrossRef]
22. Heng, Z.; Xu, X.; Xu, X.; Li, Y.; Wang, H.; Huang, W.; Yan, S.; Li, T. Integrated transcriptomic and metabolomic analysis of chili pepper fruits provides new insight into the regulation of the branched chain esters and capsaicin biosynthesis. *Food Res. Int.* **2023**, *169*, 112856. [CrossRef] [PubMed]
23. Ferrão, L.F.V.; Sater, H.; Lyrene, P.; Amadeu, R.R.; Sims, C.A.; Tieman, D.M.; Munoz, P.R. Terpene volatiles mediates the chemical basis of blueberry aroma and consumer acceptability. *Food Res. Int.* **2022**, *158*, 111468. [CrossRef]
24. Zhang, C.-R.; Khan, W.; Bakht, J.; Nair, M.G. New antiinflammatory sucrose esters in the natural sticky coating of tomatillo (*Physalis philadelphica*), an important culinary fruit. *Food Chem.* **2016**, *196*, 726–732. [CrossRef] [PubMed]
25. Yu, H.; Xie, T.; Xie, J.; Chen, C.; Ai, L.; Tian, H. Aroma perceptual interactions of benzaldehyde, furfural, and vanillin and their effects on the descriptor intensities of Huangjiu. *Food Res. Int.* **2020**, *129*, 108808. [CrossRef]

horticulturae

MDPI

Article

HS-SPME-GC–MS Profiling of Volatile Organic Compounds and Polar and Lipid Metabolites of the "Stendesto" Plum–Apricot Kernel with Reference to Its Parents

Dasha Mihaylova [1,*], Aneta Popova [2,*], Ivayla Dincheva [3] and Svetla Pandova [4]

1 Department of Biotechnology, University of Food Technologies, 4002 Plovdiv, Bulgaria
2 Department of Catering and Nutrition, University of Food Technologies, 4002 Plovdiv, Bulgaria
3 Department of Agrobiotechnologies, Agro Bio Institute, Agricultural Academy, 8 Dragan Tsankov Blvd., 1164 Sofia, Bulgaria; ivadincheva@yahoo.com
4 Department of Breeding and Genetic Resources, Fruit Growing Institute, Agricultural Academy, 4000 Plovdiv, Bulgaria; cler66@abv.bg
* Correspondence: dashamihaylova@yahoo.com (D.M.); popova_aneta@yahoo.com (A.P.)

Abstract: Plum–apricot hybrids are the successful backcrosses of plums and apricots. Plums and apricots are well-known and preferred by consumers because of their distinct sensory and beneficial health properties. However, kernel consumption remains limited even though kernels are easily accessible. The "Stendesto" hybrid originates from the "Modesto" apricot and the "Stanley" plum. Kernal metabolites exhibited quantitative differences in terms of metabolites identified by gas chromatography–mass spectrometry (GC–MS) analysis and HS-SPME technique profiling. The results revealed a total of 55 different compounds. Phenolic acids, hydrocarbons, organic acids, fatty acids, sugar acids and alcohols, mono- and disaccharides, as well as amino acids were identified in the studied kernels. The hybrid kernel generally inherited all the metabolites present in the parental kernels. Volatile organic compounds were also investigated. Thirty-five compounds identified as aldehydes, alcohols, ketones, furans, acids, esters, and alkanes were present in the studied samples. Considering volatile organic compounds (VOCs), the hybrid kernel had more resemblance to the plum one, bearing that alkanes were only identified in the apricot kernel. The objective of this study was to investigate the volatile composition and metabolic profile of the first Bulgarian plum–apricot hybrid kernels, and to provide comparable data relevant to both parents. With the aid of principal component analysis (PCA) and hierarchical cluster analysis (HCA), differentiation and clustering of the results occurred in terms of the metabolites present in the plum–apricot hybrid kernels with reference to their parental lines. This study is the first providing information about the metabolic profile of variety-defined kernels. It is also a pioneering study on the comprehensive evaluation of fruit hybrids.

Keywords: *Prunus* spp.; fruit; volatile organic compounds (VOCs); principal component analysis; metabolic chemotaxonomy; volatolomics

Citation: Mihaylova, D.; Popova, A.; Dincheva, I.; Pandova, S. HS-SPME-GC–MS Profiling of Volatile Organic Compounds and Polar and Lipid Metabolites of the "Stendesto" Plum–Apricot Kernel with Reference to Its Parents. *Horticulturae* **2024**, *10*, 257. https://doi.org/10.3390/horticulturae10030257

Academic Editors: Paola Sánchez-Bravo and Luis Noguera-Artiaga

Received: 8 February 2024
Revised: 27 February 2024
Accepted: 29 February 2024
Published: 7 March 2024

1. Introduction

Metabolomic studies are comprehensive tools used to reveal the composition of phytochemicals in various plant tissues and organs [1]. Metabolomics is an omics approach used for more than twenty years in research [2]. It can be divided into targeted and non-targeted approaches [3]. The non-targeted approach detects both known and unknown metabolites, resulting in full profiling [4]. The targeted approach uses a selective known metabolite signal [3]. Metabolomics can use different analytical platforms, including spectroscopy, chromatography, and nuclear magnetic resonance, among others [5]. Metabolomic studies can be used to detect changes during different maturity stages of growing conditions [6]. Gas chromatography–mass spectrometry (GC–MS) is a common technique applied to the identification and characterization of metabolites' composition. The production of volatile organic compounds (VOCs) is regulated via various metabolomics pathways from their precursors [7]. They are important flavor-contributing agents. Volatile compounds can be divided into primary (synthetized during maturation) and secondary (produced by tissue disruption) ones [8]. Headspace solid-phase microextraction is an effective tool for increased volatile recovery and characterization. It is commonly used with GC–MS. Solid-phase microextraction (SPME)-GC–MS is used in experimental chemistry due to its universalization [9].

The volatile composition of fruits has been widely studied, and different classes of compounds have been documented [10]. However, fruit kernels have not been thoroughly examined in terms of their volatile profile. In fact, the olfactory association of the fruit itself and its kernel widely differs. Kernel oils seldom hold the sensory feeling of the fruit. Consequently, it is of interest to define the major VOCs in kernels as well. The metabolic profile of kernels and the *Prunus* genus in particular is focused mostly on the existence of amygdalin [11]. Its presence is typical for apricot kernels and almonds, and it is known that amygdalin is enzymatically metabolized into cyanide [12]. That is why the quantity of amygdalin is important in terms of preservation of good health. Apricot seeds are also reported to contain several phenolic compounds, i.e., phenolic acids and flavonoids [13]. At present, efforts have been made to achieve sustainable exploitation of resources in every aspect of life. Food provision is viewed as a major societal challenge, and a continuous search for nutritious resources is gaining researchers' interest. The valorization of peels and fruit pomace is not new, and authors have focused on the possibility of incorporating fruit wastes in various industries knowing that they possess many health-enhancing molecules [14,15]. Several papers are now hinting that kernels are rich in metabolites and could be part of the human diet [16]. Some authors suggest that kernels could be valuable supplements in the future due to their beneficial chemical content [17]. Recently, mango seed kernels have been valorized due to their beneficial composition [18].

The plant-based diet has been recognized as having nutritive value. Fruits are generally part of the human diet, and they provide a palette of phytochemicals. Fruits are rich in phytochemicals, i.e., phenolic acids, organic acids, and sugars, among others [19]. It is known that the agrosystem changes gradually due to unfavorable meteorological conditions, the existence of pathogens, or deficiency/toxicity of minerals [20]. The agronomy sector is constantly searching for new sustainable cultivars that have better yield, need less maintenance, and use fewer resources, as well as being able to successfully thrive in the changing climate [21]. The genus *Prunus* has major representatives that are cherished worldwide, like peaches, apricots, plums, and cherries, among others. Fruit hybrids are an interesting approach towards the changing demands of consumers. They combine the most characteristic features of both their parents as well as pose an interesting niche of research with reference to their composition. Plum–apricot fruits are stone fruit like their parents, plums and apricots. Plum–apricot hybrids may result in three main types: plumcots, pluots, and apriums [22]. The "Stendesto" plum–apricot hybrid is the only successful Bulgarian one of the kind, and it is a plumcot. The plumcot is considered 50% plum and 50% apricot. Information about its composition is practically missing; the same applies for its parents, the "Modesto" apricot (father) and the "Stanley" plum (mother). The "Stendesto" plumcot was officially registered in 2013. Not many papers are available on the topic of fruit hybrid composition, which sets new research directions in the identification and application of potential biologically active sources.

A major setback in published papers is the lack of variety/cultivar identification. Not only do the geographical location, soil specificity, and local meteorology factors play an important role in the differences between fruits but also the variety/cultivar itself [23]. This makes it mandatory to pay more attention to the variety differences, especially if they occur in the same species in local and introduced lands. Fruits are providers of vitamins, minerals, and phytochemicals, but they also generate high amounts of by-products regarding their kernels. Researchers have found ways to incorporate kernels in several industries, i.e., cosmetics, biofuel, detergents, and pharmaceuticals, among others [24]. To date, information about the application of kernels in the food industry is scarce [25]. A thorough holistic approach towards their composition, beneficial compounds, and possible biological activity might set a path for their better understanding and utilization.

The objective of this study was to investigate the volatile composition and metabolic profile of the first Bulgarian plum–apricot hybrid kernels, and to provide comparable data relevant to both parents. With the aid of principal component analysis (PCA) and hierarchical cluster analysis (HCA), differentiation and clustering of the results were provided in terms of the metabolites present in plum–apricot hybrid kernels with reference to their parental lines. This work may not only represent an interesting approach for future studies in line with trending topics like zero-waste management but also be used as core information for further comparison in relative papers. Highlighting kernels as potential nutritional and functional sources will definitely aid in the utilization of this by-product.

2. Results and Discussion

The studied kernels were characterized in terms of their VOCs, polar metabolites, and lipids. Information about the compounds found in fruit kernels is scarce or missing. Thus, this is considered a first comprehensive report about the composition of apricot, plum, and plum–apricot kernels. A total of forty-three compounds were identified from the samples (Table 1). Between them, amino acids, organic acids, sugar acids and alcohols, mono- and disaccharides, phenolic acids, and hydrocarbons were identified as existing groups.

Table 1. Metabolites (mg/g dry weight) identified in studied kernels analyzed by HS-SPME-GC–MS.

RI	Class/Name	Modesto	Stanley	Stendesto
		Amino acids		
1105	Alanine	0.35 ± 0.11^{a}	0.22 ± 0.07^{a}	0.30 ± 0.13^{a}
1232	Valine	0.17 ± 0.06^{a}	0.14 ± 0.05^{a}	0.08 ± 0.02^{a}
1259	Leucine	0.76 ± 0.24^{a}	0.57 ± 0.18^{ab}	0.28 ± 0.09^{b}
1296	Isoleucine	0.30 ± 0.10^{a}	0.21 ± 0.07^{a}	0.12 ± 0.04^{a}
1302	Proline	0.93 ± 0.30^{a}	0.18 ± 0.06^{b}	0.13 ± 0.04^{b}
1343	Serine	1.07 ± 0.34^{a}	0.73 ± 0.23^{ab}	0.30 ± 0.10^{b}
1362	Threonine	0.25 ± 0.08^{a}	0.45 ± 0.15^{a}	0.36 ± 0.12^{a}
1502	Aspartic acid	0.78 ± 0.25^{a}	0.33 ± 0.11^{a}	0.48 ± 0.15^{a}
1519	Pyroglutamic acid	0.64 ± 0.20^{a}	0.42 ± 0.13^{ab}	0.11 ± 0.03^{b}
1625	Phenylalanine	0.18 ± 0.06^{ab}	0.37 ± 0.12^{a}	0.06 ± 0.02^{b}
1656	Asparagine	0.68 ± 0.22^{a}	0.39 ± 0.13^{a}	0.31 ± 0.10^{a}
1775	Glutamine	0.30 ± 0.10^{a}	0.22 ± 0.07^{a}	0.16 ± 0.05^{a}
1839	Arginine	0.88 ± 0.28^{b}	5.43 ± 1.74^{a}	4.61 ± 1.48^{a}
		Organic acids		
1119	Oxalic acid	0.73 ± 0.23^{a}	0.45 ± 0.14^{ab}	0.11 ± 0.04^{b}
1314	Succinic acid	0.19 ± 0.06^{a}	0.08 ± 0.03^{b}	0.05 ± 0.02^{b}
1330	Fumaric acid	0.11 ± 0.03^{a}	0.05 ± 0.02^{b}	0.03 ± 0.01^{b}
1475	Mallic acid	1.92 ± 0.62^{a}	0.05 ± 0.02^{b}	2.60 ± 0.83^{a}
1530	γ-Aminobutyric acid	0.13 ± 0.04^{ab}	0.16 ± 0.05^{a}	0.06 ± 0.02^{b}
1727	2-Aminoadipic acid	1.85 ± 0.59^{a}	1.36 ± 0.44^{a}	1.07 ± 0.34^{a}
1816	Isocitric acid	0.39 ± 0.12^{ab}	0.24 ± 0.08^{b}	0.74 ± 0.24^{a}
		Sugar acids and alcohols		
1264	Glycerol	1.75 ± 0.56^{a}	0.36 ± 0.11^{b}	0.47 ± 0.15^{b}
1541	Eritrreonic acid	1.19 ± 0.38^{a}	0.93 ± 0.30^{a}	0.52 ± 0.17^{a}
1611	Glutamic acid	0.43 ± 0.14^{a}	0.14 ± 0.04^{b}	0.15 ± 0.05^{b}
1695	Xylitol	2.30 ± 0.74^{a}	1.65 ± 0.53^{a}	0.87 ± 0.28^{a}
1718	Arabitol	0.68 ± 0.22^{a}	0.50 ± 0.16^{a}	0.27 ± 0.09^{a}
1801	Glyceric acid-3-phosphate	0.21 ± 0.07^{a}	0.19 ± 0.06^{a}	0.43 ± 0.14^{a}

Table 1. *Cont.*

RI	Class/Name	Modesto	Stanley	Stendesto
	Sugar acids and alcohols			
1920	Sorbitol	7.22 ± 2.32 [a]	12.31 ± 3.95 [a]	11.53 ± 3.70 [a]
2009	Gluconic acid	0.12 ± 0.04 [b]	0.30 ± 0.10 [a]	0.11 ± 0.03 [b]
2018	Glucaric acid	0.25 ± 0.08 [a]	0.15 ± 0.05 [a]	0.12 ± 0.04 [a]
2041	Myo-Inositol isomer	0.83 ± 0.27 [a]	0.53 ± 0.17 [ab]	0.29 ± 0.09 [b]
2101	Myo-Inositol isomer	2.05 ± 0.66 [a]	0.21 ± 0.07 [b]	0.74 ± 0.24 [b]
	Mono- and disaccharides			
1855	Fructose isomer	2.38 ± 0.76 [b]	6.48 ± 2.08 [ab]	7.05 ± 2.11 [a]
1869	Fructose isomer	1.60 ± 0.51 [b]	4.95 ± 1.59 [ab]	5.51 ± 1.77 [a]
1876	1-Methyl-α-D-glucopyranoside	0.10 ± 0.03 [b]	0.27 ± 0.09 [ab]	0.36 ± 0.12 [a]
1882	Glucose isomer	2.78 ± 0.89 [b]	9.90 ± 2.18 [a]	12.00 ± 2.85 [a]
1898	Glucose isomer	0.83 ± 0.27 [b]	2.21 ± 0.71 [ab]	2.60 ± 0.83 [a]
1937	Glucose 1-phosphate	0.16 ± 0.05 [b]	0.25 ± 0.08 [b]	4.05 ± 1.30 [a]
2687	Sucrose	15.54 ± 2.99 [a]	7.39 ± 1.37 [b]	10.25 ± 2.29 [ab]
	Phenolic acids			
1835	Protocatechuic acid	0.32 ± 0.10 [ab]	0.17 ± 0.06 [b]	0.60 ± 0.19 [a]
1940	*trans-p*-Coumaric acid	0.55 ± 0.18 [a]	0.41 ± 0.13 [a]	0.24 ± 0.10 [a]
2106	*trans*-Ferulic acid	0.19 ± 0.07 [a]	0.38 ± 0.12 [a]	0.26 ± 0.08 [a]
	Others			
1400	Tetradecane	0.64 ± 0.21 [a]	0.46 ± 0.15 [ab]	0.13 ± 0.04 [b]
1600	Hexadecane	0.29 ± 0.09 [a]	0.20 ± 0.06 [a]	0.12 ± 0.04 [a]

Amino acids marked in blue color are essential; RI—retention index. Different letters in the same row indicate statistically significant differences ($p < 0.05$) according to ANOVA and the Tukey test.

Amino acids were found in small quantities, yet the hybrid kernel had managed to keep all the amino acids present in both apricot and plum. The same trend applied to the other identified classes, where the hybrid had inherited all the metabolites present in its parents. A study about apricot fruit and seeds showed a resemblance in the metabolites found in the apricot kernel, although only 36 metabolites were identified [26]. Arginine was the most abundant in both hybrid and plum kernels, while serine had the highest value in the apricot kernel. Arginine is important to human nutrition since research has shown that it can increase lipolytic enzymes' activity and decrease insulin resistance [27]. Serine is also reported as exceptionally important, especially being a substrate for glucose and protein synthesis and building of phospholipids [28]. Mallic and 2-aminoadipic acids were the organic acids with the highest values in the studied kernels. The plum–apricot hybrid had the most malic acid, while the apricot had the most 2-aminoadipic acid. Aminoadipic acid is an object of scientific research due to its recent identification as a biomarker of insulin resistance and obesity [29]. Mallic acid is commonly found in fruits, and it is reported to be an antimicrobial agent combined with citric acid [30]. The current results show that organic acids also accumulate in kernels. Sorbitol had the highest values from the group of sugar acids and alcohols. The sorbitol pathway is a two-step one, where in the first step glucose is converted into sorbitol, and then in the second step sorbitol is converted into fructose [31]. The plum–apricot hybrid's kernel had lower quantities of sorbitol compared to its parents but had accumulated more glucose and fructose as isomers. Sorbitol is an alternative sweetener that is widespread in some *Prunus* spp. [32]. Sucrose and glucose and fructose isomers were predominant in the plum–apricot hybrid kernels. The amount of sucrose in the apricot kernel was 50% higher compared to the hybrid. Kernels from Turkish apricots were also high in fructose, glucose, sucrose, and maltose [33]. Protocatechuic acid was the dominant one in the plum–apricot hybrid kernels, while *trans-p*-coumaric acid had its highest values in both its parents. Protocatechuic acid is reported to possess an assortment of biological activities, i.e., antibacterial, antiviral, anticancer, antiosteoporotic, and antioxidant, among others [34].

The overall distribution of the different classes of metabolites is presented in Figure 1.

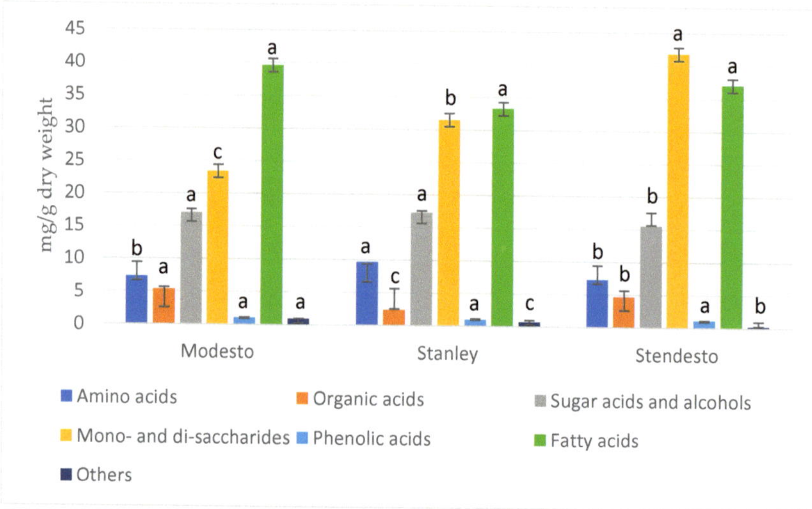

Figure 1. Distribution of metabolites in studied fruit kernels according to their chemical families. Different letters in the same chemical family indicate statistically significant differences ($p < 0.05$) according to ANOVA and the Tukey test.

The plum–apricot hybrid kernels had more mono- and disaccharides and phenolic acids compared to the parents, the same amount of amino acids as apricot kernels, and decreased sugar acids and alcohols. Considering the amount of fatty and organic acids, the amount was more similar to the apricot kernels than to the plum kernels. The distribution of the individual compounds was different, and their amounts contributed differently not only to the specific chemical family but also to the kernel variety. Due to the lack of relevant data, comparison with subject reference to other papers cannot be conducted.

The fatty acids content is presented in Table 2, where a total of twelve compounds were identified, including four fatty alcohols. Although more saturated fatty acids were discovered, their amount did not exceed the value of unsaturated fatty acids. The polyunsaturated fatty acids were more frequent, but the amount of monounsaturated fatty acids was greater. The saturated/unsaturated ratios in the studied kernels are as follows: 0.93 ("Modesto"), 1.01 ("Stanley"), and 0.96 ("Stendesto").

Table 2. Fatty acids (mg/g dry weight) identified in studied kernels analyzed by HS-SPME-GC–MS.

RI	Fatty Acids	"Modesto"	"Stanley"	"Stendesto"
1519	Lauric acid	3.93 ± 1.12^{a}	3.16 ± 1.12^{a}	2.67 ± 1.12^{a}
1572	Dodecanol	1.59 ± 0.45^{a}	1.04 ± 0.45^{a}	0.97 ± 0.45^{a}
1725	Mirystic acid	8.75 ± 2.51^{a}	8.00 ± 2.51^{a}	8.66 ± 2.51^{a}
1874	Tetradecanol	0.33 ± 0.09^{a}	0.21 ± 0.09^{a}	0.24 ± 0.09^{a}
1920	Palmitic acid	1.16 ± 0.33^{a}	0.94 ± 0.33^{a}	1.07 ± 0.33^{a}
1943	Hexadecanol	0.48 ± 0.14^{a}	0.34 ± 0.14^{a}	0.64 ± 0.14^{a}
2094	Linoleic acid	7.06 ± 2.02^{a}	4.27 ± 2.02^{a}	5.08 ± 2.02^{a}
2101	Oleic acid	12.68 ± 3.63^{a}	11.00 ± 3.63^{a}	13.39 ± 3.63^{a}
2106	Linolenic acid	0.78 ± 0.22^{ab}	1.24 ± 0.22^{a}	0.38 ± 0.22^{b}
2128	Stearic acid	0.94 ± 0.27^{b}	1.41 ± 0.27^{ab}	1.71 ± 0.27^{a}
2157	Octadecanol	0.89 ± 0.25^{a}	1.12 ± 0.25^{a}	1.44 ± 0.25^{a}
2311	Eicosanoic acid	1.00 ± 0.29^{a}	0.52 ± 0.29^{a}	0.79 ± 0.29^{a}

RI—retention index. Different letters in the same row indicate statistically significant differences ($p < 0.05$) according to ANOVA and the Tukey test.

Myristic, linoleic, and oleic acids were the predominant ones. Linoleic and oleic acids have been reported as promising anti-mycobacterial agents with high antioxidant potential [35]. It can be seen that the hybrid inherited the presence of all the identified fatty acids from its parents. Only the amounts of dodecanol and linoleic acids in the plum–apricot hybrid's kernels were less than identified in both plum and apricot. On the other hand, the plum–apricot kernel accumulated more octadecanol, oleic, and stearic acids than both its parents. Other papers investigated the composition of the oil from plum and apricot kernels, where they identified ten fatty acids with the prevalence of oleic and linoleic acids [36], demonstrating consistency with the current results.

The investigated volatile compounds are presented in Table 3. The HS-SPME-GC–MS analysis revealed the existence of thirty-five compounds profiled as aldehydes, alcohols, ketones, furans, acids, esters, and alkanes. While most metabolites were present in all the samples, alkanes were only identified in the apricot kernel. This hinted that the volatile profile of the hybrid kernel was more similar to the plum than to the apricot in terms of class distinction.

Table 3. Identified volatile compounds (% of total ion current) in studied kernels analyzed by HS-SPME-GC–MS.

RI	Name/Class	Modesto	Stanley	Stendesto
		Aldehydes		
566	2-Methylpropanal	5.43 ± 0.91^{a}	2.50 ± 0.42^{b}	6.34 ± 1.07^{a}
595	n-Butanal	1.52 ± 0.26^{ab}	1.94 ± 0.33^{a}	1.22 ± 0.20^{b}
653	3-Methylbutanal	0.61 ± 0.10^{a}	0.53 ± 0.09^{a}	0.49 ± 0.08^{a}
667	2-Methylbutanal	2.98 ± 0.50^{a}	1.33 ± 0.22^{a}	2.38 ± 0.40^{a}
698	Pentanal	4.41 ± 0.74^{b}	8.12 ± 1.36^{a}	5.53 ± 0.93^{ab}
752	(E)-2-Pentenal	0.88 ± 0.15^{a}	0.27 ± 0.05^{a}	0.70 ± 0.12^{a}
792	n-Hexanal	9.74 ± 1.64^{b}	18.61 ± 3.13^{a}	14.79 ± 2.49^{ab}
830	2-Furfural	2.61 ± 0.44^{a}	3.40 ± 0.57^{a}	2.09 ± 0.35^{a}
902	Heptanal	1.46 ± 0.25^{b}	1.67 ± 0.28^{a}	1.17 ± 0.20^{b}
961	(E)-2-Heptenal	6.15 ± 1.03^{a}	10.19 ± 1.71^{a}	4.92 ± 0.83^{a}
975	Benzaldehyde	26.99 ± 4.53^{b}	20.82 ± 3.50^{a}	22.59 ± 3.79^{b}
1011	n-Octanal	1.37 ± 0.23^{b}	2.29 ± 0.39^{a}	1.10 ± 0.18^{ab}
1073	(E)-2-Octenal	3.18 ± 0.53^{a}	5.34 ± 0.90^{a}	4.55 ± 0.76^{a}
1106	*n*-Nonanal	4.10 ± 0.69^{ab}	2.84 ± 0.48^{a}	3.28 ± 0.55^{b}
1146	(E)-2-Nonenal	0.71 ± 0.12^{b}	0.94 ± 0.16^{b}	0.57 ± 0.10^{a}
1232	(E)-2-Decenal	1.08 ± 0.18^{b}	0.80 ± 0.13^{a}	1.86 ± 0.31^{a}
		Alcohols		
500	Ethanol	0.20 ± 0.03^{b}	1.00 ± 0.17^{ab}	1.25 ± 0.21^{a}
680	1-Butanol	0.64 ± 0.11^{b}	0.95 ± 0.16^{a}	1.19 ± 0.20^{a}
689	1-Penten-3-ol	0.17 ± 0.03^{ab}	0.52 ± 0.09^{b}	0.65 ± 0.11^{a}
770	1-Pentanol	4.25 ± 0.71^{a}	2.66 ± 0.45^{b}	5.33 ± 0.90^{b}
1036	Benzyl alcohol	10.35 ± 1.74^{a}	2.30 ± 0.39^{a}	2.87 ± 0.48^{a}
1173	4-Ethylphenol	0.65 ± 0.11^{a}	0.48 ± 0.08^{a}	0.60 ± 0.10^{a}
		Ketones		
515	2-Propanone	0.95 ± 0.16^{a}	0.44 ± 0.07^{b}	0.75 ± 0.13^{ab}
691	2-Pentanone	0.75 ± 0.13^{a}	0.49 ± 0.08^{a}	0.61 ± 0.10^{a}
892	2-Heptanone	1.58 ± 0.27^{a}	1.86 ± 0.31^{a}	2.38 ± 0.40^{a}
		Furans		
995	2-Pentylfuran	1.33 ± 0.22^{b}	3.89 ± 0.65^{a}	4.86 ± 0.82^{ab}
		Acids		
741	Acetic acid	0.25 ± 0.04^{b}	2.02 ± 0.34^{a}	2.74 ± 0.46^{a}
		Esters		
617	Ethyl acetate	0.35 ± 0.06^{b}	0.57 ± 0.10^{a}	0.12 ± 0.02^{c}
1161	Benzyl acetate	0.49 ± 0.08	ND	0.69 ± 0.12
1175	Ethyl benzoate	0.63 ± 0.11	ND	0.80 ± 0.13

Table 3. *Cont.*

RI	Name/Class	Modesto	Stanley	Stendesto
		Alkanes		
900	Nonane	0.95 ± 0.16	ND	ND
1000	Decane	0.48 ± 0.08	ND	ND
1100	Undecane	0.67 ± 0.11	ND	ND
1200	Dodecane	0.29 ± 0.05	ND	ND
1300	Tridecane	0.73 ± 0.12	ND	ND

RI—retention index; ND—not detected. Different letters in the same row indicate statistically significant differences ($p < 0.05$) according to ANOVA and the Tukey test.

Figure 2 reveals the differences between the hybrid kernels and those of plum and apricot in terms of %TIC (total ion current) predominance and variety dependance.

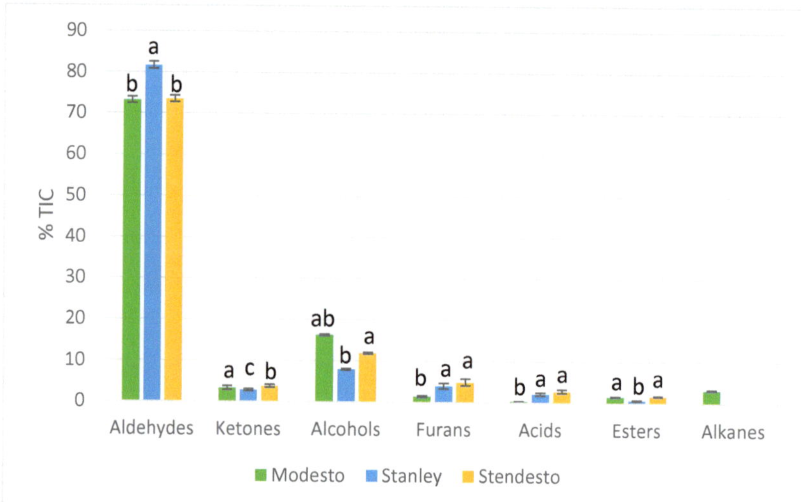

Figure 2. Distribution of volatile compounds according to their chemical families in studied fruit kernels. Different letters in the same chemical family indicate statistically significant differences ($p < 0.05$) according to ANOVA and the Tukey test.

Aldehydes contributed the most to the volatile profile of the three kernels. In terms of aldehydes, ketones, and esters, the hybrid showed more similarity to the apricot, while the %TIC of identified furans, alkanes, and acids corresponded more to the plum. The percentage of TIC of the alcohols was less than the ones in the apricot kernel and more than in the plum kernel.

Aroma has always been important not only to the food industry but also to cosmetology, pharmacology, and others. Kernel oils have a distinct smell that is not connected to the olfactory association of the fruit. The most important from the currently identified aldehydes were 2-methylpropanal ("Stendesto"); pentanal, n-hexanal, (E)-2-heptenal ("Stanley"), and benzaldehyde ("Modesto"). The hybrid kernel had the most n-hexanal and benzaldehyde. The least present from the %TIC were 3-methylbutanal and (E)-2-nonenal. Benzaldehyde is connected to the almond aroma [37]. Esters, acids, and alkanes were identified in relatively small %TIC. However, the hybrid kernel held the highest %TIC for acetic acid and ethyl benzoate. Esters contribute to the typical floral and fruity flavor of products [38]; thus, the hybrid kernel should exhibit more floral and fruity volatiles compared to its parents. Acetic acid, on the other hand, is a major odor-active component identified in fruit vinegars [39].

Benzyl alcohol dominated in the apricot kernel, while 1-pentanol dominated in the hybrid one. Benzyl alcohol is associated with a floral odor and a marzipan-like flavor [40] and 1-pentanol with a fruity odor [41]. 2-heptanone was the major ketone in the hybrid kernel. According to research, it contributes to an oxidative odor [42]. The overall volatile assessment can be linked to the threshold levels the different compounds possess. Aldehydes have lower sensory thresholds compared to alcohols [43]; thus, they might be viewed as the largest contributors to the sensory associations of the three studied kernels. Furan and its derivatives usually occur in heat-processed foods and beverages [44]. Moreover, 2-penthylfuran has been identified in a number of foods, i.e., baby food, deep-fried foods, and fruit juice [45], and 2-penthylfuran is the only furan identified in the three kernels, where the hybrid one held the highest %TIC. This compound is known for its distinct fruity flavor and caramel undertones [46].

Figure 3 is a visual presentation of the odor description of each kernel variety based on the VOCs present in them.

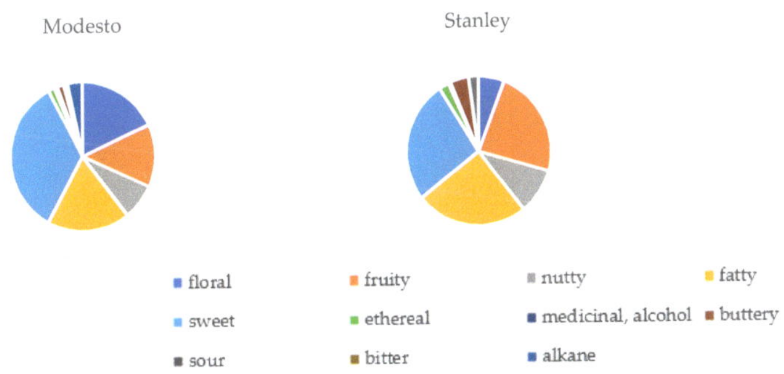

Figure 3. Odor component distribution in studied fruit kernels (according to https://foodb.ca/compounds descriptor (accessed 1 December 2023).

It can be seen that the hybrid kernel had distinct sweet (32%) and fatty (17%) sensory properties, while the plum kernel had a more fruity (23%), sweet (27%), and fatty (25%) odor description. The apricot kernel also presented a sweet (34%), fatty (19%), and floral (18%) profile. The least recognized odors in the apricot kernel were the ethereal and bitter ones, while in the plum and plum–apricot kernels those were the ethereal and sour ones. The nutty odor was evenly distributed in the hybrid and plum kernels, and more distinct in the apricot kernel. In terms of sensory perception, the plum–apricot kernel is more similar to the apricot kernel than to the plum kernel.

This research can be viewed as a pioneering study on the topic of metabolite identification of "Modesto" (apricot), "Stanley" (plum), and "Stendesto" (plum–apricot hybrid) kernels and provide a stepping stone for future evaluations and comparisons.

Principal Component and Hierarchical Cluster Analyses of HS-SPME-GC–MS Data

The chemical composition and volatile content were analyzed using principal component analysis (PCA) and further explored to distinguished separate groups with hierarchical cluster analysis (HCA). As shown in Figure 4A (metabolites), two principal components were generated in the PCA with an eigenvalue greater than 1, accounting for 65.8% (65% for PC 1 and 35% for PC2) of the total variance, whereas, in Figure 4B (VOCs), 59.7% were distributed for PC1 (52.1%) and PC2 (47.9%). Glucaric acid, glucose-1-phosphate, isocitric acid, 2-heptanone, 1-butanol, and acetic acid are positioned most positively, whereas trans-ferulic acid, threonine, (e)-2-heptenal, and 1-pentanol contributed most negatively.

As illustrated in Figure 5, the samples can be divided into two clusters for both metabolites (Figure 5A) and volatile compounds (Figure 5B).

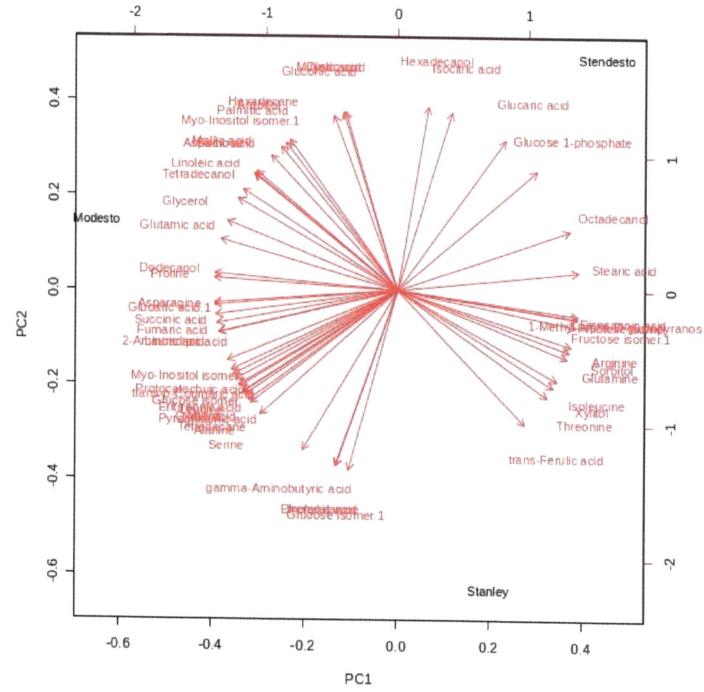

(A)

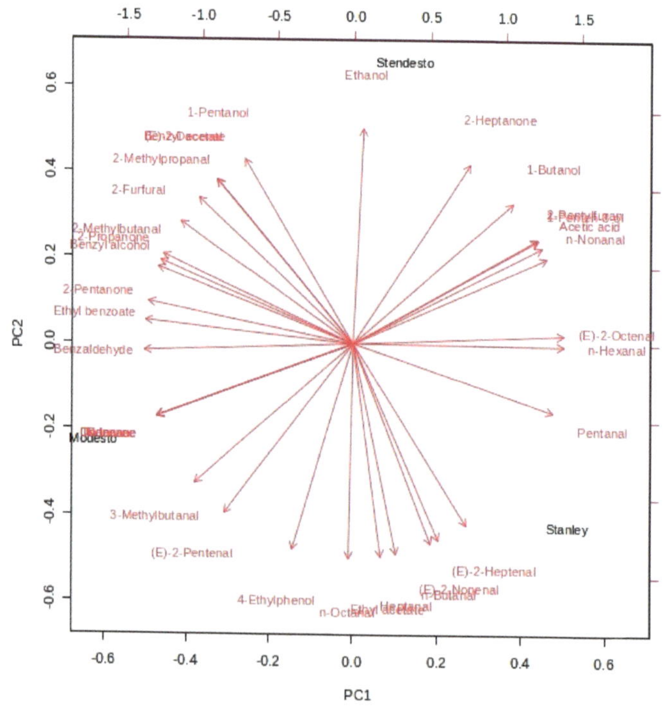

(B)

Figure 4. Principal component analysis of kernel samples. Eigenvector loading values of compounds: (**A**) primary metabolites; (**B**) VOCs.

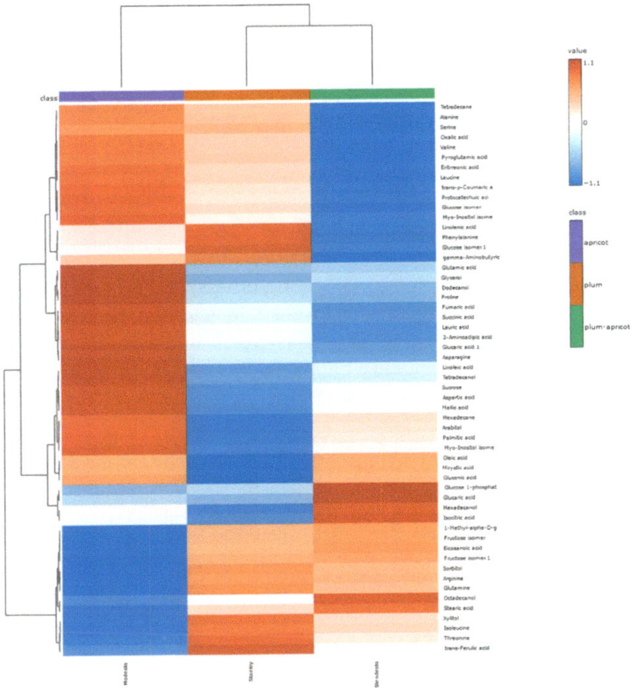

(A)

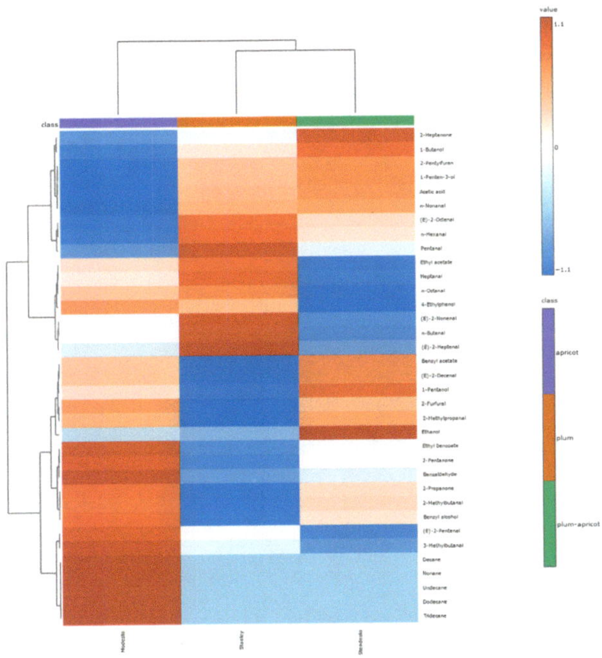

(B)

Figure 5. Clustering results of kernel samples, shown as heatmap: (**A**) primary metabolites; (**B**) VOCs. The values were normalized by $\log_{10}$ transformation.

The apricot kernel is placed separately from the plum and plum–apricot hybrid kernels. The results from the PCA and HCA were useful to preliminarily distinguish the samples. A correlation analysis of the data is presented in Figure 6. It is assumed that the positively correlated metabolite pairs have similar chemical composition, biological function, and homogeneous characteristics [47].

A positive correlation has been established between n-hexanal and pentanal, n-nonanal, acetic acid, 1-butanol, 2-pentylfuran, 1-penten-3-ol, and (E)-2-octenal. Benzaldehyde was positively correlated with sixteen structures, including ethyl benzoate, tridecane, undecane, decane, and benzyl alcohol, among others. Additionally, benzyl alcohol had a positive correlation with fifteen VOCs (2-propanone, 2-furfural, ethyl benzoate, benzyl acetate, and tridecane, among others). Serine and nineteen of the identified metabolites (alanine, valine, leucine, oxalic acid, and fumaric acid, among others) had a positive correlation. Mallic acid was positively correlated with sucrose, aspartic acid, linoleic acid, palmitic acid, oleic acid, and eight others. Sorbitol has a positive correlation with eleven metabolites, including arginine, glutamine, isoleucine, eicosanoic acid, xylitol, threonine, and others. Oleic acid was positively correlated with fifteen metabolites (myristic and gluconic acids having the highest correlation values).

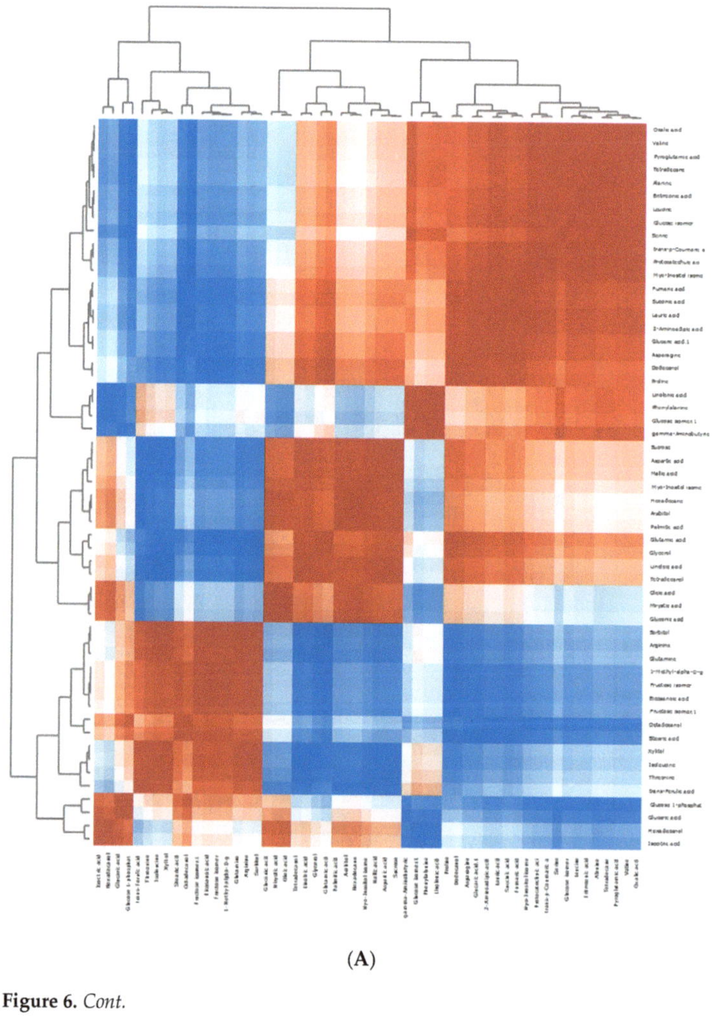

(A)

Figure 6. *Cont.*

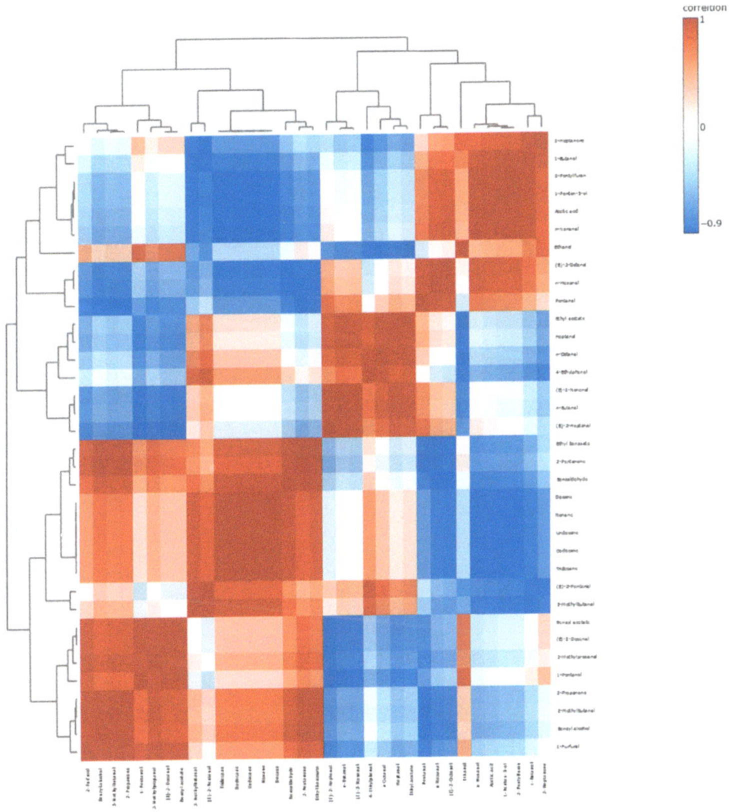

(**B**)

Figure 6. Pearson's correlation heatmaps of the differential compounds in studied kernels: (**A**) primary metabolites; (**B**) VOCs. The values were normalized by $\log_{10}$ transformation.

3. Materials and Methods

3.1. Materials

Three kernels from the following fruit varieties were objects of analysis: "Modesto" (apricot), "Stanley" (plum), and "Stendesto" (plum–apricot hybrid). The kernels were collected from several trees of the same variety. All trees were part of the same plantation. A total of sixty kernels per variety ("Modesto", "Stendesto", and "Stanley") were gathered and thoroughly cleaned from the pulp. All kernels were extracted from ripe fruits. The kernels were left to air-dry, and then were broken into pieces with a Bosch MCM3PM386 robot (Robert Bosch GmbH, Germany) and after that ground with the aid of WMF 0417070011 grinder (Tefal OBH Nordica Group AB, Sweeden). Each kernel variety was placed in a sterile container and kept in a cool, dark, and dry place prior to analysis.

3.2. Headspace Solid-Phase Microextraction (HS-SPME) and Gas Chromatography–Mass Spectrometry Analysis (GC–MS)

A 2 cm SPME fiber assembly Divinylbenzene/Carboxen/Polydimethylsiloxane (DVB/CAR/PDMS, Supelco, Bellefonte, PA, USA) was utilized for headspace sampling.

The HS-SPME extraction technique of the studied kernels followed the description of Uekane et al. [48]. An online integrated sampling procedure was automatically performed with a G1888 Network Headspace Sampler. An Agilent 7890A GC unit coupled to an Agilent 5975C MSD and a DB-5ms (30 m $\times$ 0.25 mm $\times$ 0.25 μm) column were used to analyze the volatile compounds in all samples. The oven temperature program was the following: from 40 °C (hold 1 min) to 250 °C (hold 5 min) at 2 °C/min; carrier gas: helium with flow rate: 1.0 mL/min; transfer line temperature: 270 °C; ion source temperature: 200 °C, EI energy: 70 eV, mass range: 50 to 550 m/z at 1.0 s/decade.

The extraction procedure of the polar and lipid fractions was completed as described by Ivanova et al. [49]: 0.05 g freeze-dried material was mixed with 1.0 mL methanol/water (75:25, *v*/*v*) solution and 50.0 µL of each internal standard (nonadecanoic acid methyl ester, ribitol, each in concentration 1.0 mg/mL for the quantification of metabolites of fractions A, and B, respectively), followed by heating at 70 °C for 1 h in a laboratory thermomixer (Analytik Jena AG). The solution, cooling to room temperature, was subjected to the following procedure: 500.0 mL chloroform and 200.0 mL water were added, and then the mixture was centrifuged (5 min/22 °C/13,000 rpm). The lower phase was designed for the analysis of non-polar substances (fraction A), whereas the upper phase for the polar constituents (fraction B). The two phases obtained were vacuum-dried in a centrifugal vacuum concentrator (Labconco Centrivap) at 40 °C. To the dried residue of fraction "A", 1.0 mL 2% H_2SO_4 in methanol were added and the mixture was heated on Thermo-Shaker TS-100 (1 h/96 °C/300 rpm). After cooling, the solution was extracted with n-hexane (3 × 500.0 mL). Combined organic layers were vacuum-dried in a centrifugal vacuum concentrator (Labconco Centrivap) at 40 °C.

Prior to the gas chromatography–mass spectrometry (GC–MS) analysis, fractions "A" and "B" were derivatized by the following procedures: 100.0 µL pyridine and 100.0 µL BSTFA were added to the dried residue (fraction "A"), then heated on Thermoshaker, Analytik Jena AG, Germany (45 min/70 °C/300 rpm). 1.0 µL from the solution was injected into the GC–MS.; 300.0 µL solution of methoxyamine hydrochloride (20.0 mg/mL in pyridine) was added to dried residue (fraction "B"), and the mixture was heated on Thermo-Shaker TS-100 (1 h/70 °C/300 rpm). After cooling, 100.0 µL BSTFA were added to the mixture then heated on Thermoshaker, Analytik Jena AG, Germany (40 min/70 °C/300 rpm), and 1.0 µL from the solution was injected into the GC–MS system.

The 2.73 AMDIS software (Automated Mass Spectral Deconvolution and Identification System, NIST, Gaithersburg, MD, USA) assisted in the reading of the mass spectra and the metabolite identification. The separated compounds were compared to the GC–MS spectra and retention indices (RI) of reference compounds in the Golm Metabolome Database (http://csbdb.mpimp-golm.mpg.de/csbdb/gmd/gmd.html, accessed on 1 December 2023) and NIST'08 database (NIST Mass Spectral Database, PC-Version 5.0, 2008 from National Institute of Standards and Technology, Gaithersburg, MD, USA). The 2.73 AMDIS software recorded the RIs of the compounds with a standard n-hydrocarbon calibration mixture ($C_{8–36}$, Restek, Teknokroma, Spain). Analyses were triplicated for each kernel variety.

3.3. Statistical Analysis

MS Excel software 365 was used for data analysis. Results are presented as mean ± SD (standard deviation). Additional statistical analyses of the data were presented using one-way ANOVA and a Tukey–Kramer post hoc test (α = 0.05), as described by Assaad et al. [50]. PCA and HCA of GC–MS data were conducted using MetaboAnalyst, a web-based platform (www.metaboanalyst.ca, accessed on 15 December 2023). The concentrations of the identified compounds were employed for PCA. All zero values were replaced with a value (1/2 of the minimum positive values in the original data) assumed to be the detection limit. PCA (95% confidence level) was employed to calculate the eigenvector loading values and to identify the major statistically different components among the samples. The GC–MS data were also subjected to HCA, which produced a Ward dendrogram of hierarchical clustering and Euclidean distance measurement between the analyzed samples. The values were normalized by $\log_{10}$ transformation.

4. Conclusions

No data regarding the primary metabolites and VOCs of kernels from the "Modesto" (apricot), "Stanley" (plum), and "Stendesto" (plum–apricot) varieties were present in literature, which makes this a first comprehensive evaluation regarding this important byproduct. In total, fifty-five metabolites were identified belonging to the following chemical groups: phenolic acids, hydrocarbons, organic acids, fatty acids, sugar acids and alcohols, mono- and disaccharides, and amino acids. The most abundant were the fatty acids, sugar acids and alcohols, and mono- and disaccharides. The hybrid kernel generally inherited all the metabolites present in the parental kernels. Thirty-five VOCs were identified from the three samples, with aldehydes contributing most. Considering the VOCs, the hybrid kernel had more resemblance to the plum one, bearing that alkanes were only identified in the apricot kernel.

The applied PCA placed the plum and plum–apricot kernels in the same group, leaving the apricot kernel in a separate group. The obtained results can successfully be used as a reference and stepping stone for future analyses. Focusing attention on kernels as potential nutritional and functional sources will definitely aid with the utilization of this by-product.

Author Contributions: Conceptualization, D.M. and A.P.; methodology, I.D.; software, I.D.; validation, D.M., A.P. and I.D.; formal analysis, I.D.; investigation, D.M., S.P. and A.P.; resources, S.P.; data curation, I.D.; writing—original draft preparation, A.P., D.M., I.D. and S.P.; writing—review and editing, D.M., A.P., I.D. and S.P.; visualization, A.P.; supervision, D.M.; project administration, A.P.; funding acquisition, A.P. All authors have read and agreed to the published version of the manuscript.

Funding: This research and the APC were funded by the Bulgarian National Science Fund, project no. КП-06-Н67/2.

Institutional Review Board Statement: Not applicable.

Informed Consent Statement: Not applicable.

Data Availability Statement: The data presented in this study are available on request from the corresponding author. The data are not publicly available due to privacy

Acknowledgments: The authors would like to acknowledge Argir Zhivondov for actively working on expanding plumcot varieties in Bulgaria, and registering the "Stendesto".

Conflicts of Interest: The authors declare no conflicts of interest. The funders had no role in the design of the study; in the collection, analyses, or interpretation of data; in the writing of the manuscript; or in the decision to publish the results.

References

1. Duft, R.G.; Castro, A.; Chacon-Mikahil, M.P.T.; Cavaglieri, C.R. Metabolomics and Exercise: Possibilities and Perspectives. *Mot. Rev. Educ. Física* **2017**, *23*, e101634. [CrossRef]
2. Alseekh, S.; Aharoni, A.; Brotman, Y.; Contrepois, K.; D'Auria, J.; Ewald, J.; Ewald, J.C.; Fraser, P.D.; Giavalisco, P.; Hall, R.D.; et al. Mass Spectrometry-Based Metabolomics: A Guide for Annotation, Quantification and Best Reporting Practices. *Nat. Methods* **2021**, *18*, 747–756. [CrossRef] [PubMed]
3. Dadwal, V.; Joshi, R.; Gupta, M. A Comparative Metabolomic Investigation in Fruit Sections of *Citrus medica* L. and *Citrus maxima* L. Detecting Potential Bioactive Metabolites Using UHPLC-QTOF-IMS. *Food Res. Int.* **2022**, *157*, 111486. [CrossRef] [PubMed]
4. Lippa, K.A.; Aristizabal-Henao, J.J.; Beger, R.D.; Bowden, J.A.; Broeckling, C.; Beecher, C.; Clay Davis, W.; Dunn, W.B.; Flores, R.; Goodacre, R.; et al. Reference Materials for MS-Based Untargeted Metabolomics and Lipidomics: A Review by the Metabolomics Quality Assurance and Quality Control Consortium (MQACC). *Metabolomics* **2022**, *18*, 24. [CrossRef] [PubMed]
5. Alsuhaymi, S.; Singh, U.; Al-Younis, I.; Kharbatia, N.M.; Haneef, A.; Chandra, K.; Dhahri, M.; Assiri, M.A.; Emwas, A.H.; Jaremko, M. Untargeted Metabolomics Analysis of Four Date Palm (*Phoenix dactylifera* L.) Cultivars Using MS and NMR. *Nat. Prod. Bioprospect.* **2023**, *13*, 44. [CrossRef]
6. Saini, M.K.; Capalash, N.; Varghese, E.; Kaur, C.; Singh, S.P. A Targeted Metabolomics Approach to Study Secondary Metabolites and Antioxidant Activity in 'Kinnow Mandarin' during Advanced Fruit Maturity. *Foods* **2022**, *11*, 1410. [CrossRef] [PubMed]

7. Chen, X.; Fedrizzi, B.; Kilmartin, P.A.; Quek, S.Y. Development of Volatile Organic Compounds and Their Glycosylated Precursors in Tamarillo (*Solanum betaceum* Cav.) during Fruit Ripening: A Prediction of Biochemical Pathway. *Food Chem.* **2021**, *339*, 128046. [CrossRef] [PubMed]
8. Gu, I.; Howard, L.; Lee, S.O. Volatiles in Berries: Biosynthesis, Composition, Bioavailability, and Health Benefits. *Appl. Sci.* **2022**, *12*, 10238. [CrossRef]
9. Train, A. Critical Comparison of Total Vaporization- Solid Phase Microextraction vs Headspace- Solid Phase Microextraction. Ph.D. Thesis, Indiana University, Bloomington, IN, USA, 2021. [CrossRef]
10. Yang, S.; Hao, N.; Meng, Z.; Li, Y.; Zhao, Z. Identification, Comparison and Classification of Volatile Compounds in Peels of 40 Apple Cultivars by Hs–Spme with Gc–Ms. *Foods* **2021**, *10*, 1051. [CrossRef]
11. Shirahata, T.; Kanazawa, A.; Uematsu, M.; Fuchino, H.; Kawano, N.; Kawahara, N.; Yoshimatsu, K.; Hanawa, T.; Odaguchi, H.; Kobayashi, Y. Near-Infrared Metabolic Profiling for Discrimination of Apricot and Peach Kernels. *Chem. Pharm. Bull.* **2022**, *70*, 863–867. [CrossRef]
12. Makovi, C.M.; Parker, C.H.; Zhang, K. Determination of Amygdalin in Apricot Kernels and Almonds Using LC-MS/MS. *J. AOAC Int.* **2023**, *106*, 457–463. [CrossRef]
13. Akhone, M.A.; Bains, A.; Tosif, M.M.; Chawla, P.; Fogarasi, M.; Fogarasi, S. Apricot Kernel: Bioactivity, Characterization, Applications, and Health Attributes. *Foods* **2022**, *11*, 2184. [CrossRef]
14. Leong, Y.K.; Chang, J.S. Valorization of Fruit Wastes for Circular Bioeconomy: Current Advances, Challenges, and Opportunities. *Bioresour. Technol.* **2022**, *359*, 127459. [CrossRef]
15. Teslić, N.; Kojić, J.; Đermanović, B.; Šarić, L.; Maravić, N.; Pestorić, M.; Šarić, B. Sour Cherry Pomace Valorization as a Bakery Fruit Filling: Chemical Composition, Bioactivity, Quality and Sensory Properties. *Antioxidants* **2023**, *12*, 1234. [CrossRef]
16. Jaafar, H.J. Effects of Apricot and Apricot Kernels on Human Health and Nutrition: A Review of Recent Human Research. *Tech. BioChemMed* **2021**, *2*, 139–162. [CrossRef]
17. Abd, M.; Sorour, E.; Mehanni, A.-H.E.-S.; Mahmoud, S.; Mustafa, H.; Mustafa, A.; Mustafa, M.A.; Hussien, S.M. Chemical Composition and Functional Properties of Some Fruit Seed Kernel Flours. *J. Sohag Agrisci. (JSAS)* **2021**, *6*, 184–191. [CrossRef]
18. Lebaka, V.R.; Wee, Y.J.; Ye, W.; Korivi, M. Nutritional Composition and Bioactive Compounds in Three Different Parts of Mango Fruit. *Int. J. Env. Res. Public Health* **2021**, *18*, 741. [CrossRef]
19. Lara, M.V.; Bonghi, C.; Famiani, F.; Vizzotto, G.; Walker, R.P.; Drincovich, M.F. Stone Fruit as Biofactories of Phytochemicals With Potential Roles in Human Nutrition and Health. *Front. Plant Sci.* **2020**, *11*, 562252. [CrossRef]
20. Fatima, E.A.; Moha, T.; Said, W.; Abdelilah, M.; Mohammed, R. Use of Metabolomics Data Analysis to Identify Fruit Quality Markers Enhanced by the Application of an Aminopolysaccharide. *RSC Adv.* **2021**, *11*, 35514. [CrossRef]
21. Muhie, S.H. Novel Approaches and Practices to Sustainable Agriculture. *J. Agric. Food Res.* **2022**, *10*, 100446. [CrossRef]
22. Popova, A.; Mihaylova, D.; Pandova, S.; Doykina, P. Research-Gap-Spotting in Plum–Apricot Hybrids—Bioactive Compounds, Antioxidant Activities, and Health Beneficial Properties. *Horticulturae* **2023**, *9*, 584. [CrossRef]
23. Aranha, B.C.; Hoffmann, J.F.; Barbieri, R.L.; Rombaldi, C.V.; Chaves, F.C. Untargeted Metabolomic Analysis of *Capsicum* Spp. by GC-MS. *Phytochem. Anal.* **2017**, *28*, 439–447. [CrossRef]
24. Campos, D.A.; Gómez-García, R.; Vilas-Boas, A.A.; Madureira, A.R.; Pintado, M.M. Management of Fruit Industrial By-Products—A Case Study on Circular Economy Approach. *Molecules* **2020**, *25*, 320. [CrossRef]
25. Sulaiman, M.A.; Yusoff, F.M.; Kamarudin, M.S.; Amin, S.M.N.; Kawata, Y. Fruit Wastes Improved the Growth and Health of Hybrid Red Tilapia *Oreochromis* sp. and Malaysian Mahseer, *Tor tambroides* (Bleeker, 1854). *Aquac. Rep.* **2022**, *24*, 101177. [CrossRef]
26. Farag, M.A.; Ramadan, N.S.; Shorbagi, M.; Farag, N.; Gad, H.A. Profiling of Primary Metabolites and Volatiles in Apricot (*Prunus armeniaca* L.) Seed Kernels and Fruits in the Context of Its Different Cultivars and Soil Type as Analyzed Using Chemometric Tools. *Foods* **2022**, *11*, 1339. [CrossRef]
27. McNeal, C.J.; Meininger, C.J.; Reddy, D.; Wilborn, C.D.; Wu, G. Safety and Effectiveness of Arginine in Adults. *J. Nutr.* **2016**, *146*, 2587S–2593S. [CrossRef]
28. Holeček, M. Serine Metabolism in Health and Disease and as a Conditionally Essential Amino Acid. *Nutrients* **2022**, *14*, 1987. [CrossRef]
29. Lee, H.J.; Jang, H.B.; Kim, W.H.; Park, K.J.; Kim, K.Y.; Park, S.I.; Lee, H.J. 2-Aminoadipic Acid (2-AAA) as a Potential Biomarker for Insulin Resistance in Childhood Obesity. *Sci. Rep.* **2019**, *9*, 13610. [CrossRef]
30. Baker, B.P.; Grant, J.A. *Malic Acid Profile Active Ingredient Eligible for Minimum Risk Pesticide Use*; New York State Integrated Pest Management; Cornwell University: Geneva, NY, USA, 2018.
31. Srikanth, K.K.; Orrick, J.A. *Biochemistry, Polyol Or Sorbitol Pathways*; StatPearls: Treasure Island, FL, USA, 2022.
32. Sheet, B.S.; Artik, N.; Ayed, M.A.; Abdulaziz, O.F. Some Alternative Sweeteners (Xylitol, Sorbitol, Sucralose and Stevia): Review. *Karaelmas Fen Mühendislik Derg./Karaelmas Sci. Eng. J.* **2014**, *4*, 63–70. [CrossRef]
33. Özcan, M. Short Communications Composition of Some Apricot (*Prunus armeniaca* L.) Kernels Grown in Turkey. *Acta Aliment.* **2000**, *29*, 289–293. [CrossRef]
34. Song, J.; He, Y.; Luo, C.; Feng, B.; Ran, F.; Xu, H.; Ci, Z.; Xu, R.; Han, L.; Zhang, D. New Progress in the Pharmacology of Protocatechuic Acid: A Compound Ingested in Daily Foods and Herbs Frequently and Heavily. *Pharmacol. Res.* **2020**, *161*, 105109. [CrossRef]

35. Dutta Kalita, R.; Hussain, I.; Deka, R.C.; Kumar Buragohain, A. Antimycobacterial Activity of Linoleic Acid and Oleic Acid Obtained from the Hexane Extract of the Seeds of *Mesua ferrea* L. and Their in Silico Investigation. *Indian J. Nat. Prod. Resour.* **2018**, *9*, 132–142.
36. Rodríguez-Blázquez, S.; Gómez-Mejía, E.; Rosales-Conrado, N.; León-González, M.E.; García-Sánchez, B.; Miranda, R. Valorization of Prunus Seed Oils: Fatty Acids Composition and Oxidative Stability. *Molecules* **2023**, *28*, 7045. [CrossRef]
37. Sun, Z.; Lyu, Q.; Chen, L.; Zhuang, K.; Wang, G.; Ding, W.; Wang, Y.; Chen, X. An HS-GC-IMS Analysis of Volatile Flavor Compounds in Brown Rice Flour and Brown Rice Noodles Produced Using Different Methods. *LWT* **2022**, *161*, 113358. [CrossRef]
38. Pino, J.A.; Espinosa, S.; Duarte, C. Characterization of Odor-Active Volatile Compounds of Jambolan [*Syzgium cumini* (L.) Skeels] Wine. *J. Food Sci. Technol.* **2022**, *59*, 1529–1537. [CrossRef]
39. Pino, J.A.; Pérez, J.C. Analysis of the Volatile Compounds in Acerola Vinegar by Solid-Phase Microextraction Techniques Coupled with Gas Chromatography-Mass Spectrometry. *Rev. CENIC. Cienc. Químicas* **2019**, *50*, 33–40.
40. Vichi, S.; Mayer, M.N.; León-Cárdenas, M.G.; Quintanilla-Casas, B.; Tres, A.; Guardiola, F.; Batlle, I.; Romero, A. Chemical Markers to Distinguish the Homo- and Heterozygous Bitter Genotype in Sweet Almond Kernels. *Foods* **2020**, *9*, 747. [CrossRef]
41. Franklin, L.M.; Mitchell, A.E. Review of the Sensory and Chemical Characteristics of Almond (*Prunus dulcis*) Flavor. *J. Agric. Food Chem.* **2019**, *67*, 2743–2753. [CrossRef]
42. Xi, Y.; Ikram, S.; Zhao, T.; Shao, Y.; Liu, R.; Song, F.; Sun, B.; Ai, N. 2-Heptanone, 2-Nonanone and 2-Undecanone Confer Oxidation off-Flavor in Cow Milk Storage. *J. Dairy Sci.* **2023**, *106*, 8538–8550. [CrossRef]
43. Manzano, P.; Diego, J.C.; Bernal, J.L.; Nozal, M.J.; Bernal, J. Comprehensive Two-Dimensional Gas Chromatography Coupled with Static Headspace Sampling to Analyze Volatile Compounds: Application to Almonds. *J. Sep. Sci.* **2014**, *37*, 675–683. [CrossRef]
44. Condurso, C.; Cincotta, F.; Verzera, A. Determination of Furan and Furan Derivatives in Baby Food. *Food Chem.* **2018**, *250*, 155–161. [CrossRef]
45. Tsao, W.X.; Chen, B.H.; Lin, P.; You, S.H.; Kao, T.H. Analysis of Furan and Its Derivatives in Food Matrices Using Solid Phase Extraction Coupled with Gas Chromatography-Tandem Mass Spectrometry. *Molecules* **2023**, *28*, 1639. [CrossRef] [PubMed]
46. Frankel, E.N. Frying Fats. In *Lipid Oxidation I*; Elsevier: Amsterdam, The Netherlands, 2012; pp. 355–389. [CrossRef]
47. Zheng, C.; Yang, S.; Huang, D.; Mao, D.; Chen, J.; Zhang, C.; Kong, W.; Liu, X.; Xu, Y.; Wu, Y.; et al. GC-MS Combined with Multivariate Analysis for the Determination of the Geographical Origin of *Elsholtzia rugulosa* Hemsl. in Yunnan Province. *RSC Adv.* **2022**, *12*, 21287–21296. [CrossRef]
48. Uekane, T.M.; Nicolotti, L.; Griglione, A.; Bizzo, H.R.; Rubiolo, P.; Bicchi, C.; Rocha-Leão, M.H.M.; Rezende, C.M. Studies on the Volatile Fraction Composition of Three Native Amazonian-Brazilian Fruits: Murici (*Byrsonima crassifolia* L., Malpighiaceae), Bacuri (*Platonia insignis* M., Clusiaceae), and Sapodilla (*Manilkara sapota* L., Sapotaceae). *Food Chem.* **2017**, *219*, 13–22. [CrossRef]
49. Ivanova, T.; Dincheva, I.; Badjakov, I.; Iantcheva, A. Transcriptional and Metabolic Profiling of Arabidopsis Thaliana Transgenic Plants Expressing Histone Acetyltransferase HAC1 upon the Application of Abiotic Stress—Salt and Low Temperature. *Metabolites* **2023**, *13*, 994. [CrossRef] [PubMed]
50. Assaad, H.I.; Zhou, L.; Carroll, R.J.; Wu, G. Rapid Publication-Ready MS-Word Tables for One-Way ANOVA. *Springerplus* **2014**, *3*, 474. [CrossRef]

horticulturae

MDPI

Article

Comprehensive Assessment of Coffee Varieties (*Coffea arabica* L.; *Coffea canephora* L.) from Coastal, Andean, and Amazonian Regions of Ecuador; A Holistic Evaluation of Metabolism, Antioxidant Capacity and Sensory Attributes

Raluca A. Mihai [1,*], Diana C. Ortiz-Pillajo [1], Karoline M. Iturralde-Proaño [1], Mónica Y. Vinueza-Pullotasig [1], Leonardo A. Sisa-Tolagasí [1], Mary L. Villares-Ledesma [1], Erly J. Melo-Heras [1], Nelson S. Cubi-Insuaste [1] and Rodica D. Catana [2]

[1] Army Scientific and Technological Research Center—CICTE, Department of Life Science and Agriculture, Universidad de Las Fuerzas Armadas—ESPE, Av. General Rumiñahui s/n y, Sangolqui 171103, Ecuador; dcortiz1@espe.edu.ec (D.C.O.-P.); kmiturralde@espe.edu.ec (K.M.I.-P.); myvinueza@espe.edu.ec (M.Y.V.-P.); lasisa1@espe.edu.ec (L.A.S.-T.); mlvillares@espe.edu.ec (M.L.V.-L.); ejmelo@espe.edu.ec (E.J.M.-H.); nscubi@espe.edu.ec (N.S.C.-I.)

[2] Institute of Biology Bucharest of Romanian Academy, 296 Splaiul Independentei, 060031 Bucharest, Romania; rodica.catana@ibiol.ro

* Correspondence: rmihai@espe.edu.ec

Abstract: In Ecuador, the cultivation of two main coffee species, *Coffea arabica* L. and *Coffea canephora* L., holds significant economic, environmental, social, and public health importance. *C. arabica* displays wide adaptability to diverse growing conditions, while *C. canephora* exhibits less versatility in adaptation but is superior in metabolite production in the ripe fruits (with the potential to double caffeine content). Our hypothesis revolves around the differences in the production of secondary metabolites, antioxidant capacity and sensory attributes based on the environmental conditions of the two studies species cultivated in Ecuador. The assessment of the metabolic composition of high-altitude coffee grown in Ecuador involved the determination of secondary metabolites and quantification of the antioxidant capacity through the 2,2-diphenyl-1-picrylhydrazyl assay, 2,2′-azino-bis (3-ethylbenzothiazoline-6-sulfonic acid) quenching assay, and ferric reducing antioxidant power assay. In the case of *C. arabica*, a high positive correlation was observed for total phenolic content (TPC) (4.188 ± 0.029 mg gallic acid equivalent (GAE)/g dry weight (dw)) and total flavonoid content (TFC) (0.442 ± 0.001 mg quercetin (QE)/g dw) with the antioxidant activity determined through ABTS free-radical-scavenging activity (23.179 ± 1.802 µmol Trolox (TEAC)/g dw) (R = 0.68), a medium correlation with DPPH$^{\bullet}$ radical-scavenging activity (65.875 ± 1.129 µmol TEAC/g dw) (R = 0.57), and a low correlation with ferric reducing antioxidant power assay ((100.164 ± 0.332 µmol Fe^{2+}/g dw) (R = 0.27). A high correlation (R > 90) was observed for the values evaluated in the case of *C. canephora*. The caffeine content was high in *C. arabica* beans from Los Ríos province and in *C. canephora* beans from Loja.

Keywords: biological activity; flavor; metabolites; optimal climatic conditions; coffee varieties

Citation: Mihai, R.A.; Ortiz-Pillajo, D.C.; Iturralde-Proaño, K.M.; Vinueza-Pullotasig, M.Y.; Sisa-Tolagasí, L.A.; Villares-Ledesma, M.L.; Melo-Heras, E.J.; Cubi-Insuaste, N.S.; Catana, R.D. Comprehensive Assessment of Coffee Varieties (*Coffea arabica* L.; *Coffea canephora* L.) from Coastal, Andean, and Amazonian Regions of Ecuador; A Holistic Evaluation of Metabolism, Antioxidant Capacity and Sensory Attributes. *Horticulturae* **2024**, *10*, 200. https://doi.org/10.3390/horticulturae10030200

Academic Editors: Paola Sánchez-Bravo and Luis Noguera-Artiaga

Received: 10 January 2024
Revised: 14 February 2024
Accepted: 14 February 2024
Published: 21 February 2024

1. Introduction

Ecuador stands out as one of the most biodiverse countries globally, benefiting from optimal climatic conditions that support a wide array of crops. The Coastal region experiences a tropical, arid, and dry humid climate, with the rainy season occurring in April and May, and a dry season (characterized by low rainfall) from June to November/December. The average temperature ranges from 22 to 25 °C, and rainfall levels fluctuate between 60 and 2000 mm, influenced by the Humboldt current [1]. Moving to the Sierra region, extreme rainfall is observed in March, April, and October due to the sun's position. The average annual rainfall in this area ranges from 700 to 1200 mm, and the temperature, varying between 13 and 16 °C, is contingent on altitude (ranging from 500 to 6000 m above sea level (m.a.s.l.). In the eastern region of Ecuador, the average altitude is 1000 m.a.s.l., with consistent humidity throughout the year, as the annual rainfall hovers between 2000 and 3000 mm.

The cultivation of coffee in Ecuador holds significant economic importance, contributing substantially to income generation and foreign exchange. Moreover, it plays a crucial role in the social fabric, involving diverse communities and ethnic groups across various provinces in Ecuador. From an environmental perspective, coffee cultivation is predominantly practiced within agroforestry systems, thereby aiding in preserving native flora and fauna within diverse soil and climatic conditions. Additionally, the institutional and health aspects of coffee production are noteworthy, as coffee consumption has been linked to a reduced incidence of health risks such as type 2 diabetes, liver damage, and neurodegenerative diseases [2,3]. According to Duicela-Guambi et al. [4], 68% of coffee plantations in Ecuador yield arabica coffee, while the remaining 32% produce robusta coffee.

Coffea is indeed a diverse genus, comprising around 80 different species, with *C. arabica* L. and *C. canephora* L. being the most economically significant. Species within the *Coffea* genus typically manifest as trees or saplings with dense, horizontally spreading branches [5]. Their hermaphroditic flowers are arranged in intricate inflorescences, featuring generally white corollas, occasionally pale pink, with exposed anthers and an elongated style [6]. The fruits take the form of drupe-like berries, exhibiting a color range from yellowish to reddish at maturity, with the coffee kernel or seed covered by a double layer and characterized by a ventral invagination [7]. The leaves of *Coffea* plants are shiny and dark green, with visible veining, and grow in an opposite arrangement with the next pair of leaves [8].

In Ecuador, *C. arabica* is indeed the most economically significant coffee species. It is a seasonal crop that exhibits wide adaptability, thriving in altitudes ranging between 1500 and 2000 m.a.s.l. and temperatures of 17–23 °C, while also demonstrating resistance to drought [9]. The ideal soil conditions for coffee cultivation in Ecuador are described as loam, sandy loam, or clay loam with a granular texture, a deep "A" horizon, good drainage, adequate organic matter content, and a pH level ranging from 5.5 to 6.5 [10,11].

C. canephora, also known as robusta coffee, contains about twice as much caffeine as *C. arabica*. It is less fragrant compared to arabica and exhibits superior resistance to heat, diseases, and pests. This species thrives in regions with a yearly temperature range of 22–26 °C, annual rainfall between 2000–3500 mm, and humidity levels of 85–90%. Its optimal growing range is at altitudes of 0–700 m.a.s.l., and it produces round beans [12] with oval-shaped seeds [13]. Robusta coffee is typically cultivated at altitudes below 600 m.a.s.l., primarily in humid tropical forest areas prevalent in the Amazon and Coastal regions. The climatic conditions of *C. arabica* and *C. canephora* are described in Figure 1.

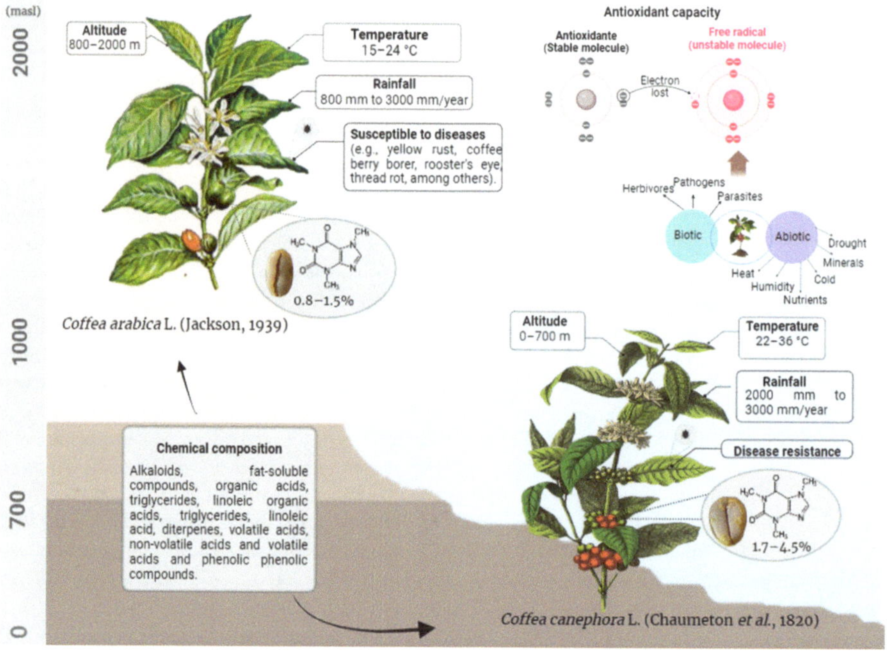

Figure 1. Summary of general characteristics of coffee plants *Coffea arabica* L. and *Coffea canephora* L. The optimal height for *C. arabica* is 800 to 2000 m above sea level, while *C. canephora* develops better from sea level to 700 m (Adapted from Gómez-Merino et al. [14], Starovicova & Hartemink [15], Cornelissen [16]).

According to Gotteland & De Pablo [2], coffee has various applications beyond being a popular beverage due to its rich content of bioactive compounds, including phenolics, in particular, chlorogenic acids, which act as antioxidants in the body. These antioxidants play a role in combating oxidative stress and neutralizing free radicals, potentially contributing to protection against chronic diseases such as cardiovascular disease, type 2 diabetes, and certain types of cancer. Additionally, coffee is beneficial in protecting against oxidative damage and skin aging [17].

Metabolites with high biological activity can be found in plants of the *Coffea* genus, including phenolic compounds (chlorogenic acids), diterpenes, and melanins [18]. The production of these compounds is influenced by various internal and external factors such as soil composition, water demand, climate, growing conditions, and ripening stage [19]. Caffeine and chlorogenic acids are known to provide several health benefits, primarily related to their antioxidant properties, which help protect against damage from reactive oxygen species (ROS) and oxidative stress through the donation of hydrogen atoms [20].

According to Patay et al. [21], the most abundant phenolic compound in *Coffea* species is 5-caffeoylquinic acid and its isomers, and the most important alkaloid is caffeine. The characteristic aroma of coffee is attributed to the presence of terpenoids such as 4-vinylguaiacol, α-2-furfurylthiol, and 3-methylbutane yrosin [22]. Additionally, other secondary metabolites found in the mature leaves and fruits of *Coffea* plants include trigonelline (alkaloid), 2-ethyl-3,5-dimethylpyrazine, 3-methylbutanal, 2-furfurylthiol, methylpropanal, 3-mercapto-3-methylbutyl formate, and methanethiol [23]. Robusta coffee also contains 16-*O*-methylcaffeophestol [24], as described in Figure 2.

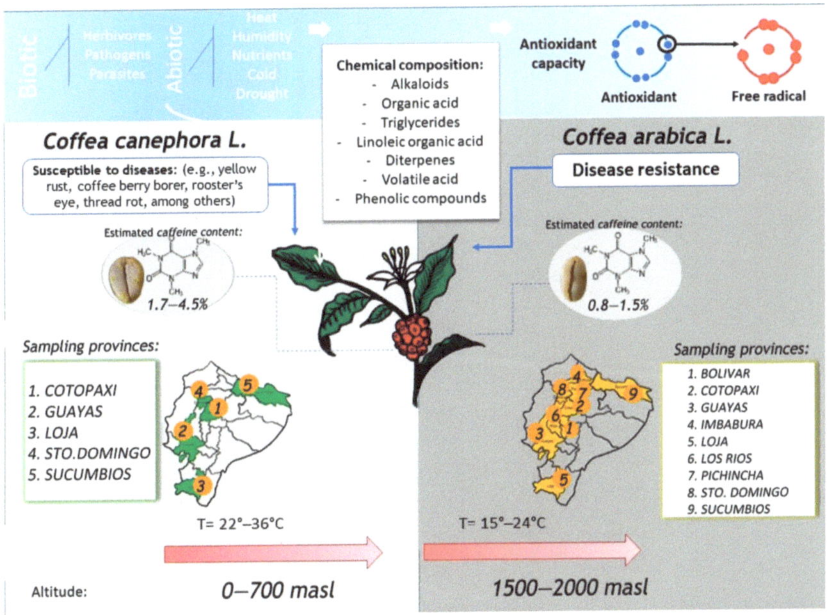

Figure 2. *Coffea arabica* L. and *Coffea canephora* L. sampling of crops cultivated in different Ecuadorian Provinces, stating the common chemical compounds and the role that antioxidant compounds play against free radicals, as well as the optimal growth conditions for each species, the species *C. canephora* is cultivated in the provinces of Cotopaxi, Guayas, Loja, Santo Domingo y Sucumbios while the species *C. arabica* is cultivated in the provinces of Bolivar, Cotopaxi, Guayas, Imbabura, Loja, Los Rios, Pichincha, Santo Domingo, Sucumbios.

The present study aims to evaluate the metabolic composition, biological activity, and sensorial attributes of coffee varieties grown in Ecuador. The research emphasizes the significance of investigating the production of secondary metabolites not only at the bean level but also within the plant to achieve a high antioxidant capacity. This investigation considers differences in the production of secondary metabolites, antioxidant capacity, and sensorial attributes according to the environmental conditions of the two species *C. arabica* and *C. canephora* cultivated in Ecuador. This research will provide valuable insights into the influence of environmental factors on the production of bioactive compounds in coffee plants, particularly at high altitudes, and may have implications for the development of coffee cultivation practices to optimize the antioxidant potential of the beans.

2. Materials and Methods

2.1. Chemicals

For the determination of secondary metabolites, Folin—Ciocalteu reagents (for phenolics), aluminum chloride ($AlCl_3$), and sodium acetate (CH_3COONa) (for flavonoids) were used. For the quantification of the antioxidant capacity, 2,2-diphenyl-1-picrihydrazyl (DPPH), 2,2-azinobis (3-ethylbenzothiazoline-6-sulfonic acid) diammonium salt (ABTS), 2,4,6-Tri(2-pyridyl)-triazine (TPTZ), and iron (III) chloride hexahydrate were used. Calibration curves were constructed using standard solutions of 6-hydroxy-2,5,7,8-tetramethylchroman-2-carboxylic acid (Trolox), 3,4,5-trihydroxybenzoic acid (gallic acid), potassium persulfate, sulfate iron (III) heptahydrate, quercetin, and ethanol 95%. All of the chemicals were purchased from Sigma-Aldrich Chemical Co. (St. Louis, MO, USA).

2.2. Sample Collection and Processing

Leaves and fruits of both species were collected. *C. arabica* samples were collected from certified coffee-producing farms from Bolívar, Cotopaxi, and Guayas provinces in the rainy season, Imbabura in the dry season, Pichincha in the rainy season, Santo Domingo de los Tsáchilas in the rainy season, and Loja, Los Ríos, and Sucumbíos, in the rainy season as well. The *C. canephora* samples were collected from the Cotopaxi, Guayas, Loja, Santo Domingo de los Tsáchilas, and Sucumbíos provinces. Random sampling with 5 replicates was carried out at different points of the crop to guarantee the homogeneity of the samples of the *C. canephora* and *C. arabica* species. For fruits, the physical aspect was taken into account. Priority was given to ripe beans that exhibited a characteristic intense red color, and absence of rotten spots, or bruises, indicating the optimum state for harvesting. In contrast, for leaves, ripe leaves with a uniform color were taken, these were wrapped in paper towels and stored with silica gel, for transport in a cooler for analysis to the CICTE laboratory of the Universidad de las Fuerzas Armadas in Cantón Rumiñahui.

2.3. Extraction of Active Ingredients

Prior to the determination of the antioxidant character and phytochemical compounds produced by coffee plants, leaves, and fruits were washed with distilled water to remove impurities. Ground leaves (5 g) were placed in 25 mL of 95% ethanol (v/v), used as a solvent, for 24 h at 4 °C in the dark. Fresh coffee fruits were ground, using 30 g of the raw material with 100 mL of 95% ethanol for extraction in a 24 h period. The supernatants were filtered through Whatman no. 1 filter paper, vacuum evaporated, and kept at 4 °C until use [25]. Then, for phenolic and flavonoid content, and antioxidant capacity assays, three replicates from each sample were used for analysis.

2.4. Determination of Total Phenol Content

The method described by Madaan et al. [26], with modifications, was used to determine the total phenol content (TPC). Following the reaction between the phenolic groups and the phosphomolybdic and phosphotungstic acids from the Folin-Ciocalteu reagent, green-blue complex detectable at 710 nm is formed. Briefly, an aliquot of the extract was dissolved in the solvent extraction, diluted to 5 mL of Milli-Q water, and added to 1.5 mL of the Folin-Ciocalteu reagent. To carry out the reaction, the mixture was left for 5 min at room temperature (25 °C), then 2 mL of a 100 g/L solution of Na_2CO_3 was added. The absorbance was measured after 30 min with a spectrophotometer against a control without extract. Gallic acid was used for the calibration curve using standard solutions at a concentration between 0–250 mg/L, obtaining the equation, $y = 0.0112x + 0.1759$ with a correlation factor of $R^2 = 0.9794$.

2.5. Determination of Flavonoid Content

The determination of the total content of flavonoids from our extracts was carried out spectrophotometrically according to the $AlCl_3$ colorimetric method or the Dowd method [27], a method based on the formation of a flavonoid-aluminum complex with maximum absorption at 430 nm. A 1 mL aliquot of the extract solution was mixed with 0.3 mL of 10% (v/v) $AlCl_3$ solution in methanol, 0.2 mL (1 M) potassium acetate, and 5.6 mL of distilled water. The mixture was incubated for 10 min at room temperature before measuring the absorbance of the reaction mixture at 430 nm. The standard calibration curve was performed at a concentration between 0–1.5 mg/L using quercetin as the standard, obtaining through linear regression the equation, $y = 1.4566x + 0.0265$ with a correlation factor of $R^2 = 0.9935$.

2.6. HPLC Quantification of Caffeine Content

Quantification of the caffeine content present in coffee beans was performed by high-performance liquid chromatography (HPLC) using the methodology proposed by Wanyika et al. [28]. An Agilent 1100/1200 Series HPLC (Santa Clara, CA, USA) instrument with a binary pump (G1312A), column oven (G1316A), and auto-injector (G1329A) managed by Chemstation software LTS 01.11 (Agilent Technologies, Santa Clara, CA, USA) was used for caffeine content quantification. An Agilent Zorbax SB C_{18} column 150 × 3.9 mm^2, internal diameter, particle size, 4 µm, at a flow rate of 1.4 mL/min and a temperature of 22 °C, using as the mobile phase a 20% (v/v) methanol solution in deionized water was used for the separation phase. The caffeine standard solution was prepared at a concentration of 100 mg/mL to match the mobile phase. The calibration curve was performed using the standard solutions prepared by serial dilution of the stock solution with the mobile phase at concentrations of 0, 0.02, 0.04, 0.06, and 0.08 mg/mL. Coffee pericarp was freeze-dried and a solution was prepared with distilled water at a concentration of 20 mg/mL. The sample was prepared with 5 mL of the freeze-dried coffee solution, and calibrated with the mobile phase to 50 mL. Standards and sample were placed in the HPLC system under the conditions: column, reversed phase, ODS detector, 250 × 4.6 nm^2, a flow rate of 1 mL/min, wavelength of 278 nm, pressure of 150 khf/cm^2, mobile phase of water, acetic acid, and methanol (79.9:0.1:20) and injection volume of 10 µL. The results (average values ± standard deviations) were expressed as mg/g dw.

2.7. Determination of Antioxidant Capacity

2.7.1. Free Radical Scavenging by Use of the DPPH Radical

The DPPH free radical method was based on the protocol established by Guo et al. [29] with modifications. A volume of 2 mL DPPH stock solution (0.1 mM) was added in each of the test tubes containing 0.1 mL of the crude extract sample and incubated in the dark, at room temperature, for 30 min. The absorbance was measured at a wavelength of 517 nm and the scavenging ability as a percentage was calculated as DPPH scavenging ability = (A control − A sample/A control) × 100. A Trolox standard solution was used for the calibration curve determination with concentrations ranging from 0–0.625 mM, obtaining the following equation, y = 158.07x − 1.6766 with a correlation factor of R^2 = 0.9955.

2.7.2. Free Radical Scavenging through the Use of the ABTS Radical Cation

The method described by Loizzo et al. [30] was used to determine the antioxidant activity based on the ABTS free radical assay. A mixture of ABTS (2 mM) and potassium persulfate (70 mM) was allowed to stand overnight at room temperature in the dark to form the ABTS radical cation before use. The ABTS solution was then diluted with 80% methanol to obtain an absorbance of 0.700 ± 0.005 at 734 nm. A total of 100 µL of appropriately diluted samples was added to 2 mL of ABTS solution and the absorbance was recorded at 734 nm after 1 min of incubation at room temperature. The standard calibration curve was built using Trolox standard solutions with concentrations ranging from 0 to 2.5 mM, producing the equation, y = 31.995x + 3.9568 with a correlation factor of R^2 = 0.9697.

2.7.3. Ferric Reducing Antioxidant Power Assay FRAP

The method used for determining the reducing capacity [31] uses antioxidants as reductants in a redox-related colorimetric method, using a slightly reduced oxidant, the ferric ion (Fe^{3+}), present in stoichiometric excess. The ferric tripyridyltriazine complex (Fe^{3+}-TPTZ) is reduced (in the presence of low pH) by the antioxidants in the sample to the ferrous form (Fe^{2+}-TPTZ), which appears as a deep blue color. The reaction is monitored at 593 nm. The FRAP reagent was prepared by mixing the acetate buffer (300 mM, pH 3.6), a solution of 10 mM TPTZ in 40 mM HCl and 20 mM$FeCl_3$ at 10:1:1 ($v/v/v$). The sample was incubated with 2 mL of FRAP solution (prepared by mixing 25 mL acetate buffer solution, 5 mL TPTZ solution, and 10 mL $FeCl_3{\cdot}6H_2O$ solution) at 37 °C, for 30 min in the dark. All solutions were prepared on the day of use. A standard curve was constructed by using different concentrations of $FeSO_4 \times 7H_2O$ ranging from 0–5 mM, following a linear regression to obtain the equation, y = 0.3654x − 0.0532 with a correlation factor of R^2 = 0.9686.

2.8. Sensory and Organoleptic Evaluation

In this study, a panel of 36 coffee enthusiasts, who are regular consumers of coffee and possess a strong knowledge of its sensory attributes, was selected to participate in a sensory evaluation of fresh coffee fruits. Before the evaluation, the panelists received training to familiarize themselves with the specific characteristics of fresh coffee fruits and the lexicon used for sensory evaluation. The lexicon for sensory evaluation of fresh coffee fruits (in a standardized manner) includes descriptive terms related to aroma (e.g., floral, fruity, roasted), flavor (e.g., nutty, chocolatey, acidic), after-flavor (e.g., lingering, bitter), acidity (e.g., bright, sharp), sourness (e.g., mild, pronounced), body (e.g., full-bodied, light), color (e.g., dark, medium, light), and texture (e.g., smooth, gritty). Each panelist received a cup with 3 coffee beans from each province, for both *C. arabica* and *C. canephora* species. A total of 27 samples were tested for *C. arabica*, originating from the provinces of Bolívar, Cotopaxi, Guayas, Imbabura, Loja, Los Ríos, Pichincha, Santo Domingo, and Sucumbíos, while each panelist tested 15 samples from Cotopaxi, Guayas, Loja, Santo Domingo, and Sucumbíos for *C. canephora*. Before the evaluation, the samples were meticulously prepared to offer a comprehensive sensory experience. The procedure begins with a cleaning of the samples to exclude physical defects and impurities. After the first step, all samples must pass through sensory evaluations such as olfactory and visual; additionally, an examination of defects is prepared followed by the analysis of the physical variables represented by fragrance, flavor, residual flavor, acidity, body, bitterness, color, and texture [27]. The panelists received information about the specific characteristics of fresh coffee fruits and underwent training to familiarize themselves with the sensory attributes. Using a standardized questionnaire, panelists scored various attributes including aroma, flavor, after-flavor, acidity, sourness, body, color, and texture of the fresh coffee fruits. The data obtained from the sensory evaluation was then aggregated and analyzed to identify trends and preferences related to the characteristics of the fresh coffee fruits. All variables receive a rating from 0 to 10, where 9–10 is classified as very good, 6–8 as good, 4–5 as intermediate, 2–3 as low, and 0–1 as bad, according to the Specialty Coffee Association of Africa protocol [4].

2.9. Statistical Analysis

For statistical analysis, the R Studio statistic program was used. Data was analyzed by a three-way ANOVA test under a $p < 0.05$ value of significance. The assays were conducted in triplicate and presented as mean ± standard deviation. Permutational multivariate analysis of variance (PERMANOVA), which extends the univariate factorial linear model to multiple dimensions without requiring a known probability distribution of the dependent variables [32], was used to perform the statistical analysis of non-parametric data. The correlation between total phenolic content, total flavonoid content, and antioxidant capacity was evaluated by using Pearson's correlation coefficient, while the correlation matrix was depicted as a scatter plot matrix. A negative value indicates a negative linear correlation, a positive value indicates a positive linear correlation, and 0 indicates no linear correlation [33]. The correlation ranking (values as *** high correlation, ** medium correlation, * low correlation) was established based on Maurage et al. [34]

3. Results

3.1. Total Phenolic Content and Total Flavonoid Content

The phenolic concentration was evaluated in leaves and fruit for each of the species, from different provinces of Ecuador. In the evaluated samples from *C. arabica*, the top concentration in fruit was recorded in the province of Cotopaxi (4.188 ± 0.029 mg GAE/g dw, while the lowest concentration was recorded in the province of Bolivar (0.996 ± 0.066 mg GAE/g dw). In the case of the leaf samples, the top phenolic content was in the province of Guayas (10.869 ± 0.002 mg GAE/g dw), while the lowest concentration was observed in the province of Los Ríos (1.124 ± 0.181 mg GAE/g dw). For the fruits of *C. canephora*, the province with the uppermost phenolic content was in Cotopaxi (3.036 ± 0.317 mg GAE/g dw), and the lowest concentration in the province of Loja (1.298 ± 0.101 mg GAE/g dw). In the case of the leaf samples, it was determined that the peak phenolic content was in Guayas Province (10.782 ± 0.004 mg GAE/g dw), and the lowest concentration in the province of Loja (2.533 ± 0.112 mg GAE/g dw). An analysis of variance (ANOVA) was carried out, determining significant differences between the samples (p-value < 0.0001) as shown in Figure 3a.

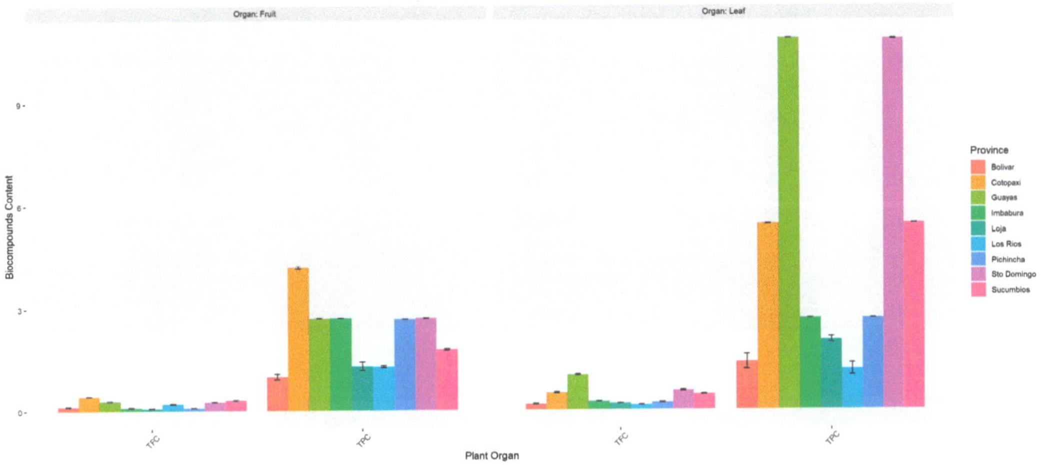

(**a**)

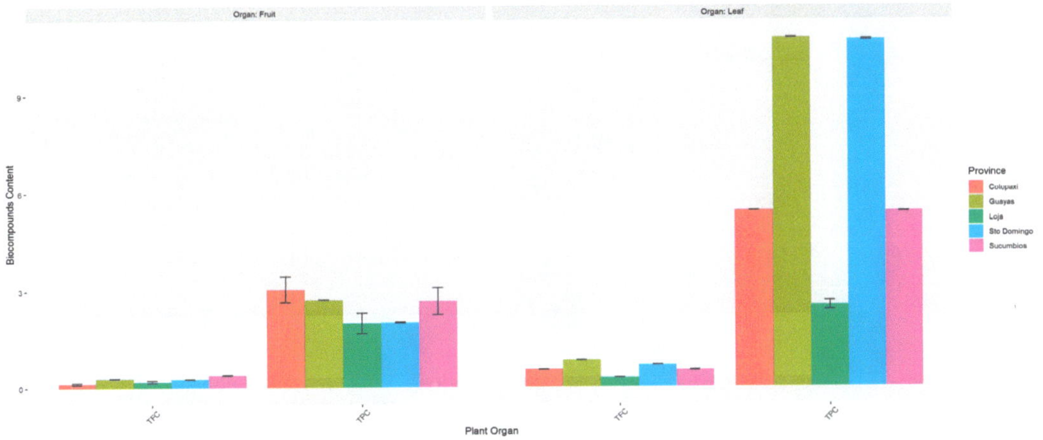

(**b**)

Figure 3. Total phenolic and flavonoid content from leaves and fruits for *C. arabica* (**a**) and *C. canephora* (**b**) collected from different regions of Ecuador. Legend: TFC—total flavonoid content, TPC—total phenolic content.

The province with the upper value of total flavonoid content in *C. arabica* fruit samples was Cotopaxi (0.442 ± 0.001 mg QE/g dw), while the lowest content was recorded in the province of Loja (0.076 ± 0.006 mg QE/g dw). In leaf samples, the total flavonoid content with the highest result was recorded in Guayas (1.028 ± 0.014 mg QE/g dw), and the lowest in Bolivar (0.129 ± 0.008 mg QE/g dw). In the case of the fruit of *C. canephora*, the top total flavonoid content was recorded in Sucumbios (0.385 ± 0.008 mg QE/g dw) and the lowest in Cotopaxi (0.128 ± 0.02 mg QE/g dw). An analysis of variance was carried out (p-value < 0.0001) with results reported in Figure 3b.

3.2. Determination of Caffeine Content

The highest value of caffeine content in *C. arabica* samples was recorded in the province of Los Ríos (15.09 mg/g dw); while the lowest value was found in the province of Imbabura (0.91 ± 0.00629 mg/g dw). Regarding the *C. canephora* species, the highest value was recorded in the province of Loja (18.05 ± 0.15803 mg/g dw), and the lowest value was recorded in the province of Santo Domingo de los Tsáchilas (1.97 ± 0.03557 mg/g dw), as shown in Table 1.

Table 1. Total Caffeine Content in *C. arabica* and *C. canephora* collected from different regions of Ecuador.

Species	Province	Caffeine Content (mg/g dw)
	Pichincha	1.76 ± 0.0729
	Imbabura	0.91 ± 0.00629
	Sucumbios	12.23 ± 0.52208
	Cotopaxi	13.96 ± 0.27289
Coffea arabica	Loja	9.29 ± 0.03744
	Bolívar	13.87 ± 0.15794
	Los Ríos	15.09 ± 0.15803
	Guayas	13.38 ± 0.23911
	Santo Domingo de los Tsáchilas	3.26 ± 0.09705
	Sucumbios	10.32 ±0.02242
	Cotopaxi	16.31 ± 0.10919
Coffea canephora	Loja	18.05 ± 0.21232
	Guayas	11.61 ± 0.12294
	Santo Domingo de los Tsáchilas	1.97 ± 0.03557

3.3. Antioxidant Capacity

When evaluating the antioxidant capacity through the ABTS test, for the *C. arabica* species, the best antioxidant capacity in fruit samples was obtained in the province of Cotopaxi (23.179 ± 1.802 µmol TEAC/g dw) and the lowest value in Imbabura (1.885 ± 0.05 µmol TEAC/g dw), while in leaf samples, the highest result was found in Santo Domingo de los Tsáchilas Province (100.085 ± 0.817 µmol TEAC/g dw), and the lowest was recorded in the province of Imbabura (2.488 ± 0.045 µmol TEAC/g dw). For the fruit of *C. canephora* species, the best antioxidant capacity was found in the province of Guayas (24.533 ± 0.202 µmol TEAC/g dw), and the lowest was recorded in Santo Domingo de los Tsáchilas (8.331 ± 0.127 µmol TEAC/g dw), while for leaf samples, the highest antioxidant capacity was found in Guayas (24.533 ± 0.202 µmol TEAC/g dw) and the lowest in Loja (23.932 ± 0.416 µmol TEAC/g dw).

In the DPPH test, for *C. arabica* fruit, the highest antioxidant capacity was obtained in the province of Imbabura (65.875 ± 1.129 µmol TEAC/g dw) and the lowest value in Los Ríos (8.225 ± 1.138 µmol TEAC/g dw), however, for the leaf, the highest capacity was from the province of Loja (304.876 ± 25.455 µmol TEAC/g dw) and the lowest capacity was found in the province of Bolivar (21.028 ± 1.252 µmol TEAC/g dw). In the case of *C. canephora*, fruit samples from the Loja province showed the highest antioxidant capacity (85.869 ± 1.727 µmol TEAC/g dw), while the lowest was recorded in Sucumbios (24.124 ± 1.203 µmol TEAC/g dw). Regarding the antioxidant capacity of the leaf samples, the highest result was recorded in Guayas (196.956 ± 0.279 µmol TEAC/g dw) and the lowest in Loja (71.149 ± 9.519 µmol TEAC/g dw).

In the FRAP test, peak radical reduction power in *C. arabica* fruit was obtained in the province of Imbabura (100.164 ± 0.332 µmol Fe^{2+}/g dw), and the lowest was registered from Los Ríos (8.225 ± 1.138 µmol Fe^{2+}/g dw). Analyses of leaf samples resulted in a higher reduction power in the province of Pichincha (102.705 ± 0.447 µmol Fe^{2+}/g dw). In the case of *C. canephora*, fruit samples from the Loja Province registered a higher radical reduction power (36.567 ± 1.127 µmol Fe^{2+}/g dw), than the results recorded in Santo

Domingo de los Tsáchilas (8.336 ± 0.127 μmol Fe^{2+}/g dw). As for the leaf samples, the top capacity was recorded in Guayas (100.286 ± 0.114 μmol Fe^{2+}/g dw) and the lowest in Loja (35.928 ± 2.164 μmol Fe^{2+}/g dw). An analysis of variance was performed (p-value < 0.0001) and the results are presented in Figures 4 and 5. Data were expressed in TEAC/g dw for ABTS and DPPH and μmol Fe^{2+}/g dw for FRAP.

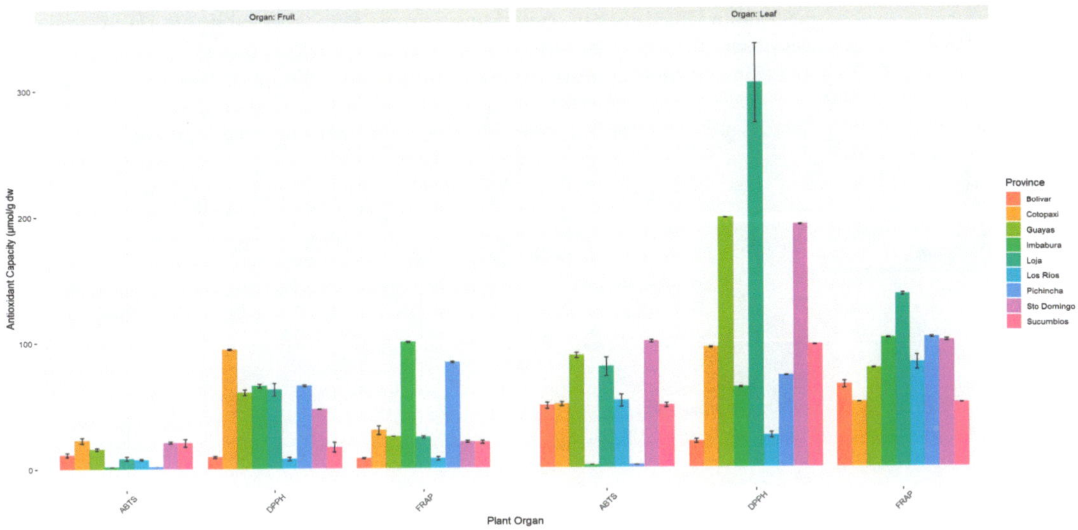

Figure 4. Antioxidant activity (ABTS, DPPH, and FRAP) of *Coffea arabica* fruits and leaves from all the tested provinces. Legend: ABTS—Free radical scavenging using ABTS radical cation assay, DPPH—Free radical scavenging using DPPH radical assay, FRAP—Ferric reducing antioxidant power assay).

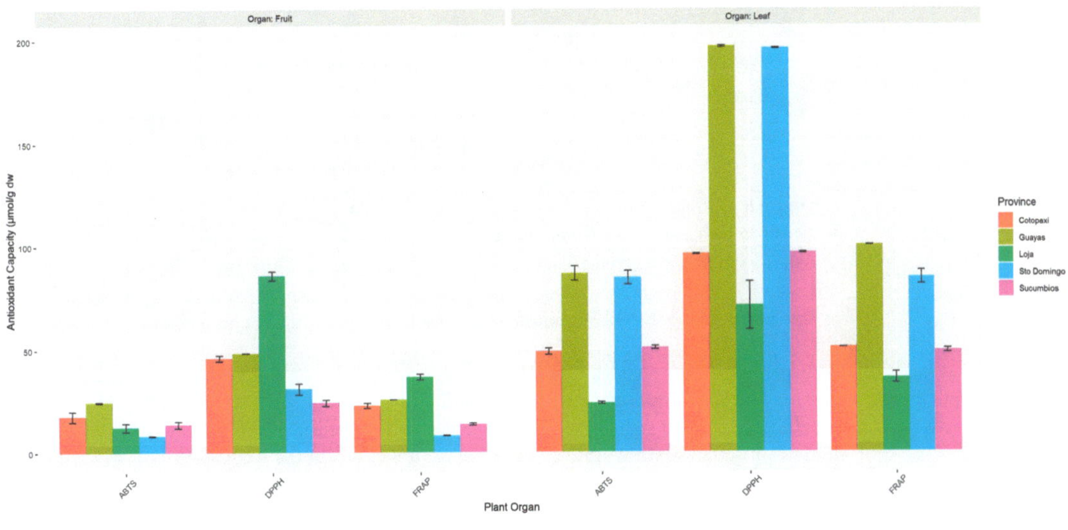

Figure 5. Antioxidant activity (ABTS, DPPH and FRAP) of *Coffea canephora* fruits and leaves from all the tested provinces. Legend: ABTS—Free radical scavenging using ABTS radical cation assay, DPPH—Free radical scavenging using DPPH radical assay, FRAP—Ferric reducing antioxidant power assay).

3.4. Correlation

The correlations between the total content of secondary metabolites, phenolics, and flavonoids and the three antioxidant capacity methods (ABTS, DPPH, and FRAP) were evaluated for each species of *Coffea* sp. In the case of *C. canephora*, a high positive correlation (R > 90) was observed for all of the values evaluated. For *C. arabica*, the correlation of all of the values evaluated resulted in a high positive correlation for TPC in ABTS (R = 0.68), medium in DPPH (R = 0.57), and low in FRAP (R = 0.27), while for TFC a similar trend was observed with values for ABTS (R = 0.64) and DPPH (R = 0.47) of medium positive character and a low positive value for FRAP (R = 0.05) (Figures 6 and 7).

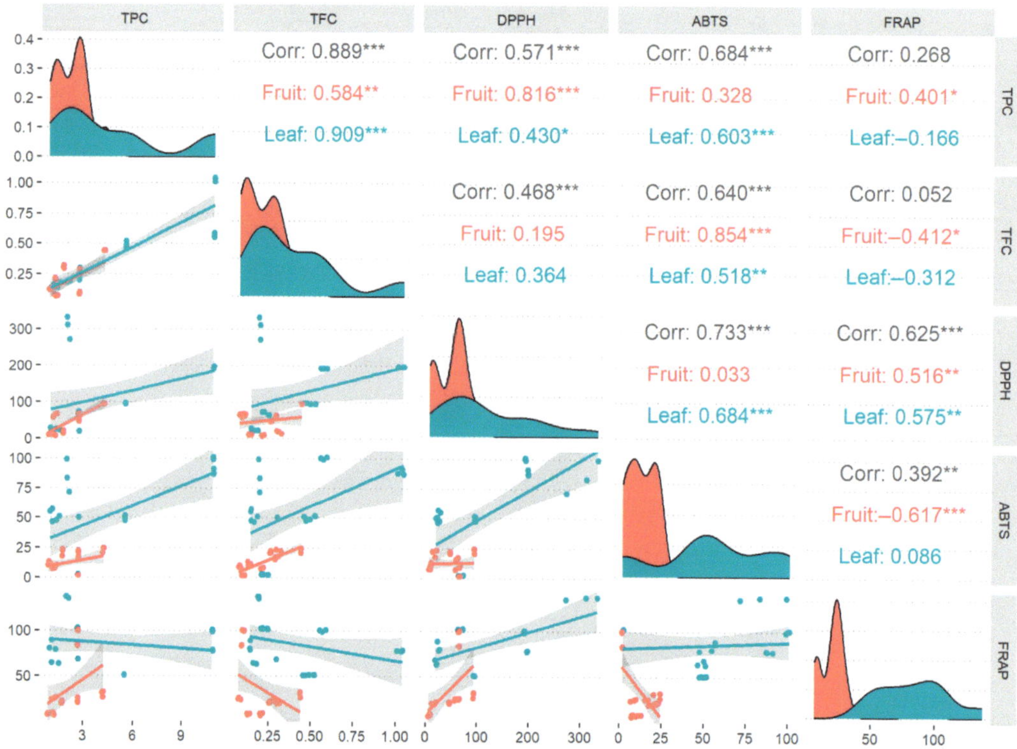

Figure 6. Correlation of total flavonoid content and total phenolic content to ABTS, DPPH, and FRAP assays in the *C. arabica* between plant organs leaf and fruit. A high positive correlation is observed. Asterisks indicate the type of correlation, *** high correlation, ** medium correlation, * low correlation. Legend: ABTS—Free radical scavenging using the ABTS radical cation assay, DPPH—Free radical scavenging using the DPPH radical assay, FRAP—Ferric reducing antioxidant power assay).

3.5. Sensory and Organoleptic Evaluation of Raw Coffee Beans

The *C. arabica* species predominates in fragrance with a mean score of 8.05 ± 1.234, while the lowest value was for bitterness with a mean of 6 ± 2.534. *C. canephora* predominates in bean texture with a mean score of 8.8, while the lowest score was for bitterness 5.8 ± 2.201. Flavor, residual flavor, acidity, body, and coloring are comparable in the two species (Figure 8).

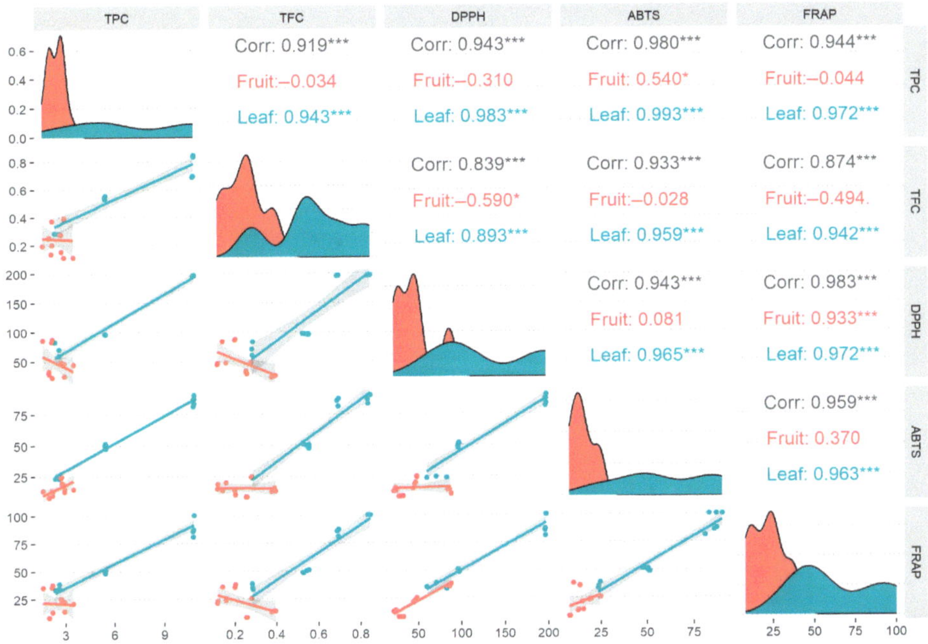

Figure 7. Correlation of total flavonoid content and total phenolic content to ABTS, DPPH, and FRAP assays in the *C. robusta* between plant organs leaf and fruit. A high positive correlation is observed. Asterisks indicate the type of correlation, *** high correlation, * low correlation. Legend: ABTS—Free radical scavenging using the ABTS radical cation assay, DPPH—Free radical scavenging using the DPPH radical assay, FRAP—Ferric reducing antioxidant power assay).

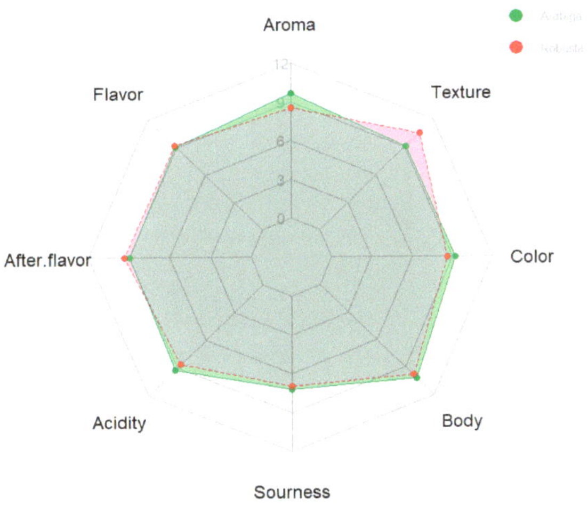

Figure 8. Sensory and organoleptic evaluation of raw coffee beans.

4. Discussion

When comparing the results obtained for both *Coffea* species, it is not possible to establish a relationship between the province and the highest phenolic content, as both recorded higher values in different provinces, but when the lowest values were recorded, a tendency was established in the province of Loja for *C. arabica* and *C. canephora*. For C.

canephora, Guayas province is among the provinces with the highest content. According to Carvalho Neto et al. [35], the beans contain high concentrations of caffeine, phenolic compounds, flavonoids, and triacylglycerols, bioactive compounds with high antioxidant and antimicrobial activity. Hudáková et al. [36] support these claims by stating that coffee is an exceptional source of antioxidants, particularly phenolic compounds, which play a significant role in neutralizing free radicals.

Mature coffee plants contain phenolic compounds such as catechins, carotenoids, anthocyanins, and chlorogenic acids [37]. The high content of phenolic compounds in leaves compared to coffee beans is congruent with Patay et al. [21], who reported that leaves have a higher concentration of chlorogenic acid compared to other parts of the plant because its presence is related to the level of exposure to UV light.

Chlorogenic acids comprise hydroxycinnamic esters with quinic acid, including caffeoylquinic acid, dicaffeoylquinic acid, feruloylquinic acid, and coumaroylquinic acid, compounds that show antioxidant capacity [38]. Chlorogenic acids, present in coffee, have attracted the attention of several researchers as it has antioxidant, antibacterial, antiviral, antibiotic, anti-inflammatory, and neuroprotective properties. Clinical applications are being developed thanks to their cardiovascular properties in the treatment of hypertension, as its role in plants is an intermediate for the synthesis of lignin [39].

The results regarding the neutralization of DPPH radicals show that for each gram of sample, a Trolox equivalent of 65.875 ± 1.129 μmol TEAC/g dw is inhibited in the fruits of the *C. arabica* species originating in the province of Imbabura, while the lowest concentration is observed in Los Ríos. As for leaves, the highest concentration was recorded in the province of Loja, while the lowest value was found in Bolivar. Abdulmajid [40] and Guambi et al. [4] mention that the antioxidant character of coffee varies depending on various pre-harvest factors such as altitude, soil type, geographical location, and plant genetics. Concerning the fruits of the *C. canephora* species, the highest antioxidant concentration was recorded in the province of Loja, in contrast to Sucumbios (lowest value). For leaves, the highest concentration was found in the province of Guayas, while the lowest was recorded in Loja. Lemos et al. [41] reported that the diversity present in the environments where coffee is grown can influence the chemical composition depending on the genetic variability of the species, harvesting processes, and other agronomic conditions. The highest antioxidant activity determined through the ABTS method was observed in *C. arabica* (23.179 ± 1.802 μmol TEAC/g dw) and *C. canephora* (24.533 ± 0.202 μmol TEAC/g dw) ripe fruits from Cotopaxi and Guayas provinces, respectively. Acidri et al. [42] reported on the antioxidant capacity of ripe coffee beans, highlighting that the fruits are the most crucial component of the plant and that as they ripen, approximately 7% of the total content of chlorogenic acids is lost, increasing the antioxidant character. Additionally, Platzer et al. [43] mention that absorbance readings can be affected, since the ABTS method measures hydrophilic antioxidants, leading to overestimates due to the thermodynamics of the reaction.

In the FRAP assay, Pacheco-Coello & Rabottini-Villamizar [44] reported the reducing potential of the ferric ion (Fe^{3+}) with a value of 178.32 ± 0.99 μmol Fe^{2+}/g, which differs significantly from the results obtained here, where the best readings were given in fruits and leaves of *C. arabica* species originating from the provinces of Imbabura (100.164 ± 0.332 μmol Fe^{2+}/g dw) and Pichincha (102.705 ± 0.447 μmol Fe^{2+}/g dw).

Echegaray et al. [45] point out that FRAP, being an assay based on redox reactions, is susceptible to any substance capable of yielding electrons, causing measurement errors due to possible interference. This assay is also limited when certain substances are found in the sample used through the assay, such as fluorides, phosphates, citrates, tartrates, the presence of lipids, and other intracellular molecules which can interfere with phenolics determination [46,47]. Ahmed et al. [23] mention in their study that they found both increases and decreases in secondary metabolites and sensory characteristics that define coffee quality in response to changes in environmental conditions. It is also influenced by the microenvironment in which the compound is located, which could result in interactions with each other, producing synergistic or inhibitory effects [48].

It is important to consider that the antioxidant capacity present in food originating from plants is not determined solely by the sum of the individual antioxidant capacities of each component [49]. Our results showed a positive correlation for *C. arabica* between the DPPH assay and TPC (0.571 ***), TFC (0.468 ***), and between ABTS and TPC (0.684 ***) and TFC (0.640 ***). In the case of *C. canephora*, significant correlations were observed between antioxidant activity determined through all assays used and secondary metabolites: DPPH and TPC (0.943 ***) and TFC (0.839 ***), ABTS and TPC (0.980 ***) and TFC (0.933 ***). In contrast to *C. arabica*, *C. canephora* showed significant correlation between FRAP and TPC (0.944 ***) and TFC (0.874 ***).

A representative metabolite of coffee is caffeine, contained in both green and roasted coffee beans and also in coffee beverages; the average caffeine content in green beans in *C. arabica* ranged from 0.7 to 1.3% [50]. In our tested samples, the caffeine content was higher in *C. canephora* collected from 2 of 5 provinces, compared with *C. arabica;* but lower than that described in the literature [51]. The caffeine content was found in higher percentages in the fruit of *C. canephora* from the province of Loja. According to Koshiro et al. [52], the caffeine content increases related to the stage of growth, being higher when the fruit is green or not completely reddened. Perdani et al. [53] noted that coffee fruits have a high content of phenolics, with chlorogenic acids being the highest phytochemical compound found (90%). In the same study, it was determined that the geographical position of the crop for the plant has a significant influence on the content of both phenolics and caffeine. According to Vega et al. [54], the amount of caffeine present in coffee plants is related to its role as a defense mechanism against herbivores, while for Stevenson et al. [55], the concentration of this metabolite could be related to the need to stimulate pollinators.

Sensory analysis of pulped coffee fruit showed differences between the parameters measured by Cotacallapa et al. [56], the organoleptic quality of the fruit depends on the cultivation practices, as well as the genetic origin or similarity of the variables. For Alves et al. [57], the organoleptic properties of coffee beans are affected by the amount of direct light received by the plant. The highest score was received for aroma. Ludwig et al. [58] mention that green coffee is characterized by mild aromas related to plant parts, such as flowers, fruits, and vegetables; the desirable fragrance associated with coffee beverages develops during roasting. According to Velásquez & Banchón [59], *C. arabica* is considered a high-altitude coffee and is characterized by a weak body, somewhat acidic, and strong fragrance due to its low caffeine content, while the altitude of cultivation of *C. canephora* is lower and gives it a less acidic sensory characteristic.

5. Conclusions

Coffee grown in the provinces in the Coastal region, with optimal temperatures ranging from 22 to 26 °C, has been found to yield better concentrations of phenols and flavonoids, along with enhanced antioxidant capacity. Notably, the highest antioxidant capacity for DPPH and ABTS tests was observed in Cotopaxi province for *C. arabica* and Loja province for *C. canephora*. Furthermore, *C. canephora* exhibited the highest caffeine content, which is a typical characteristic of this species. In sensory tests, *C. arabica* obtained the best score, while *C. canephora* scored the highest in terms of texture. These findings underscore the influence of environmental conditions on the production of bioactive compounds, biological capacity, and the sensory attributes of coffee, highlighting the potential for optimizing cultivation practices to enhance the quality and antioxidant properties of coffee beans.

Author Contributions: Conceptualization, R.A.M.; methodology, D.C.O.-P., K.M.I.-P. and M.Y.V.-P.; software, L.A.S.-T. and M.L.V.-L.; formal analysis, D.C.O.-P., K.M.I.-P., M.Y.V.-P., L.A.S.-T., M.L.V.-L. and N.S.C.-I.; writing—original draft preparation, R.A.M., E.J.M.-H. and R.D.C.; writing—review and editing, R.A.M., E.J.M.-H. and R.D.C.; supervision, R.A.M.; project administration, R.A.M.; funding acquisition, R.A.M. All authors have read and agreed to the published version of the manuscript.

Funding: This research was funded by Universidad de las Fuerzas Armadas-ESPE, grant number CV-GNP-0066-2020, and the Institute of Biology, Bucharest, Romanian Academy, grant number RO1567-IBB08/2023. The APC was funded by Universidad de las Fuerzas Armadas-ESPE.

Data Availability Statement: Data are contained within the article.

Acknowledgments: The authors gratefully acknowledge CICTE from the University of Army Forces-ESPE.

Conflicts of Interest: The authors declare no conflicts of interest. The funders had no role in the design of the study; in the collection, analyses, or interpretation of data; in the writing of the manuscript; or in the decision to publish the results.

References

1. Varela, L.A.; Ron, S.R. Geografía y Clima del Ecuador. BIOWEB. Pontificia Universidad Católica del Ecuador. 2018. Available online: https://bioweb.bio/geografiaClima.html/ (accessed on 19 December 2023).
2. Gotteland, M.; De Pablo, S. Some truths about coffee. *Rev. Chil. Nutr.* **2007**, *34*, 105–115. [CrossRef]
3. Ponce-Vaca, L.A.; Orellana-Suarez, K.D.; Acuña-Velásquez, I.R.; Alfonso-Alemán, J.L.; Fuentes Figueroa, T. Situación de la caficultura ecuatoriana: Perspectivas. *Rev. Estud. Desarro. Soc. Cuba. América Lat.* **2018**, *6*, 307–325.
4. Guambi, L.A.D.; Cedeño, S.d.R.V.; Talledo, D.S.F. Organoleptic Quality of Arabica Coffees in Relation to the Varieties and Altitudes of the Cultivation Zones, Ecuador. *Rev. Iberoam. Tecnol. Postcosecha* **2017**, *18*, 67–77.
5. Arreaga Ronquillo, E.; Quezada Campoverde, J.; Barrezueta Unda, S.; Cervantes Alava, A.; Prado Carpio, E. Economic impact generated by coffee production in Ecuador in the period 2016–2019. *593 Digit. Publ. CEIT* **2021**, *6*, 83–91. [CrossRef]
6. Davis, A.P.; Govaerts, R.; Bridson, D.M.; Stoffelen, P. An annotated taxonomic conspectus of the genus *Coffea* (Rubiaceae). *Bot. J. Linn. Soc.* **2006**, *152*, 465–512. [CrossRef]
7. Carvajal Herrera, J.J.; Aristizábal Torres, I.D.; Oliveros Tascón, C.E.; Mejía Montoya, J.W. Colorimetry of the Coffee Fruit (*Coffea arabica* L.) during its Development and Ripening. *Rev. Fac. Nac. Agron. Medellín* **2011**, *64*, 6229–6240.
8. Farah, A.; Ferreira dos Santos, T. Chapter 1—The Coffee Plant and Beans: An Introduction. In *Coffee in Health and Disease Prevention*; Preedy, V.R., Ed.; Academic Press: Cambridge, MA, USA, 2015; pp. 5–10. [CrossRef]
9. Beining, A.; Burkhardt, J.; Fetene, M. Water relations of Ethiopian wild coffee populations: Genetic fixation and phenotypic plasticity. In Proceedings of the 22nd International Scientific Colloquium on Coffee, Campinas, International Scientific Association on Coffee, CDROM-A105, Paris, France, 14–19 September 2008.
10. INIAP. 2017 Guide to Facilitate Learning in the Management of Robusta Coffee (*Coffea canephora* P.). Learning Guide N° 008. Available online: https://repositorio.iniap.gob.ec/bitstream/41000/4788/7/iniapeecaga008.pdf (accessed on 2 December 2023).
11. Vizuete-Montero, M.O.; Figueroa-Saavedra, H.F.; Barbaru-Grajales, A.D.; Zapata- Mayorga, H.A.; Herrera-Ocaña, H.R.; Moya, W. Physio-edaphoclimatic factors show optimal soil suitability for three tropical crops in the Ecuadorian Amazon. *Sci. Agric.* **2024**, *81*, e20220214. [CrossRef]
12. Dethlefsen & Balk. Plant, Varieties and Growing Areas. 2018. Available online: https://www.dethlefsen-balk.de/ESP/10889/Kaffeepflanze.html (accessed on 20 December 2023).
13. Clifford, M.N.; Willson, K.C. *Coffee: Botany, Biochemistry and Production of Beans and Beverage*, 1st ed.; Clifford, M.N., Ed.; Springer: Berlin, Germany; Croom Helm;: London, UK; Sydney, Australia; Westport, CT, USA, 1985; pp. 13–47.
14. Gómez-Merino, F.; Trejo-Téllez, L.I.; Morales-Ramos, V.; Marín-Garza, T.; Crosby-Galván, M. Nutritional assessment of Robusta coffee (*Coffea canephora*) beans of different origins processed in Mexico. *Agroproductividad* **2018**, *11*, 25–29.
15. Starovicova, M.; Hartemink, R. Food-Info.net: La planta de café. Food-Info: La Planta de Café. 2021. Available online: https://www.food-info.net/es/products/coffee/plant.htm (accessed on 20 November 2023).
16. Cornelissen, P. Pictura Antique Prints: Antique Prints, Art, Maps and Books. Cafe Cofee Tree Coffea Arabica. 2021. Available online: https://pictura-prints.com/product/antique-print-pl-85-cafe-coffee-tree-coffea-arabica-turpin-chaumeton-1814/ (accessed on 20 November 2023).
17. Vilaplana, M. Antioxidantes presentes en los alimentos. Vitaminas, minerales y suplementos. *Ámbito Farmac. Nutric.* **2007**, *26*, 79–86. Available online: https://www.elsevier.es/es-revista-offarm-4-articulo-antioxidantes-presentes-alimentos-vitaminas-minerales-13112893 (accessed on 20 December 2023).
18. Wang, X.; Meng, Q.; Peng, X.; Hu, G.; Qiu, M. Identification of new diterpene esters from green Arabica coffee beans, and their platelet aggregation accelerating activities. *Food Chem.* **2018**, *263*, 251–257. [CrossRef] [PubMed]
19. Good Kitzberger, C.S.; dos Santos Scholz, M.B.; de Toledo Benassi, M. Bioactive compounds content in roasted coffee from traditional and modern *Coffea arabica* cultivars grown under the same edapho-climatic conditions. *Food Res. Int.* **2014**, *61*, 61–66. [CrossRef]
20. Mitraka, G.-C.; Kontogiannopoulos, K.N.; Batsioula, M.; Banias, G.F.; Assimopoulou, A.N. Spent Coffee Grounds' Valorization towards the Recovery of Caffeine and Chlorogenic Acid: A Response Surface Methodology Approach. *Sustainability* **2021**, *13*, 8818. [CrossRef]

21. Patay, É.B.; Sali, N.; Kőszegi, T.; Csepregi, R.; Balázs, V.L.; Németh, T.S.; Németh, T.; Papp, N. Antioxidant potential, tannin and polyphenol contents of seed and pericarp of three *Coffea* species. *Asian Pac. J. Trop. Med.* **2016**, *9*, 366–371. [CrossRef]
22. Buffo, R.A.; Cardelli-Freire, C. Coffee flavor: An overview. *Flavour Frag. J.* **2004**, *19*, 99–104. [CrossRef]
23. Ahmed, S.; Brinkley, S.; Smith, E.; Sela, A.; Theisen, M.; Thibodeau, C.; Warne, T.; Anderson, E.; Van Dusen, N.; Giuliano, P.; et al. Climate Change and Coffee Quality: Systematic Review on the Effects of Environmental and Management Variation on Secondary Metabolites and Sensory Attributes of *Coffea arabica* and *Coffea canephora*. *Front. Plant Sci.* **2021**, *12*, 708013. [CrossRef]
24. Buzzo Mori, A.L.; Kalschne, D.L.; Gava Ferrão, M.A.; Almeida da Fonseca, A.F.; Gava Ferrão, R.; de Toledo Benassi, M. Diterpenes in *Coffea canephora*. *J. Food Compos. Anal.* **2016**, *52*, 52–57. [CrossRef]
25. Kaisangsri, N.; Selammasakul, O.; Sonklin, C. Phenolic Compounds and Biological Activities of Coffee Extract for Cosmetic Product. *J. Sci. Engin.* **2019**, *1*, 71–76. Available online: https://www.jstage.jst.go.jp/article/sjse/1/1/1_71/_pdf/-char/ja (accessed on 20 December 2023).
26. Madaan, R.; Bansal, G.; Kumar, S.; Sharma, A. Estimation of total phenols and flavonoids in extracts of *Actaea spicata* roots and antioxidant activity studies. *Indian J. Pharm. Sci.* **2011**, *73*, 666–669. [CrossRef]
27. Arvouet-Grand, A.; Vennat, B.; Pourrat, A.; Legret, P. Standardization of propolis extract and identification of principal constituents. *J. Pharm. Belg.* **1994**, *49*, 462–468.
28. Wanyika, H.N.; Gatebe, E.G.; Gitu, L.M.; Ngumba, E.K.; Maritim, C.W. Determination of caffeine content of tea and instant coffee brands found in the Kenyan market. *Afr. J. Food Sci.* **2010**, *4*, 353–358. [CrossRef]
29. Guo, D.J.; Cheng, H.L.; Chan, S.W.; Yu, P.H.F. Antioxidative activities and the total phenolic contents of tonic Chinese medicinal herbs. *Inflammopharmacology* **2008**, *16*, 201–207. [CrossRef]
30. Loizzo, M.R.; Pacetti, D.; Lucci, P.; Núñez, O.; Menichini, F.; Frega, N.G.; Tundis, R. *Prunus persica* var. platycarpa (Tabacchiera Peach): Bioactive compounds and antioxidant activity of pulp, peel and seed ethanolic extracts. *Plant Foods Hum. Nutr.* **2015**, *70*, 331–337. [CrossRef]
31. Benzie, I.F.F.; Szeto, Y.T. Total antioxidant capacity of teas by the ferric reducing/antioxidant power assay. *J. Agric. Food Chem.* **1999**, *47*, 633–636. [CrossRef]
32. Garrido-Martín, D.; Calvo, M.; Reverter, F.; Guigó, R. A fast non-parametric test of association for multiple traits. *Genome Biol.* **2023**, *24*, 230. [CrossRef]
33. Osman, M.A.; Mahmoud, G.I.; Shoman, S.S. Correlation between total phenols content, antioxidant power and cytotoxicity. *Biointerface Res. Appl. Chem.* **2020**, *11*, 10640–10653. [CrossRef]
34. Maurage, P.; Heeren, A.; Pesenti, M. Does Chocolate Consumption Really Boost Nobel Award Chances? The Peril of over-Interpreting Correlations in Health Studies. *J. Nutr.* **2013**, *143*, 931–933. [CrossRef] [PubMed]
35. Carvalho Neto, D.P.D.; Gonot-Schoupinsky, X.P.; Gonot-Schoupinsky, F.N. Coffee as a Naturally Beneficial and Sustainable Ingredient in Personal Care Products: A Systematic Scoping Review of the Evidence. *Front. Sustain.* **2021**, *2*, 697092. [CrossRef]
36. Hudáková, J.; Marcinčáková, D.; Legáth, J. Study of antioxidant effects of selected types of coffee. *Folia Vet.* **2016**, *60*, 34–38. [CrossRef]
37. Maxiselly, Y.; Anusornwanit, P.; Rugkong, A.; Chiarawipa, R.; Chanjula, P. Morpho-Physiological Traits, Phytochemical Composition, and Antioxidant Activity of Canephora Coffee Leaves at Various Stages. *Int. J. Plant Biol.* **2022**, *13*, 11. [CrossRef]
38. Tajik, N.; Tajik, M.; Mack, I.; Enck, P. The potential effects of chlorogenic acid, the main phenolic components in coffee, on health: A comprehensive review of the literature. *Eur. J. Nutr.* **2017**, *56*, 2215–2244. [CrossRef] [PubMed]
39. Behne, S.; Franke, H.; Schwarz, S.; Lachenmeier, D.W. Risk Assessment of Chlorogenic and Isochlorogenic Acids in Coffee By-Products. *Molecules* **2023**, *28*, 5540. [CrossRef] [PubMed]
40. Abdulmajid, A.M. Sensory Evaluation of beverage characteristics and biochemical components of *Coffee* Genotypes.. *Agric. Food Sci.* 2015. Available online: https://api.semanticscholar.org/CorpusID:212469690 (accessed on 20 November 2023).
41. Lemos, M.F.; de Andrade Salustriano, N.; de Souza Costa, M.M.; Lirio, K.; da Fonseca, A.F.A.; Pacheco, H.P.; Endringer, D.C.; Fronza, M.; Scherer, R. Chlorogenic acid and caffeine contents and anti-inflammatory and antioxidant activities of green beans of conilon and arabica coffees harvested with different degrees of maturation. *J. Saudi Chem. Soc.* **2022**, *26*, 101467. [CrossRef]
42. Acidri, R.; Sawai, Y.; Sugimoto, Y.; Handa, T.; Sasagawa, D.; Masunaga, T.; Yamamoto, S.; Nishihara, E. Phytochemical Profile and Antioxidant Capacity of Coffee Plant Organs Compared to Green and Roasted Coffee Beans. *Antioxidants* **2020**, *9*, 93. [CrossRef]
43. Platzer, M.; Kiese, S.; Herfellner, T.; Schweiggert-Weisz, U.; Miesbauer, O.; Eisner, P. Common Trends and Differences in Antioxidant Activity Analysis of Phenolic Substances Using Single Electron Transfer Based Assays. *Molecules* **2021**, *26*, 1244. [CrossRef]
44. Pacheco-Coello, F.; Rabottini-Villamizar, T. *Cafeína, Compuestos Fenólicos y Actividad Antioxidante en Granos de café Orgánico (Coffea arabica var. caturra) Sometidos a Tres Tiempos de Tostado, Nirgua, Estado Yaracuy, Venezuela*; Saber, Universidad de Oriente: Cumaná, Venezuela, 2022; Volume 34, pp. 20–27. [CrossRef]
45. Echegaray, N.; Pateiro, M.; Munekata, P.E.S.; Lorenzo, J.M.; Chabani, Z.; Farag, M.A.; Domínguez, R. Measurement of Antioxidant Capacity of Meat and Meat Products: Methods and Applications. *Molecules* **2021**, *26*, 3880. [CrossRef]
46. Rajurkar, N.; Gaikwad, K.; Razavi, M.S. Evaluation of Free Radical Scavenging Activity of Justicia Adhatoda: A Gamma. *Int. J. Pharm. Pharm. Sci.* **2012**, *4*, 1–4. Available online: https://web.archive.org/web/20180413024655id_/http://ijppsjournal.com/Vol4Suppl4/3734.pdf (accessed on 20 November 2023).

47. Tuncay, M.E. Comparison of Spectrophotometric Assays Determining Total Antioxidant Capacity Toplam Antioksidan Kapasiteyi Belirleyen Spektrofotometrik Yöntemlerin Karşılaştırılması. *Türk Klin. Biyokim. Derg.* **2019**, *17*, 111–119. Available online: https://tkb.dergisi.org/pdf/pdf_TKB_350.pdf (accessed on 20 November 2023).
48. Mendoza Isaza, N.A.; Hoyos-Arbeláez, J.A.; Peláez-Jaramillo, C.A. Antioxidant capacity and total polyphenol content of *Stevia rebaudiana* stem extracts in several in vitro models. *Rev. EIA* **2020**, *17*, 1–9. [CrossRef]
49. Rivas, D.; Dueñas, A.; Rodriguez, J. Secondary metabolites and antioxidant activity of wild tomatillo (*Solanum pimpinellifolium* L.). *Rev. Ténica La Fac. Ing.* **2020**, *2*, 26–32.
50. Demianova, A.; Bobkova, A.; Lidiková, J.; Jurcaga, L.; Bobko, M.; Belej, L.; Kolek, E.; Polakova, K.; Iriondo-DeHond, A.; del Castillo, M. Volatiles as chemical markers suitable for identification of the geographical origin of green *Coffea arabica* L. *Food Controñ* **2022**, *136*, 3–7. [CrossRef]
51. Teixeira, A.L.; Rocha, R.B.; Espindula, M.C.; Ramalho, A.R.; Júnior, J.R.V.; Alves, E.A.; Lunz, A.M.P.; Souza, F.d.F.; Costa, J.N.M.; Fernandes, C.d.F. Amazonian Robustas: New *Coffea canephora* coffee cultivars for the Western Brazilian Amazon. *Crop Breed. Appl. Biotechnol.* **2020**, *20*, e323420318. [CrossRef]
52. Koshiro, Y.; Zheng, X.-Q.; Wang, M.-L.; Nagai, C.; Ashihara, H. Changes in content and biosynthetic activity of caffeine and trigonelline during growth and ripening of *Coffea arabica* and *Coffea canephora* fruits. *Plant Sci.* **2006**, *171*, 242–250. [CrossRef]
53. Perdani, C.G.; Pranowo, D.; Qonitatilah. Total phenols content of green coffee (*Coffea arabica* and *Coffea canephora*) in East Java. *IOP Conf. Ser. Earth Environ. Sci.* **2019**, *230*, 012093. [CrossRef]
54. Vega, F.E.; Ziska, L.H.; Simpkins, A.; Infante, F.; Davis, A.P.; Rivera, J.A.; Barnaby, J.Y.; Wolf, J. Early growth phase and caffeine content response to recent and projected increases in atmospheric carbon dioxide in coffee (*Coffea arabica* and *C. canephora*). *Sci. Rep.* **2020**, *10*, 5875. [CrossRef] [PubMed]
55. Stevenson, P.C.; Nicolson, S.W.; Wright, G.A. Plant secondary metabolites in nectar: Impacts on pollinators and ecological functions. *Funct. Ecol.* **2017**, *31*, 65–75. [CrossRef]
56. Cotacallapa, R.E.; Lima-Medina, I.; Cornejo-Condori, G.B. Comparative organoleptic quality of coffee (*Coffea arabica* L.) in Puno-Peru and La Paz-Bolivia. *Rev. Investig. Altoandinas* **2019**, *21*, 283–292. [CrossRef]
57. Alves, A.L.; Pessoa, M.S.; de Souza, P.E.N.; Partelli, F.L.; Moscon, P.S.; da Silva, E.C.; Guimarães, A.O.; Muniz, E.P.; Pinheiro, P.F.; Borém, F.M.; et al. Influence of Environmental and Microclimate Factors on the Coffee Beans Quality (C. 65 canephora): Correlation between Chemical Analysis and Stable Free Radicals. *Agric. Sci.* **2018**, *9*, 9082. [CrossRef]
58. Ludwig, I.A.; Clifford, M.N.; Lean, M.E.J.; Ashihara, H.; Crozier, A. Coffee: Biochemistry and potential impact on health. *Food Funct.* **2014**, *5*, 1695–1717. [CrossRef] [PubMed]
59. Velásquez, S.; Banchón, C. Influence of pre- and post-harvest factors on the organoleptic and physicochemical quality of coffee: A short review. *J. Food Sci. Technol.* **2022**, *60*, 1–13. [CrossRef] [PubMed]

horticulturae

Article

Physicochemical Properties of Geographical Indication (GI) Sweet Cherries in China and Their Influencing Factors of Cultivar, Climate Type, and Soil Condition

Ying Nie, Jiazhang Huang *, Rui Liu, Pei Wang, Peng Liu, Man Lu and Junmao Sun *

Institute of Food and Nutrition Development, Ministry of Agriculture and Rural Affairs, Beijing 100081, China; nieying01@caas.cn (Y.N.); liurui@caas.cn (R.L.); wangpei_929@126.com (P.W.); liupeng@caas.cn (P.L.); luman@caas.cn (M.L.)
* Correspondence: huangjiazhang@caas.cn (J.H.); sunjunmao@caas.cn (J.S.)

Abstract: As one of the fruits widely planted in China, the quality of sweet cherries is affected by various factors. This study aims to investigate the characteristics of geographical indication (GI) sweet cherries grown in China and to analyze the effects of cultivars, climate types, and soil conditions on their quality traits. Twenty-two parameters of nine cherry samples and their planted soil properties were analyzed through a descriptive analysis and correlation analysis. There were significant differences in the physiochemical traits. Notable positive correlations between the fruit weight and its size, rate of edibility, and flavonoid content were shown. The Univariate-General Line Model exhibited that weight, soluble solids content (SSC), titratable acidity (TA), and total phenolic content (TPC) were mainly influenced by both the cultivar and climate type, while only the cultivar affected the maturity index (MI). Soil condition parameters were significantly different for each sample. Based on the established linear regression models, it was found that soil P had a positive impact on SSC and TA, but a negative effect on TPC ($p < 0.05$). On the other hand, soil K had a negative effect on TA but a positive impact on TPC ($p < 0.05$).

Keywords: sweet cherry; physiochemical properties; influencing factors; geographical indication

Citation: Nie, Y.; Huang, J.; Liu, R.; Wang, P.; Liu, P.; Lu, M.; Sun, J. Physicochemical Properties of Geographical Indication (GI) Sweet Cherries in China and Their Influencing Factors of Cultivar, Climate Type, and Soil Condition. *Horticulturae* **2023**, *9*, 1118. https://doi.org/10.3390/horticulturae9101118

Academic Editors: Paola Sánchez-Bravo and Luis Noguera-Artiaga

Received: 1 September 2023
Revised: 27 September 2023
Accepted: 3 October 2023
Published: 10 October 2023

1. Introduction

Sweet cherry (*Prunus avium* L.) is a commercially valuable fruit in the global market, due to its appealing appearance, delicious taste, and nutritional traits [1,2]. Large fruit size is considered to be a priority aim for breeding programs and is highly associated with commercial value [3]. The size-related parameters of width and weight are widely applied as grading indexes in North American, European, and Asian countries [4–6]. The levels of soluble solids, total sugar, and titratable acidity affect cherries' taste, both sweetness and sourness. These factors impact the overall perception of the taste of cherries and can determine whether consumers enjoy them or not [7,8]. Cherries are rich in polyphenolics that offer functional benefits to our health, making them a remarkable supplement to our diet. Anthocyanins, flavonoids, and hydroxycinnamic acids are dominant compounds for polyphenolics in cherry fruits [9]. Studies have shown that cherries contain phenolic compounds that can protect the nervous system, alleviate diabetes, inhibit chronic inflammation, and reduce serum uric acid levels. These benefits are attributed to the strong antioxidant properties of the compounds found in cherries [10–13].

The quality of cherries is influenced by various factors such as their genotypes, ripening stage, soil conditions, agronomic practices, and how they are treated after being harvested [14–16]. Sweet cherries grown in the same region exhibited noticeable variances in size, weight, SSC, TA, and phenolic composition [7,17]. When the same sweet cherry cultivars were introduced to separate experiment sites with different climate conditions, they showed varying levels of adaptability in terms of tree growth and fruit size [18,19].

With the soil amendment of sweet cherry trees, the changes in soil nutrition indicators have had an influence on their yielding and fruit characterizations [20,21]. Potassium plays an important role in carbon assimilation and transportation; metabolism regulation; affecting fruit size, color, soluble solids, and titratable acidity [22,23]; and increasing the ability of plants to resist diseases, and cold [24]. Phosphorus is one of the vital constitutional elements for cell membranes, primarily participating in energy transfer, enzyme activation/inactivation, and photosynthesis. It significantly contributes to fruit yield and modulating the production of metabolites and soluble solids [25].

The sweet cherry (*P. avium*) was first introduced in China more than 150 years ago and has become the dominant species with popular cultivars, such as Tieton, Hondeng, Brooks, Summit, Russia NO.8, Van, and Santina [26,27]. According to the USDA's annual report, China has emerged as the top consumer of sweet cherries in recent times. Cherry production in China was 650,000 tons in 2022/2023, followed by Turkey (980,000 tons) and the European Union (727,000 tons). By 2022, cherry planting areas in China reached about 170,000 hectares, ranking first in the world [27]. With greatly increasing demand in the Chinese market, sweet cherries are not only grown in the originally introduced place Bohai Bay but also widely expanded in various provinces in the northwestern and southwestern parts of China. Since there is a lack of reasonable evaluation for growers, the newly introduced varieties cannot fully reflect the high quality in the new planting area as in the original places.

A geographical indication (GI) identifies a good that originated in a specific region, where given quality, reputation, or characteristics are essentially attributed to its geographical origin [28]. By 2022, there were 39 pieces of cherry GIs (mostly belonging to sweet cherries) registered in the Ministry of Agriculture and Rural Affairs of the People's Republic of China (PRC). Each GI cherry product has unique characteristics due to its variations in cultivar and planting conditions. With regard to GI products (cherries included), researchers paid more attention to establishing traceability models to distinguish the authenticity of agricultural products by detecting multiple mineral elements and volatile compounds [29–31], rather than analyzing differences in product quality traits contributing to their cultivar and unique planting conditions [32,33].

Though many researchers have studied the impact of growing conditions (weather, fertilization, and irrigation) on sweet cherries in major producing countries [34–36], most of them were more likely to focus on vegetative growth and to study the effects of the specific soil nutrients on growth [37,38]; the influencing factors on fruit quality parameters have also been rarely involved. Also, a few surveys focused on cherry fruit characteristics in China—an emerging producer. In order to make suggestions for growers to select suitable sweet cherry cultivars in the appropriate planting environment and to manage orchards effectively, it is necessary to investigate the physicochemical properties of GI cherries in different regions and identify the factors (cultivar, climate type, and soil conditions) that affect them.

2. Materials and Methods

2.1. Fruit and Soil Materials

Nine cherry samples were obtained from nine GI products registered in the Ministry of Agriculture and Rural Affairs of the PRC (Table 1). For each GI product, one presentative commercial mature sweet cherry cultivar was selected. Three orchards located relatively scattered were chosen within the protective range for a GI product. The same weight of cherry fruits (500 g) was randomly picked and collected in an orchard. Cherries from three orchards were fully mixed as a whole, regarded as one GI cherry sample (about 1.5 kg for each). The samples were immediately packed with ice bags and transported to the laboratory within 24 h. All parameters of fresh fruits, including weight, size, juice yield, edible rate, soluble solids content, titratable acidity, maturity index, sugar composition, total sugar, total phenolics, total anthocyanins, total flavonoids, procyanidin, β-carotene, ascorbic acid, and minerals (potassium, magnesium, calcium, and iron), were analyzed immediately.

Table 1. Name, origin, variety, climate type, and maturation time for GI cherries in this study.

GI Name	Origin	Variety	Climate Type	Picking Time
Baqiao	Xi'an, Shanxi	Hongdeng	Temperate monsoon	Mid May
Tianbao	Taian, Shandong	Huangmi	Temperate monsoon	Mid May
Jiangxian	Yuncheng, Shanxi	Tieton	Temperate monsoon	Mid May
Wenchuan	Aba Tibetan and Qiang Autonomous Prefecture, Sichuan	Hongdeng	Subtropical monsoon	Mid May
Xiehu	Lianyungang, Jiangsu	Brooks	Subtropical monsoon	Mid May~Early June
Tongzhou	Tongzhou, Beijing	Hongdeng	Temperate monsoon	Early June
Qinzhou	Tianshui, Gansu	Tieton	Temperate monsoon	Mid June
Ledu	Haidong, Qinghai	Tieton	Plateau mountain	Mid June
Dalian	Dalian, Liaoning	Russian NO.8	Temperate monsoon	Late June

Taking the trunk as a circular dot and extending outward to the edge of the crown projection for soil collection, each tree took two points symmetrically, and 0–40 cm of soil was drilled with stainless steel soil. The soil samples collected from the three planting areas were thoroughly blended to ensure homogeneity. In each orchard, 300 g of a soil sample was collected, and for each GI product, a total of 900 g of a soil sample was obtained.

2.2. Chemicals and Reagents

All chemicals applied were an analytical grade, and the mobile phase ingredient and solution for the determination of sugar composition, total anthocyanins, β-carotene, and ascorbic acid were of an HPLC grade. Sugar standards included glucose (≥99.5%), fructose (≥99.0%), lactose (≥98.0%), sucrose (≥99.5%), and maltose (≥99.0%), as well as L (+)-ascorbic acid (≥99.0%), all purchased from Sigma (St. Louis, MO, USA). Gallic acid (≥99.0%), cyanidin-3-O-rutinoside (≥95.0%), procyanidin (≥95.0%), β-carotene (≥98.0%), delphinidin (≥97.0%), cyanidin-O-glucoside (≥95.0%), petunidin chloride (≥95.0%), pelargonium chloride (≥96.0%), peonidin (≥97.0%), and malvidin (≥97.0%) were from Yuanye Bio-Technology (Shanghai, China). The purities of potassium, magnesium, calcium, and iron standards were all 1000 μg/mL, all of which were purchased from the National Nonferrous Metals and Electronic Materials Analysis and Testing Center (Beijing, China).

2.3. Weight and Size Measurements

Twenty-five cherries of each GI sample were randomly chosen for each batch. Their weights were measured with the ME204/02 digital weight scale (METTLER TOLEDO, Shanghai, China). The width and length of fruits were determined using a digital vernier caliper (GELIXI, Hangzhou, China). The shape index was calculated as

$$\text{shape index} = \text{length}/\text{width}$$

The results were expressed as average values of 25 cherry fruits.

2.4. Juice Yield and Edible Rate Measurements

Fruit juice was obtained from 500 g of cherry flesh pulped with an electronic juicer. The juice sample was then centrifuged at 6149× g for 5 min using a high-speed centrifuge and the supernatant was collected and weighed [39]. The edible rate was calculated by randomly selecting 15 cherries, then weighing the whole fruit, stone, and stalk, respectively. These two parameters were measured in triplicate.

$$edible\ rate = \frac{\text{M(whole fruit)} - \text{M(stone)} - \text{M(stalk)}}{\text{M(whole fruit)}} \times 100\%$$

2.5. Soluble Solids Content, Titratable Acidity, Maturity Index, Sugar Composition, and Total Sugar

These parameters were detected with an independent homogenate (n = 2), extracted from 25 pitted cherries. Soluble solids content (SSC) was measured with a digital refractometer (MDS-R500, Shanghai, China) and the result was expressed as Brix°. Titratable acidity (TA) was determined with an FE28 pH meter (METTLER TOLEDO, Shanghai, China), titrating the sample with 0.1 M NaOH up to pH 8.2. The result was expressed as g malic acid 100 g^{-1} fresh weight (FW). The maturity index (MI) was the ratio of SSC and TA. Sugar composition was determined with a CM5400 HPLC system (HITACHI, Tokyo, Japan) equipped with a ZOBRAX NH2 column (4.6 mm × 250 mm × 5 μm) and Chromaster 5430 refractive index detector (HITACHI, Tokyo, Japan). Each sample was injected twice. Total sugar values were calculated with the sum of individual sugar contents [40].

2.6. *Total Phenolics, Total Anthocyanins, Total Flavonoids, and Procyanidin*

Total phenolic content (TPC) was assayed according to the Folin–Ciocalteu method [41] with a T6 UV spectrophotometer (PERSEE, Beijing, China). Gallic acid was used as a standard for quantification. The results were expressed as gallic acid mg 100 g^{-1} FW. Total anthocyanins content (TAC) was determined with an HPLC system equipped with a ZORBAX SB C18 column (4.6 mm × 250 mm × 5μm) and chromaster 5100 UV detector (HITACHI, Japan), measured at 530 nm. The column was eluted with the mobile phase (A: 1% formic acid solution; B: 1% formic-acid–acetonitrile solution). Anthocyanins were ultrasonically extracted from the cherry homogenate combined with the ethanol-water-HCl solution (*v*:*v*:*v* = 2:1:1) for 30 min, and then hydrolyzed with boiling water for 1 h [42]. Quantification was carried out by comparing the peak area of the sample solution with that of the mixed standard solution, which was dissolved with a HCl-methanol solution (*v*:*v* = 10:90). The anthocyanin content was the sum of six anthocyanins, and expressed as mg kg^{-1} FW. According to Marinella De Leo [1], total flavonoid content (TFC) was measured with the aluminum chloride colorimetric method. The solution mixture was measured with a T6 UV spectrophotometer (PERSEE, Beijing, China) at 510 nm. The results were expressed as quercetin (mg100 g^{-1} FW). Procyanidin content was determined using Monika with a few modifications [43]. A T6 UV spectrophotometer (PERSEE, Beijing, China) was applied to detect anthocyanidin ions generated by being treated with a 5% (*v*/*v*) HCl-n-butanol solution and a 2% (*w*/*v*) $NH_4Fe(SO_4)_2$ solution (prepared with 2 M HCl) in boiling water for 40 min. The absorbance was measured at 546 nm. The results were expressed as mg 100 g^{-1} FW.

2.7. *β-Carotene and Ascorbic Acid*

The β-carotene content was measured with an a1260 HPLC system (Agilent, Santa Clara, CA, USA) equipped with an Ultimate XB-C18 column (4.6 mm × 250 mm × 5 μm) and G1315D VL DAD detector (Agilent, Santa Clara, CA, USA). Homogenized samples were saponified to release carotene into a free state, and dichloromethane was extracted with petroleum ether to constant volume, and then separated using reversed-phase chromatography with quantification using an external standard method. Mobile phase A and B were methanol–acetonitrile–water (*v*:*v*:*v* = 73.5:24.5:2) and methyl tert-butyl ether (MTBE), respectively. The results were expressed as ug100 g^{-1} FW [44]. The mixture of cherry flesh and metaphosphoric acid dissolution was homogenized as tested samples for measuring ascorbic acid [45], which was evaluated with a CM5110 HPLC system (HITACHI, Japan) equipped with an Ultimate XB-C18 column (4.6 mm × 250 mm × 5 μm) and G1315D VL DAD detector (Agilent, Santa Clara, CA, USA). The column was diluted with the mobile phase (A: KH_2PO_4-Cetyltrimethylammonium bromide water solution; B: 100% methanol). The results were reported as mg100 g^{-1} FW.

2.8. *Mineral Content of Cherry Fruits*

The mineral content was detected with an iCAP6300 inductively coupled plasma emission spectrometer (Thermo, MA, USA) with homogenized samples, prepared with the wet digestion method according to Matos-Reyes [46]. The sample was added with a 10 mL mixture solution of nitric acid and perchloric acid (*v*:*v* = 10:1). Then, the mixture was heated at a high temperature until white smoke appeared and the digestion solution was transparent and colorless. The solution was cooled down and water was added to 25 mL for measuring. The results were expressed as mg kg^{-1} FW.

2.9. Parameters of Planted Soil

All soil samples were dried and passed through a 2 mm sieve before measurements. Soil pH was measured with a PHS-3C pH meter (JINGHONG, Shanghai, China) with an electrode immersed in standard buffers of pH 4.01, pH 6.87, and pH 9.18 in turn for calibration. The soil–water suspension samples (w:v = 1:2.5) were prepared with water removing CO_2. The electrode should be washed with clean water, and be absorbed using filter paper strips after each test [47]. Soil organic matter was evaluated with an excessive 0.4 M potassium-dichromate–sulfuric-acid solution to oxidize organic carbon in the air-dried soil sample, and the excess potassium dichromate was titrated with a ferrous sulfate standard solution. The amount of organic carbon was calculated according to the oxidation correction coefficient from the consumed amount of potassium dichromate and then multiplied by the constant of 1.724, which was the content of soil organic matter [47]. Available phosphorus was determined with a UV-1800 spectrophotometer (SHIMADZU, Kyoto, Japan) with the tested sample, prepared with a hydrochloric acid and sulfuric acid solution to dissolve and release iron phosphate and aluminum salt. The phosphorus standard series chromogenic solution was made with a 5.0 mL aluminum ladder anti-developer at 20 °C (30 min) and then measured at 700 nm [47]. Available potassium was detected with a 5800 ICP-OES inductively coupled plasma optical emission spectrometer (Agilent, Santa Clara, CA, USA) with 1 M acetic acid extraction [48]. Available calcium and available iron were measured with ammonium-hydrogen-carbonate–diethylene-triaminepentaacetic-acid (AB-DTPA) extraction and DTPA extraction, respectively. Then, the prepared samples were detected with a 5800 ICP-OES inductively coupled plasma optical emission spectrometer (Agilent, Santa Clara, CA, USA) [49].

2.10. Statistics Analysis

All experiment data were detected twice and expressed as the mean ± SD using Microsoft Excel 2019 (Microsoft, Redmond, WA, USA), except for fruit weight, length, width, and shape index, which were expressed as the mean ± SD of 25 determinations. An analysis of variance (ANOVA) was performed on all the physicochemical parameters. A significant difference analysis of all parameters between groups was performed with Duncan's test and $p < 0.05$ was regarded as the significant level. A correlation analysis of physiochemical and soil parameters was determined with Pearson's correlation coefficient at a significant level for $p < 0.05$, $p < 0.01$, and $p < 0.001$, respectively. The correlation analysis of physiochemical parameters was completed with the Omicshare online platform (https://www.omicshare.com/tools/Home/Task/taskdetail?tasknum=reporticawg5e8d0d, accessed on 10 August 2023). The univariate general linear model was applied to analyze the influencing factors of main quality traits. The relationship between quality and soil condition was established with the multiple linear regression model after a collinearity analysis. All these analysis methods were completed with SPSS Statistics V17.0 (IBM Inc., Chicago, IL, USA).

3. Results and Discussions

3.1. Variability in Physicochemical Properties of GI Cherries

The external properties of GI cherries are shown in Table 2. The size and weight of cherry fruits are important commercial attributes for market acceptance [3,50]. The width and length of cherry samples varied from 21.61 mm to 34.78 mm, and 20.33 mm to 31.15 mm, respectively. The fruit weight ranged from 5.00 g to 17.92 g. Both the size and weight of cherries exhibited significant differences ($p < 0.05$). Dalian showed significantly bigger size and weight than the others, while Tianbao was the smallest among all samples. The result of fruit weight exhibited as higher than that of Italian cherries (3.85~12.97 g) and Portuguese cherries (4.87~11.75 g) [7,51]. The shape index of GI cherries ranged from

0.79 to 0.99, and the average value was 0.89. Although Tongzhou had the same variety as Baqiao and Wenchuan, it had a distinct shape that set it apart from the others. Both juice yield and edible rate are important factors influencing the taste of commercial cherries. The edible rates of cherry samples ranged from 90.73 ± 0.15% to 96.23 ± 0.25% with a CV of 1.92%. However, the juice yield had a wider range, varying from 39.70 ± 0.61% to 71.10 ± 0.82% with a CV of 16.61%. This result is in accord with Jia et al. [39], as the juice yield is comprehensively affected by genotype, planting patterns, and maturity [52,53].

Table 2. External properties of GI cherry samples.

Sample	Width (mm)	Length (mm)	Shape Index	Weight (g)	Juice Yield (%)	Edible Rate (%)
Baqiao	28.66 ± 0.98 d	24.53 ± 0.83 c	0.86 ± 0.03 bc	10.24 ± 0.09 d	60.47 ± 1.29 c	92.50 ± 0.36 b
Tianbao	21.61 ± 1.65 a	21.38 ± 0.95 ab	0.99 ± 0.04 e	5.00 ± 0.23 a	63.63 ± 1.10 d	90.93 ± 0.75 a
Jiangxian	24.08 ± 1.24 b	21.05 ± 5.67 ab	0.91 ± 0.18 bcd	11.00 ± 0.04 e	66.08 ± 1.63 e	93.37 ± 0.31 c
Wenchuan	24.23 ± 1.35 b	20.33 ± 1.20 a	0.84 ± 0.04 ab	10.40 ± 0.01 d	49.09 ± 1.30 b	92.93 ± 0.54 bc
Xiehu	24.41 ± 0.68 b	22.23 ± 0.67 b	0.91 ± 0.04 d	6.71 ± 0.11 b	68.20 ± 0.23 f	90.73 ± 0.15 a
Tongzhou	26.93 ± 2.68 c	21.27 ± 1.48 ab	0.79 ± 0.04 a	6.97 ± 0.21 b	39.70 ± 0.61 a	92.67 ± 0.42 bc
Qinzhou	31.77 ± 1.34 e	29.45 ± 1.05 d	0.93 ± 0.03 d	14.76 ± 0.24 f	71.10 ± 0.82 g	95.20 ± 0.27 d
Ledu	26.80 ± 1.43 c	23.60 ± 0.90 c	0.88 ± 0.03 bcd	8.22 ± 0.25 c	66.17 ± 0.35 e	92.50 ± 0.36 b
Dalian	34.78 ± 1.26 f	31.15 ± 1.24 e	0.90 ± 0.03 cd	17.92 ± 0.10 g	59.97 ± 0.35 c	96.23 ± 0.25 e
Mean	27.03	23.89	0.89	10.14	60.49	93.01
Std.	4.15	3.88	0.06	4.10	10.05	1.78
CV%	15.37	16.27	6.41	40.47	16.61	1.92

Note: Std. and CV represent standard deviation and coefficient of variation, respectively. Different letters in each column stand for mean values significantly different according to Duncan's test at $p < 0.05$.

Soluble solids content (SSC) and titratable acidity (TA) represent the cherry composition of soluble content (vitamin, mineral, amino acids, sugar, etc.) and organic acids, respectively [54]. SSC and TA are also essential parameters reflecting cherry flavor with sweetness and sourness, and their levels gradually increase during the ripening process for cherries. Ultimately, MI as a maturity index is often analyzed to evaluate the optimum harvest date and to determine consumer preference [52,54,55]. From Table 3, the average SSC was 16.54 ± 3.15 Brix°, ranging from 12.64 ± 0.10 Brix° to 23.67 ± 1.00 Brix°. Significant differences among samples were exhibited in terms of TA, which varied widely from 0.33 ± 0.01% to 0.99 ± 0.01%. Both the ranges of SSC and TA were in general accordance with the worldwide level (SSC: 13.5 to 24.5 Brix°; TA: 0.5% to 1.3%) [54]. Previous studies reported that SSC values above 15~18 Brix° could be regarded as commercially ideal [56,57]. Tianbao (12.64 ± 0.10 Brix°) and Qinzhou (15.69 ± 0.06 Brix°) have lower soluble solid content compared to others. However, their MI values are significantly higher, with Tianbao at 38.56 ± 1.82 and Qinzhou at 35.22 ± 0.30, which can be attributed to their lower titratable acidity. A series of studies confirmed that the SSC and TA of cherries mainly depend on cultivars [58,59]. Therefore, it can be seen that taking SSC, TA, and MI into account is more comprehensive and reasonable than only detecting the SSC of cherries to measure the fruit's degree of ripening. Compared to SSC, total sugar can be more exact to confirm the sugar content of fruits and to reflect the sweet flavor of cherries. Table 3 exhibits that the average total sugar was 13.81 ± 3.65 g/100 g, varying greatly from 9.09 ± 0.16 g/100 g (Baqiao) to 19.70 ± 0.22 g/100 g (Tongzhou). With regard to sugar composition, the results are similar to the previous report that glucose and fructose are the two main sugars found in GI cherries, and the glucose content is bigger than that of fructose [60,61]. For fructose, it ranged from 4.20 ± 0.01 g/100 g FW (Baqiao) to 6.37 ± 0.41 g/100 g FW (Wenchuan), in accordance with the sweet cherries grown in Canada and Italy (4.4 g/100 g FW to 6.7 g/100 g FW) [9,62]. Meanwhile, glucose of GI cherries owned a broader range (4.89 ± 0.09 g/100 g FW to 13.79 ± 0.50 g/100 g FW) than that of cherries in other countries (5.2 g/100 g FW to 10 g/100 g FW).

Table 3. Taste properties of GI cherry samples.

Sample	Soluble Solids (Brix°)	Titratable Acidity (g/kg)	Maturity Index	Total Sugar (g/100 g)	Glucose (g/100 g)	Fructose (g/100 g)
Baqiao	12.65 ± 0.49 a	4.9 ± 0.00 c	25.65 ± 1.10 b	9.09 ± 0.16 a	4.89 ± 0.09 a	4.20 ± 0.01 a
Tianbao	12.64 ± 0.10 a	3.3 ± 0.01 a	38.56 ± 1.82 e	9.83 ± 0.01 a	5.33 ± 0.26 a	4.50 ± 0.01 a
Jiangxian	17.47 ± 1.15 c	5.4 ± 0.00 c	32.59 ± 2.16 d	15.69 ± 0.05 e	10.30 ± 0.45 d	5.39 ± 0.11 b
Wenchuan	23.67 ± 1.00 e	8.0 ± 0.00 e	29.52 ± 0.96 c	18.75 ± 0.36 f	12.38 ± 0.64 e	6.37 ± 0.41 d
Xiehu	15.29 ± 0.17 b	8.1 ± 0.00 e	18.82 ± 0.10 a	13.85 ± 0.79 d	9.42 ± 0.40 c	4.43 ± 0.18 a
Tongzhou	19.08 ± 0.61 d	9.9 ± 0.01 f	19.21 ± 0.84 a	19.70 ± 0.22 g	13.79 ± 0.50 f	5.91 ± 0.27 c
Qinzhou	15.69 ± 0.06 b	4.5 ± 0.00 b	35.22 ± 0.30 e	12.79 ± 0.13 c	7.28 ± 0.34 b	5.51 ± 0.18 bc
Ledu	17.72 ± 0.28 c	6.9 ± 0.02 d	25.72 ± 0.37 b	12.73 ± 0.02 c	6.83 ± 0.08 b	5.90 ± 0.08 c
Dalian	17.22 ± 0.28 c	5.5 ± 0.01 c	31.13 ± 0.78 cd	11.86 ± 0.06 b	6.67 ± 0.09 b	5.19 ± 0.11 b
Mean	16.54	6.10	28.49	13.81	8.11	5.24
Std.	3.15	2.00	6.77	3.65	3.08	0.68
CV%	19.01%	32.94%	23.77%	26.45%	37.93%	12.93%

Notes: Std. and CV represent standard deviation and coefficient of variation, respectively. Different letters in each column stand for mean values significantly different according to Duncan's test at $p < 0.05$.

The nutritional properties of GI cherries showed a significant difference ($p < 0.05$) in this study (Table 4). Polyphenols including anthocyanins, flavonoids, and phenolic acids [54] are the dominant phytochemicals for sweet cherries with strong antioxidant activity in vitro and in vivo [63–65]. The formation and accumulation of total phenolic compounds are influenced by various factors, such as cultivars, agronomic conditions, maturity degree, and postharvest preservation [7,50,66]. Regarding TPC, the highest value was 197.72 ± 3.94 mg/100 g FW for Dalian, while the lowest value of 100.04 ± 3.88 mg/100 g FW was observed in Wenchuan. The TPC range of GI cherries in this study is higher than that of twenty-four sweet cherry cultivars planted in Italy and Turkey (84.96 ± 3.37 mg/100 g FW to 156 ± 1.30 mg/100 g FW; 58.31 ± 10.56 mg/100 g FW to 115.41 ± 7.98 mg/100 g FW) [60,67]. During the ripening stages of sweet cherries, phenolic compounds, particularly anthocyanins, play a significant role in producing the red color [2]. TPC and TAC of Dalian exhibited as significantly higher than others (197.72 ± 3.94 mg/100 g FW and 53.82 ± 2.31 mg/kg FW, respectively) as the Dalian cherry sample owned a dark red color mainly due to genotype and ripening stage [2]. Flavonoids also contribute to the biological activities of sweet cherries. TFC of GI cherries ranged from 7.86 ± 0.49 mg/100 g FW (Tianbao) to 35.64 ± 2.22 mg/100 g FW (Dalian), which generally represented the total flavonoids content of sweet cherries grown in China (7.79 ± 0.26 mg/100 g FW to 25.03 ± 0.26 mg/100 g FW) [68]. However, the result is lower than that of cherry fruits planted in Poland and Spain [64,65]. Procyanidin is a type of colorless chemical compound that is not present in all cherry cultivars [7]. However, most sour cherry cultivars contain high levels of procyanidin, which is known to have strong antioxidant properties and can effectively scavenge free radicals [69]. The content of procyanidin greatly varied from 2.98 ± 1.34 mg/100 g FW (Xiehu) to 61.77 ± 1.85 mg/100 g FW (Dalian). Carotenoids are a group of pigment compounds primarily composed of β-carotene. This yellow lipophilic pigment is responsible for converting into vitamin A in the human body [70]. Baqiao was significantly the highest (followed by Tianbao), while Jiangxian, Ledu, Qinzhou, and Wenchuan were under 50 μg/100 g. The β-carotene levels of Baqiao, Wenchuan, and Tongzhou, despite belonging to the same cultivar, displayed noteworthy differences ($p < 0.05$). This result could also be explained with their growing conditions [70,71]. Ascorbic acid is one of the metabolites with antioxidant activity, besides phenolic compounds and carotenoids [8]. Generally, the ascorbic acid of sweet cherries and sour cherries ranged from 6~10 mg/100 g FW and 5~22 mg/100 g FW, respectively [9,69]. The content of ascorbic acid in the samples varied, with Qinzhou showing the highest at 7.61 ± 0.42 mg/100 g FW and Dalian exhibiting the lowest at 3.69 ± 0.24 mg/100 g FW. The mean value was 6.09 ± 1.80 mg/100 g FW, which was higher than Van, Noir De Guben, and 0–900 Ziraat grown in Turkey [72]. The mineral properties of GI cherries varied significantly ($p < 0.05$). The mineral composition of cherries not only depends on fruit cultivars but also on soil conditions [69,73]. When it comes to potassium levels, Tongzhou (3090.44 ± 114.08 mg/kg FW) was significantly the highest while Baqiao (1840.91 ± 77.73 mg/kg FW) was the lowest. The mean value for all samples was 2268.84 ± 402.47 mg/kg FW. Cherries are considered to be a supplemental intake source of potassium [7,9]. The 100 g of GI cherries analyzed in this study could provide 9.20~15.45% for AI (2000 mg/d) of potassium. Compared to the abundant content of potassium, lower concentrations of magnesium (92.31 ± 0.85~184.10 ± 2.91 mg/kg FW), calcium (70.13 ± 4.89~155.01 ± 5.40 mg/kg FW), and iron (3.01 ± 0.14~9.14 ± 0.38 mg/kg FW) were also found [9].

A variance analysis of physiochemical parameters in Tables 2–4 suggests that high variability existed for most of the characteristics in cherry fruits due to various influencing factors. Dalian is characterized as a big size with high fruit weight, and phenolics enrichment. Wenchuan is outstanding for its high values of soluble solids content, while presenting a low value of total phenolic compounds. Qinzhou owned high values of juice yield and maturity index. Tongzhou is notable for mineral enrichment (K, Mg, and Ca) with a high content of titratable acidity, total sugar, and glucose. Baqiao was enriched in β-carotene with a low value regarding soluble solids content. Though Baqiao, Wenchuan, and Tongzhou belonged to the same cultivar (Hongdeng), these samples exhibited different quality characterizations, which maybe contributed to planting environment conditions. Moreover, 8 out of 22 parameters showed smaller CVs than 20% [74,75], which indicated low variability among GI cherries for width, length, shape index, juice yield, edible rate, fructose, and potassium. Breeding programs focus on several important characteristics that are desirable to consumers, including fruit size, juice yield, edible rate, and soluble solids content. These traits can be easily noticed and appreciated by consumers [50,76,77]. All nutritional trait parameters had a coefficient of variation (CV) greater than 20% (except for K) with values ranging from 21.16% for TPC to 113.46% for β-carotene. In recent times, consumers have come to recognize the importance of the health benefits associated with the nutrient-rich properties of cherries, in addition to their appearance and taste. As a result, breeding efforts should also aim to improve nutritional parameters alongside size and taste.

Table 4. Nutritional properties of GI cherry samples.

Sample	TPC (mg/100 g)	TAC (mg/kg)	TFC (mg/100 g)	Procyanidin (mg/100 g)	β-Carotene (ug/100 g)	Ascorbic Acids (mg/100 g)	Potassium (mg/kg)	Magnesium (mg/kg)	Calcium (mg/kg)	Iron (mg/kg)
Baqiao	123.04 ± 2.23 bc	5.79 ± 0.27 a	11.59 ± 0.67 bc	16.35 ± 0.86 c	187.86 ± 4.93 f	6.62 ± 0.13 c	1840.91 ± 77.73 a	104.22 ± 4.41 b	105.66 ± 3.01 d	9.14 ± 0.38 f
Tianbao	143.10 ± 0.60 d	6.78 ± 0.42 a	7.86 ± 0.49 a	2.98 ± 1.34 a	147.85 ± 9.44 e	5.48 ± 0.13 b	1883.01 ± 117.06 a	114.47 ± 1.89 c	126.93 ± 3.90 e	3.66 ± 0.10 ab
Jiangxian	125.65 ± 1.44 c	17.60 ± 0.57 de	12.23 ± 0.56 c	28.70 ± 1.74 e	42.94 ± 2.91 bc	5.59 ± 0.17 b	2144.99 ± 65.46 bc	100.27 ± 1.56 b	91.85 ± 2.51 c	3.01 ± 0.14 a
Wenchuan	100.04 ± 3.88 a	15.76 ± 1.06 cd	14.54 ± 0.11 d	57.91 ± 1.67 g	17.00 ± 0.95 a	5.55 ± 0.17 b	2459.53 ± 33.57 d	128.82 ± 1.48 e	125.23 ± 2.04 e	5.05 ± 0.34 d
Xiehu	104.71 ± 5.68 a	10.75 ± 0.36 b	9.54 ± 0.08 ab	4.31 ± 0.06 a	115.07 ± 7.86 d	5.74 ± 0.23 b	2591.51 ± 47.16 d	112.76 ± 1.46 c	79.10 ± 2.96 b	6.50 ± 0.30 e
Tongzhou	115.10 ± 7.18 b	14.74 ± 0.06 c	16.06 ± 0.44 de	11.36 ± 0.28 b	127.26 ± 8.60 d	3.81 ± 0.04 a	3090.44 ± 114.08 e	184.10 ± 2.91 g	155.01 ± 5.40 f	4.82 ± 0.33 d
Qinzhou	132.65 ± 0.14 c	18.81 ± 1.00 e	21.80 ± 0.90 f	22.28 ± 1.46 d	30.94 ± 1.06 b	7.61 ± 0.42 d	2016.18 ± 31.19 ab	92.31 ± 0.85 a	89.83 ± 0.39 c	3.89 ± 0.27 bc
Ledu	149.06 ± 6.31 d	19.04 ± 0.95 e	17.82 ± 0.76 e	39.69 ± 2.24 f	40.00 ± 1.59 bc	6.96 ± 0.26 c	2529.14 ± 0.81 d	138.92 ± 2.21 f	100.93 ± 0.82 d	4.25 ± 0.11 cd
Dalian	197.72 ± 3.94 e	53.82 ± 2.31 f	35.64 ± 2.22 g	61.77 ± 1.85 h	52.12 ± 2.42 c	3.69 ± 0.24 a	2268.05 ± 131.92 c	120.96 ± 3.83 d	70.13 ± 4.89 a	3.03 ± 0.02 a
Mean	131.5	18.44	18.69	25.43	125.94	6.09	2268.84	123.77	108.32	4.82
Std.	27.83	13.49	10.86	21.35	142.89	1.8	402.47	26.52	27.31	1.95
CV%	21.16%	73.15%	58.10%	83.97%	113.46%	29.56%	17.74%	21.42%	25.21%	40.57%

Notes: Std. and CV represent standard deviation and coefficient of variation, respectively. Different letters in each column stand for mean values significantly different according to Duncan's test at $p < 0.05$. TPC, total phenolic content; TAC, total anthocyanins content; TFC, total flavonoid content.

3.2. Correlation Analysis of Quality Parameters for GI Cherries

Correlations of physicochemical properties exhibited at significant levels (Figure 1). Cherry fruit weight showed remarkably positive correlations with width (r = 0.856 **), length (r = 0.834 **), and edible rate (r = 0.956 **), which were supported with the previous literature studying sweet cherries [55,59,74]. Additionally, fruit weight showed noticeably high correlation with total flavonoids content (r = 0.855 **). Edible rate showed a positive correlation with flavonoids (r = 0.897 **) and anthocyanin (r = 0.805 **). It can be explained that the weight of the fruit increases as phenolic compounds accumulate during the maturation of sweet cherries [78–80]. The content of soluble solids in fruits is closely linked to both total sugar and fructose content, with strong positive correlations (r = 0.824 ** and r = 0.896 **, respectively). Soluble solids content is a key indicator of fruit maturity and sweetness, making them important parameters to consider [79,80]. The amount of sugar in the sweet cherries was mostly made up of total sugar and had a strong correlation with glucose (r = 0.993 **). According to Figure 1, there was a strong positive correlation (r = 0.945 **) between anthocyanin and total flavonoid content. It has been reported that pre-synthesis genes of anthocyanin play a dominant role in the biosynthesis of flavonoids during sweet cherries' development [79]. Titratable acidity showed positive correlations with glucose (r = 0.821 **) and potassium (r = 0.949 **). Accumulation of organic acids happened in the same ripening stage as the synthesis of sugar. Both the content of organic acids and sugar increased to a summit point at harvest time for sweet cherries [81,82]. Previous literature found that the titratable acidity of sweet cherries enhanced with increasing doses of potassium fertilization [38,83], simultaneously exerting positive effects on the content of potassium in cherry fruits [8]. A significantly positive correlation between potassium and magnesium (r = 0.851 **) was also detected in this study. Currently, there is insufficient data on the correlation between mineral composition in cherries. Only a few researchers have found that applying potassium fertilization led to a decrease in magnesium levels in sweet cherry fruits and leaves [84,85]. The result from Figure 1 shows the negative correlation between Ca and juice yield (r = −0.765 *); however, no current literature reported a direct relationship between them. It is possible to explain that Ca in fruit determines the functionality of sweet cherry xylem development. This contributes to increasing fruit firmness, reducing fruit water absorption, and avoiding cracking [35,86,87].

The cherry properties can be described with external parameters (width, length, shape index, fruit weight, juice yield, and edible rate), taste parameters (soluble solids content, titratable acidity, maturity index, total sugar, glucose, and fructose), and nutrition parameters (total phenolics, total flavonoids, total anthocyanins, procyanidin, β-carotene, ascorbic acid, potassium, magnesium, calcium, and iron). After taking into account the connections between various external factors and ease of detection, the fruit weight was selected as a representation of the exterior qualities of cherries [3,5,74]. When evaluating the taste of cherries, the balance of sweet and sour flavors is considered the most important factor. The flavor is influenced by factors such as the soluble solids content, titratable acidity, and maturity index, which are also helpful for measuring the maturity of the fruit. Polyphenol compounds are the dominant bioactive constituents. Both anthocyanins and flavonoids belong to phenolic compound groups. Total phenols, anthocyanins, and flavonoids show significant positive correlations, as seen in Figure 1. Therefore, fruit weight, soluble solids content, titratable acidity, maturity index, and total phenols are regarded as typical quality parameters of GI cherries to study their influencing factors.

3.3. Effect of Cultivar and Climate Type on the Properties of GI Cherries

The Univariate-General Line Model presents the average values of quality parameters and the effects of variety and climate types in Table 5. Significant effects of variety and climate on fruit weight, acidity, and phenols exist ($p < 0.01$). Russia NO.8 had significantly higher fruit weight than other cultivars. Cherries grown in temperate monsoon climates were also heavier than those in subtropical monsoon and plateau mountain climates. Different cultivars exhibit a range of fruit weights due to the correlation between fruit weight and cell numbers as determined with the quantitative trait locus (QTL) [3,88]. Also, cherry fruit weight was significantly influenced by precipitation, evapotranspiration, and insolation [89]. The difference in soluble solids content between cultivars was significant ($p < 0.05$), but it was more sensitive to climate types ($p < 0.01$). Researchers have confirmed that the same cherry fruit cultivars exhibit diverse ranges of soluble solids content depending on weather parameters in different planting sites [90–92]. The soluble solids content of cherries was dominantly influenced by weather conditions, compared with the cultivar factor. Iryna et al. [93] already mathematically substantiated that weather condition parameters (especially the humidity and temperature) greatly affect the accumulation of SSC for cherry fruits. Titratable acidity was determined with the cultivar and climate type, as supported with previous studies [89,94]. Table 5 shows significant differences in titratable acidity among the three climate types, which can be attributed to the dilution effect of precipitation conditions. The maturity index of Russia NO.8 was significantly higher (35.64 ± 2.33) than that of other cultivars. The maturity index was highly related to soluble solids content and titratable acidity, which was influenced by both factors of genotype and ripening stage [81] to present various sensory characteristics. However, there was no significant effect of climate type on the maturity index (calculated from SSC and TA) ($p > 0.05$). Previous studies have shown that different cherry cultivars have varying levels of total phenol content. Russia NO.8 had the highest, while Tieton had the lowest. The total phenol content of GI cherries grown in different climate types was also significantly different from each other ($p < 0.01$). Tables 2–4 shows no significant difference in TPC between Jiangxian and Qinzhou, both grown in temperate monsoon climates with the same cultivar (Tieton) ($p > 0.05$). However, their TPC levels were noticeably lower than that of Ledu (Tieton) planted in the climate of a plateau mountain ($p < 0.05$).

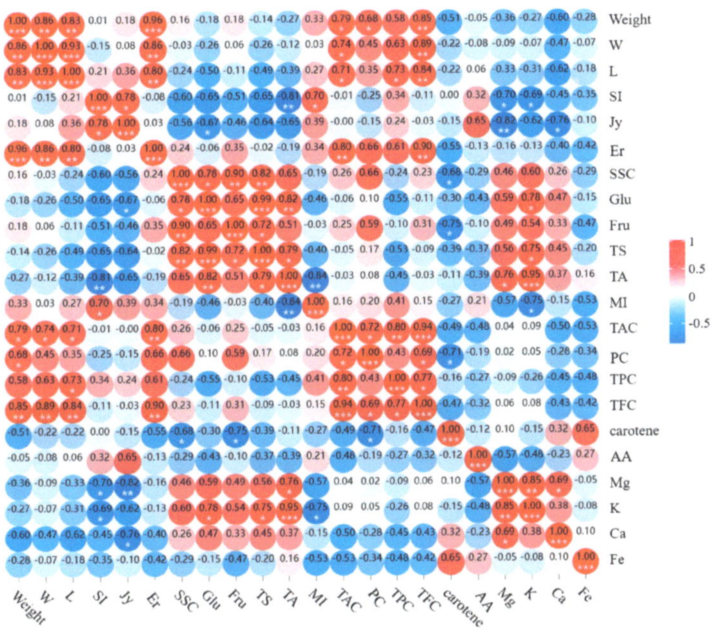

Figure 1. Pearson's correlation coefficients of physiochemical parameters for GI cherries. Note: ***, **, * Correlation is significant, respectively. W, weight; L, length; SI, shape index; Jy, juice yield; Er, edible rate; SSC, soluble solids content; Glu, glucose; Fru, fructose; TS, total sugar; TA, titratable acidity; MI, maturity index; TAC, total anthocyanins content; PC, procyanidins; TPC, total phenolic content; TFC, total flavonoid content. AA, ascorbic acids; Mg, magnesium; K, potassium; Ca, calcium; Fe, iron.

Table 5. Effects (mean ± SE) of cultivars and climate types on typical quality parameters of GI cherries.

Factors	Fruit Weight	Soluble Solids Content	Titratable Acidity	Maturity Index	Total Phenols
Cultivars					
Honegdeng	9.50 ± 0.49 b	19.77 0.90 b	19.06 ± 0.30 b	14.18 ± 1.43 ab	112.72 ± 11.11 a
Huangmi	5.00 ± 0.793 a	12.64 ± 1.47 a	27.50 ± 0.49 d	7.86 ± 2.33 a	143.10 ± 0.60 b
Tieton	10.55 ± 0.49 c	17.15 ± 0.90 b	21.34 ± 0.30 c	17.42 ± 1.43 b	135.79 ± 5.68 b
Brooks	6.71 ± 0.79 a	15.29 ± 1.47 ab	15.64 ± 0.49 a	9.55 ± 2.33 a	104.71 ± 3.94 a
Russia NO.8	17.93 ± 0.79 d	17.22 ± 1.47 b	18.86 ± 0.49 b	35.64 ± 2.33 c	197.72 ± 28.67 c
Climate types					
Temperate monsoon	11.10 ± 0.34 b	15.58 ± 0.64 a	22.55 ± 0.21 c	18.58 ± 1.01 b	147.26 ± 1.53 b
Subtropical monsoon	8.56 ± 0.56 a	19.48 ± 1.04 b	18.15 ± 0.35 b	12.04 ± 1.65 a	102.37 ± 2.94 a
Plateau Mountain	8.22 ± 0.79 a	17.72 ± 1.47 ab	16.29 ± 0.49 a	17.82 ± 2.33 b	149.06 ± 3.53 c
ANOVA *p*-value Main Effects					
Cultivars	0.000 ***	0.011 **	0.000 ***	0.000 ***	0.000 ***
Climate types	0.000 ***	0.004 ***	0.000 ***	NS	0.000 ***

Note: *** and ** represent the significance levels at $p < 0.01$ and $p < 0.05$, respectively. NS means non-significant. Means followed by the same letters within each column represent no significant difference (with Duncan's test, $\alpha < 0.05$).

3.4. The Relationship between Soil Conditions and Properties of GI Cherries

The results of soil conditions for GI cherries are presented in Table 6. Besides the factors of cultivar and climate type, pollination success, reproductive maximization, and high-quality cherry fruits are related to proper soil conditions [95,96]. The average pH value of the soil was 7.53 ± 0.59 (pH 6.0~7.0 was recommended), ranging from 6.61 ± 0.01 (Xiehu) to 8.26 ± 0.01 (Ledu) [97]. According to Cristóbal and Melakeberhan, improper pH would lead to inhibiting plant growth by disturbing mineral absorption [97,98]. Soil organic matter is comprised of multiple fractions from well-decomposed living organisms that can provide nutrients for plant growth. It is often used to assess soil quality measurements [99]. The mean organic matter value was 24.14 ± 12.84 g/kg, shown in Table 6. The levels of soil organic matter from Tongzhou (35.60 ± 0.21 g/kg), Dalian (42.00 ± 0.35 g/kg), and Wenchuan (44.10 ± 0.49 g/kg) are similar to Watsona's report; however, the others are significantly lower than previous studies [20,100]. Both P and K of soil samples varied significantly ($p < 0.05$), and the coefficients of variation were 120.62% and 97.07%, respectively. The SSC level of Wenchuan was the highest, which may be explained with that P supply is a crucial factor that affects SSC modulation for cherries during the preharvest time, and biosynthesis of sugar is negatively influenced by P deficiency [97,101]. Soil K of samples ranged from 128.00 ± 0.71 mg/kg (Tianbao) to 1760.00 ± 3.54 (Dalian). According to AĞLAR and Bustamante, higher values of fruit weight and size were obtained with potassium fertilization treatment [83,102], which can illustrate the phenomenon of Dalian cherry fruit with the biggest weight and size, while Tianbao is with the smallest among all samples. Though soil K of Dalian was the highest, the lowest content of Ca was found in the fruit sample (70.13 ± 4.89 mg/kg) from Tables 2–4. It was supported with a previous report that high-dose K application promotes the decrease in the fruit Ca level [84]. Soil Fe and soil Ca are also necessary for sweet cherry growth and fruit quality promotion [103,104]. Previous researchers discovered that soil pH determined both the levels of Fe and Ca in planted soil. Soil Ca enrichment and soil Fe reduction are related to elevated soil pH [105], which are in accordance with the results in this study (Table 6). However, the adverse result was also found that enrichment for soil Fe was stimulated with higher soil pH [106]. That may be attributed to different cultivars of fruits [95,106].

Table 6. Parameters of soil conditions for GI cherries.

Samples	pH	Organic Matter g/kg	Available P mg/kg	Available K mg/kg	Available Fe mg/kg	Available Ca mg/kg
Baqiao	7.85 ± 0.01 f	14.50 ± 0.14 c	13.90 ± 0.42 a	237.00 ± 0.00 d	20.90 ± 0.00 b	335.00 ± 1.41 b
Tianbao	6.65 ± 0.00 b	18.50 ± 0.57 d	40.00 ± 0.85 b	128.00 ± 0.71 a	28.10 ± 0.00 e	392.00 ± 0.71 e
Jiangxian	7.98 ± 0.00 g	19.30 ± 0.06 e	71.60 ± 0.90 c	419.00 ± 1.41 f	29.70 ± 0.21 f	333.00 ± 2.12 b
Wenchuan	7.21 ± 0.01 c	44.10 ± 0.49 h	542.00 ± 22.63 f	837.00 ± 2.12 h	25.50 ± 0.00 d	427.00 ± 2.83 f
Xiehu	6.61 ± 0.01 a	13.40 ± 0.14 b	49.90 ± 2.55 b	187.00 ± 1.41 c	54.60 ± 0.07 i	291.00 ± 1.41 a
Tongzhou	7.77 ± 0.00 e	35.60 ± 0.21 f	248.00 ± 4.95 d	752.00 ± 9.19 g	34.40 ± 0.00 h	371.0 ± 02.12 d
Qinzhou	7.98 ± 0.00 g	19.30 ± 0.07 e	60.84 ± 2.31 bc	358.00 ± 0.00 e	23.00 ± 0.00 c	344.00 ± 1.41 c
Ledu	8.26 ± 0.01 h	10.60 ± 0.14 a	14.35 ± 0.52 a	173.00 ± 2.12 b	5.37 ± 0.01 a	438.00 ± 2.83 g
Dalian	7.43 ± 0.01 d	42.00 ± 0.35 g	308.50 ± 13.44 e	1760.00 ± 3.54 i	31.80 ± 0.35 g	461.00 ± 2.12 h
Mean	7.53	24.14	149.90	539.00	28.15	376.89
Std.	0.59	12.84	180.81	523.23	13.04	56.70
CV (%)	7.90%	53.18%	120.62%	97.07%	46.30%	15.05%

Note: Std. and CV represent standard deviation and coefficient of variation, respectively. Different letters in each column stand for mean values significantly different according to Duncan's test at $p < 0.05$.

The correlation analysis of soil condition parameters was performed before the establishment of multiple linear regression models (Table 7). Organic matter exhibited significantly positive correlations with available P (r = 0.939) and available K (r = 0.849), respectively (not shown). A collinearity diagnosis was carried out based on the correlation result. Regression models characterizing the dependence of cherry quality traits on soil condition factors are exhibited in Table 7. SSC was only significantly impacted by soil P ($R^2 = 0.648$, $p < 0.05$), and the regression equation was $Y_{SSC} = 0.016X_1 + 14.488$. The equation between SSC and soil available P substantiated that phosphorus played an important role in accumulating the sugar of cherry fruits [68,98]. The highest SSC value of Wenchuan was related to its high level of soil P (Table 6). Researchers have also found that phosphorus fertilization could improve SSC levels in the Rosaceae family [107,108]. The phenomenon may be caused by two reasons: (1) higher P content caused in leaf tissue by phosphorus fertilization helps to enhance the photosynthetic rate [109], and then to improve the accumulation of carbohydrates, and (2) sugar transportation and allocation from roots and shoots are affected by P content in the plant [99,110]. High P content in cherries is obtained by improving sugar transportation from leaf to fruit [111]. The regression equation for TA with soil parameters was $Y_{TA} = 0.008X_1 - 0.008X_2 + 0.455X_3 + 0.081X_4 + 5.709X_5 - 76.980$ ($R^2 = 0.861$). From Table 7, TA was positively influenced by soil P, Fe, Ca, and pH; however, a significantly negative influence was exerted with soil K ($p < 0.05$). Researchers also found that TA change was caused by phosphorus and potassium fertilization, while little literature discovered the effects of soil Fe and Ca on TA [112,113]. Phosphorus serves as a vital constituent of nucleic acid, adenosine triphosphate, and cell membrane structures [114]. High P content in the fruit increased with P fertilization enhances the requirement of carbohydrates (like TA) for modulating important processes related to energy transfer and membrane synthesis [115]. Previous literature reported that potassium supply to fruits has a positive impact on their TA levels [115,116], but negative effects [38,117] or non-significant effects on TA and organic acids were also found [118,119]. There are two explanations for the relationship between potassium and TA change. (1) In most cases, additional potassium takes part in the upregulation of the tricarboxylic acid pathway to promote the synthesis of organic acids. (2) Potassium also affects tonoplastic transport regulation, and is essential for organic acids to vacuole storage or release by modulating cell membrane permeability [22,120]. This process cannot have an effect on the TA level, as no protons are formed to affect it. However, a definite illustration of the negative effect of potassium on TA has not been studied at present. The role of potassium in TA change depends on the kinds of fruit, fertilization dose, and physiological status of fruits [22,116,119]. A specific relationship between TPC and soil factors was also obtained in this study. The regression model was $Y_{TPC} = -0.009X_3 - 0.002X_1 + 0.001X_2 + 1.409$

(R^2 = 0.917). A significantly positive impact on TPC was shown with soil Fe and K, while it was significantly negatively influenced by soil P ($p < 0.05$). The TPC positively affected by soil K may contribute to the enhancement of the carbohydrates. Potassium plays a vital role in promoting photosynthesis activity and stimulating carbohydrate translocation [118], which indirectly increases the TPC content in the plant with potassium fertilization [121]. The result of positive correlations between the carbohydrates and total phenolics under high K application has already been studied in previous research [121,122]. It is noteworthy that though P supply is a critical factor in improving cherry quality during preharvest periods, P deficiency promotes total phenol compound accumulation and flesh-browning inhibition in fruit cells [97,113].

Table 7. Multiple linear regression equations of typical quality properties dependent on soil condition parameters.

Quality Parameters	Influencing Factors	Regression Equation	F	Sig.	Modified R^2
SSC	available P	$Y_{SSC} = 0.016X_1 + 14.488$	15.731	0.005	0.648
TA	available P, available K, available Fe, available Ca, pH	$Y_{TA} = 0.008X_1 - 0.008X_2 + 0.455X_3 + 0.081X_4 + 5.709X_5 - 76.980$	10.882	0.039	0.861
TPC	available Fe, available P, available K	$Y_{TPC} = -0.009X_3 - 0.002X_1 + 0.001X_2 + 1.409$	30.542	0.001	0.917

Note: X_1, available P; X_2, available K; X_3, available Fe; X_4, available Ca; X_5, pH.

4. Conclusions

The present work studied the physicochemical properties of GI cherries grown in China and investigated the factors affecting fruit quality. Peculiar quality traits of cherry samples exhibited significant differences in each physicochemical parameter, attributed to their cultivars and specific environmental conditions. Dalian presented the largest size and weight with the highest content of bioactive compounds. The highest SSC was found in Wenchuan, and Baqiao presented the highest content of β-carotene. Tongzhou was rich in minerals (K, Mg, and Ca). Moreover, the influencing factors of both cultivars and climate types presented the main effects on weight, SSC, TA, and TPC ($p < 0.05$ and $p < 0.01$). However, MI was only influenced by cultivars. Significant differences in soil conditions for GI cherries were achieved. This would help make suggestions for growers and breeding scientists to select cultivars planted in places with appropriate climate types. The multiple linear regressions showed that soil P positively affected SSC and TA, while exerting a negative influence on TPC. The soil K showed a negative effect on TA, with a positive effect on TPC. This would help orchard operators improve cherry fruit quality by modulating fertilization plans. Even so, further research should be carried out by investigating the influence mechanisms of the specific climatical parameters (like humidity and precipitation) or the combined soil nutrients on fruit quality traits, in order to help growers specifically determine suitable growing environments for different cultivars, and to obtain optimum fertilization formulas.

Author Contributions: Conceptualization and writing—original draft, Y.N.; methodology, Y.N. and J.H.; project administration and supervision, J.H. and J.S.; software, R.L. and P.W.; validation, J.H., R.L. and P.L.; data curation and formal analysis, P.W. and M.L. All authors have read and agreed to the published version of the manuscript.

Funding: This research was funded by the Government Procurement Service Program of the Department of Agricultural Products Quality and Safety Supervision, Ministry of Agriculture and Rural Affairs in China (Grant Number: 14225045).

Data Availability Statement: The data of this article can be available from the corresponding author.

Acknowledgments: Thanks to nine cherry geographical indication product management departments for providing samples and sample information in time. Thanks to the Geographical Indications Management Office of China Green Food Development Center for the support of sampling coordination.

Conflicts of Interest: The authors declare no conflict of interest.

References

1. Marinella, D.L.; Anna, M.I.; Maria, P.G.; Valeria, D.; Fabiano, C.; Filippo, S.; Gianfranco, D.; Nunziatina, D.T.; Alessandra, B. Comparative chemical analysis of six ancient Italian sweet cherry (*Prunus avium* L.) varieties showing antiangiogenic activity. *Food Chem.* **2021**, *360*, 129999.
2. Manuel, J.S.; Alberto, M.; Santiago, R.M.; Alejandro, H.; Margarita, L.C.; María, G.C. Physicochemical and sensorial characterisation of four sweet cherry cultivars grown in Jerte Valley (Spain). *Food Chem.* **2012**, *133*, 1551–1559.
3. Szilágyi, S.; Horváth-Kupi, T.; Desiderio, F.; Bekefi, Z. Evaluation of sweet cherry (*Prunus avium* L.) cultivars for fruit size by FW_G2a QTL analysis and phenotypic characterization. *Sci. Hortic.* **2022**, *292*, 110656. [CrossRef]
4. *07-1537;* United States Standards for Grades of Sweet Cherries. Agricultural Marketing Service. United States Department of Agriculture: Washington, DC, USA, 2007.
5. *GB/T 26901-2011;* Grades of Cherries. State Administration for Market Regulation. Standardization Administration: Beijing, China, 2011.
6. *Recommended Quality Guide for Export Cherries.* Chilean Cherry Committee: Chile, 2021. Available online: https://fructidor.com/newsdetail.aspx?idn=58303 (accessed on 23 August 2023).
7. Ana, C.G.; Gonçalo, C.; Gilberto, A.; Cristina, G.V.; Diego, A.M.; Luís, R.S. Physical and phytochemical composition of 23 Portuguese sweet cherries as conditioned by variety (or genotype). *Food Chem.* **2021**, *335*, 127637.
8. Kubilay, V.; Hasim, K.; Serkan, S. A study on some chemical and physico-mechanic properties of three sweet cherry varieties (*Prunus avium* L.) in Turkey. *J. Food. Eng.* **2006**, *74*, 568–575.
9. Federica, B.; Oomah, B.D. Sweet and sour cherries: Origin, distribution, nutritional composition and health benefits. *Trends Food Sci. Technol.* **2019**, *86*, 517–529.
10. Kim, D.O.; Heo, H.J.; Kim, Y.J.; Hyun, S.Y.; Chang, Y.L. Sweet and Sour Cherry Phenolics and Their Protective Effects on Neuronal Cells. *J. Agric. Food Chem.* **2005**, *53*, 9921–9927. [CrossRef]
11. Guillermo, C.; Francisco, L.; Maria, P.G.; Carine, S.; Víctor, L. Bioactive and functional properties of sour cherry juice (*Prunus cerasus*). *Food Funct.* **2016**, *7*, 4675–4682.
12. Fatih, M.Y. Sour Cherry By-products: Compositions, Functional Properties and Recovery Potentials. *Crit. Rev. Food Sci. Nutr.* **2019**, *59*, 3549–3563.
13. Pissard, A.; Lateur, M.; Baeten, V.; Magein, H.; Dupont, P.; Tabart, J.; Pincemail, J.; Kevers, C. Determination of total phenolic compound content and antioxidant activity in cherry species and cultivars. *J. Berry Res.* **2016**, *6*, 81–91. [CrossRef]
14. Ewa, S.; Tomasz, K.; Karolina, M.K.; Sebastian, P. Fruit quality and contents of some bioactive compounds in selected Czech sweet cherry (*Prunus avium* L.) cultivars under conditions of Central Poland. *Agriculture* **2022**, *12*, 1859.
15. Sofia, C.; Filipa, Q.; Carlos, R.; Alice, V.; Alfredo, A.; Ana, I.B.; Rob, S.; Ana, P.S.; Berta, G. Effects of calcium and growth regulators on sweet cherry (*Prunus avium* L.) quality and sensory attributes at harvest. *Sci. Hortic.* **2019**, *248*, 231–240.
16. Giménez, M.J.; María, S.; Juan, M.V.; Domingo, M.R.; Salvador, C.; Daniel, V.; Fabián, G. Preharvest salicylic acid and acetylsalicylic acid treatments preserve quality and enhance antioxidant systems during postharvest storage of sweet cherry cultivars. *J. Sci. Food Agric.* **2017**, *97*, 1220–1228. [CrossRef] [PubMed]
17. Mehmet, A.O.; Yusuf, C.G.; Husnu, D.; Leyla, D.; Idris, M.; Gul, C.O. Comparative evaluation of phenolic profile and antioxidant activity of new sweet cherry (*Prunus avium* L.) genotypes in Turkey. *Phytochem. Anal.* **2022**, *33*, 564–576.
18. Thouraya, A.G.; Tahar, S.; Youssef, A. Behavior and morphometric characterization of local and introduced cultivars of sweet cherries (*Prunus avium*), tested in a multi-site trial in Tunisia. *Sci. Hortic.* **2020**, *270*, 109455.
19. Li, B.; Xie, Z.; Zhang, A.; Xu, W.; Zhang, C.; Liu, Q.; Liu, C.; Wang, S. Tree growth characteristics and flower bud differentiation of sweet cherry (*Prunus avium* L.) under different climate conditions in China. *Hortic. Sci.* **2010**, *37*, 6–13. [CrossRef]
20. Tristan, T.W.; Louise, M.N.; Denise, N.; Gerry, H.N.; Tom, A.F. Soil amendments influence *Pratylenchus penetrans* populations, beneficial rhizosphere microorganisms, and growth of newly planted sweet cherry. *Appl. Soil Ecol.* **2017**, *117*, 212–220.
21. Anna, B.; Zdzisław, K.; Jan, K.; Jerzy, Z. Yielding and fruit quality of Lithuanian sweet cherry cultivars grown under the climatic and soil conditions of Warmia. *Folia Hortic.* **2011**, *23*, 101–106.
22. Mohd, H.I.; Hawa, Z.E.J.; Ehsan, K.; Ali, G. Primary, secondary metabolites, photosynthetic capacity and antioxidant activity of the malaysian herb kacip fatimah (*Labisia Pumila* Benth) exposed to potassium fertilization under greenhouse conditions. *Int. J. Mol. Sci.* **2012**, *13*, 15321–15342.
23. McGuire, R.G. Reporting of objective color measurements. *HortScience* **1992**, *27*, 1254–1255. [CrossRef]
24. Christian, Z.; Mehmet, S.; Edgar, P. Potassium in agriculture—Status and perspectives. *J. Plant Physiol.* **2014**, *171*, 656–669.
25. Bai, Q.; Shen, Y.; Huang, Y. Advances in mineral nutrition transport and signal transduction in Rosaceae fruit quality and postharvest storage. *Front. Plant Sci.* **2021**, *12*, 620018. [CrossRef] [PubMed]
26. Zhang, K.; Yan, G.; Zhang, X.; Wang, J.; Duan, X. Sweet cherry growing in China. *Acta Hortic.* **2019**, *1235*, 133–140. [CrossRef]

27. FAS Beijing. Stone Fruit Annual (Report No. CH2022-0078). USDA Foreign Agriculture Service. 2022. Available online: https://www.fas.usda.gov/data/china-stone-fruit-annual-6 (accessed on 23 August 2023).
28. World Trade Organization. Uruguay Round Agreement: Trips. 2016. Available online: https://www.wto.org/english/docs_e/legal_e/27-trips_04b_e.htm#3 (accessed on 23 August 2023).
29. Angela, G.P.; Antonio, F.M.; Maria, R.F.; Giuseppe, S.; Giacomo, D.; Vincenzo, L.T.; Rosaria, C.F.C.; Marcella, D.B.; Giuseppa, D.B. Multielement and chemometric analysis for the traceability of the Pachino Protected Geographical Indication (PGI) cherry tomatoes. *Food Chem.* **2022**, *386*, 132746.
30. Longobardi, F.; Ventrella, A.; Bianco, A.; Catucci, L.; Cafagna, I.; Gallo, V.; Mastrorilli, P.; Agostiano, A. Non-targeted ^{1}H NMR fingerprinting and multivariate statistical analyses for the characterization of the geographical origin of Italian sweet cherries. *Food Chem.* **2013**, *141*, 3028–3033. [CrossRef] [PubMed]
31. Fu, J.; Liu, R.R.; Chen, Y.F.; Xing, J.G. Discrimination of geographical indication of Chinese green teas using an electronic nose combined with quantum neural networks: A portable strategy. *Sens. Actuators B-Chem.* **2023**, *375*, 132946. [CrossRef]
32. Shi, Q.Q.; Han, G.; Liu, Y.; Jiang, J.J.; Jia, Y.Y.; Li, X.G. Nutrient composition and quality traits of dried jujube fruits in seven producing areas based on metabolomics analysis. *Food Chem.* **2022**, *385*, 132627. [CrossRef]
33. Su, S.; Wu, J.; Peng, X.Y.; Li, B.; Li, Z.J.; Wang, W.; Ni, J.W.; Xu, X.Q. Genetic and agro-climatic variability in seed fatty acid profiles of *Akebia trifoliata* (Lardizabalaceae) in China. *J. Food Compos. Anal.* **2021**, *102*, 104064. [CrossRef]
34. Palma, M.; Sepúlveda, Á.; Yuri, J.A. Effect of plastic roof and high tunnel on microclimate, physiology, vegetative growth and fruit characteristics of 'Santina' sweet cherry. *Sci. Hortic.* **2023**, *317*, 112037. [CrossRef]
35. Rutkowski, K.; Łysiak, G.P. Weather conditions, orchard age and nitrogen fertilization influences yield and quality of 'Łutówka' sour cherry fruit. *Agriculture* **2022**, *12*, 2008. [CrossRef]
36. Victor, B.; Pedro, J.B.R.; Roque, T.S.; Rafael, D. Influence of regulated deficit irrigation and environmental conditions on reproductive response of sweet cherry trees. *Plants* **2020**, *9*, 94.
37. Uçgun, K. Effects of nitrogen and potassium fertilization on nutrient content and quality attributes of sweet cherry fruits. *Not. Bot. Horti Agrobot.* **2019**, *47*, 114–118. [CrossRef]
38. Hüseyin, Y.; Ömer, A. Effects of potassium fertilization on leaf nutrient content and quality attributes of sweet cherry fruits (*Prunus avium* L.). *J. Plant Nutr.* **2020**, *44*, 946–957.
39. Jia, C.S.; Geoffrey, I.N.; Sun, D.X.; Sun, Y.G.; Wu, P. Variety compound quality relationship of 12 sweet cherry varieties by HPLC-chemometric analysis. *Int. J. Food Sci. Technol.* **2019**, *54*, 2897–2914. [CrossRef]
40. Elena, C.C.; Mireia, C.; Alfonso, M.; Dolores, H.; Carla, M.S.; Paula, M.B.; Antonio, J.M.M. Effect of regulated deficit irrigation on commercial quality parameters, carotenoids, phenolics and sugars of the black cherry tomato (*Solanum lycopersicum* L.) 'Sunchocola'. *J. Food Compos. Anal.* **2022**, *105*, 104220.
41. Singleton, V.L.; Orthofer, R.; Lamuela–Raventos, R.M. Analysis of total phenols and other oxidation substrates and antioxidants by means of Folin–Ciocalteau reagent. *Methods Enzymol.* **1999**, *299*, 152–178.
42. Kim, M.K.; Kim, H.; Kim, H.S.; Youngsang, L.; Yongho, K. Identification and quantification of anthocyanin pigments in colored rice. *Nutr. Res. Pract.* **2008**, *2*, 46–49. [CrossRef]
43. Olszewska, M.A. Variation in the phenolic content and in vitro antioxidant activity of *Sorbus aucuparia* leaf extracts during vegetation. *Acta Pol. Pharm.* **2011**, *68*, 937–944.
44. *GB 5099.83-2016*; National Food Safety Standards, Determination of Carotene in Food China. Food and Drug Administration, National Health and Family Planning Commission of the PRC: Beijing, China, 2017.
45. *GB 5099.86-2016*; National Food Safety Standards, Determination of Ascorbic Acid in Food. China Food and Drug Administration, National Health and Family Planning Commission of the PRC: Beijing, China, 2017.
46. Matos-Reyes, M.N.; Simonot, J.; Lopez-Salazar, O.; Cervera, M.L.; Guardia, M. Authentication of Alicante's mountain cherries protected designation of origin by their mineral profile. *Food Chem.* **2013**, *141*, 2191–2197. [CrossRef]
47. Yu, Y.; Yang, J.; Zeng, S.; Wu, D.; Jacobs, D.F.; Sloan, J.L. Soil pH, organic matter, and nutrient content change with the continuous cropping of *Cunninghamia lanceolata* plantations in South China. *J. Soils Sediments* **2017**, *17*, 2230–2238. [CrossRef]
48. Skjemstad, J.O.; Krull, E.S.; Swift, R.S.; Szarvas, S. Mechanisms of protection of soil organic matter under pasture following clearing of rainforest on an Oxisol. *Geoderma* **2008**, *143*, 231–242. [CrossRef]
49. Tekalign, M.; Ritcher, C.; Heiligtag, B. Comparison of extractants for the determination of available phosphorus, potassium, calcium, magnesium and sodium in some Ethiopian and German soils. *Commun. Soil Sci. Plant Anal.* **1996**, *27*, 2197–2212.
50. Ballistreri, G.; Continella, A.; Gentile, A.; Amenta, M.; Fabroni, S.; Rapisarda, P. Fruit quality and bioactive compounds relevant to human health of sweet cherry (*Prunus avium* L.) cultivars grown in Italy. *Food Chem.* **2013**, *140*, 630–638. [CrossRef] [PubMed]
51. Kalyoncu, I.H.; Ersoy, N.; Yilmaz, M. Contents of sweet cherry (*Prunus avium* L.) type grown in Konya. *Afr. J. Biotechnol.* **2009**, *8*, 2744–2749.
52. Marlon, J.R.S.; Ana, P.M.P.; Adilson, P.J.; Camilo, A.P.C.; Daniel, C.; Mara, F.M.; Sarita, L.; Marco, A.T. Yield performance of new juice grape varieties grafted onto different rootstocks under tropical conditions. *Sci. Hortic.* **2018**, *241*, 194–200.
53. Suwimol, C.; John, B.G.; Quan, V.V.; Konstantinos, P.; Costas, E.S. Sweet cherry: Composition, postharvest preservation, processing and trends for its future use. *Trends Food Sci. Technol.* **2016**, *55*, 72–83. [CrossRef]
54. Ozgur, C.; Cemile, E.O.; Atakan, G.; İsa, E.; İsmail, D. Determination of optimum harvest date of sweet cherry cv. Lapins grown in Isparta. *Turk. J. Agric. For.* **2014**, *2*, 1905–1910.

55. Wani, A.A.; Singh, P.; Gul, K.; Wani, M.H.; Langowski, H.C. Sweet cherry (*Prunus avium*): Critical factors affecting the composition and shelf life. *Food Packag. Shelf Life* **2014**, *1*, 86–99. [CrossRef]
56. Kappel, F.; Granger, A.; Hrotkó, K.; Schuster, M. *Fruit Breeding*; Springer: Berlin, Germany, 2011; pp. 459–504.
57. Ooraikul, B.; Stiles, M.E. *Modified Atmosphere Packaging of Food*; Ellis Horwood Ltd.: Chichester, UK, 1991; pp. 169–242.
58. Sakshi, P.; Rachel, A.I.; Carlos, R.C.; Luis, F.L.; Veronica, C.S.; Juan, C.C.; Dario, J.C. Fruit Characterization of *Prunus serotina* subsp. capuli. *Horticulturae* **2022**, *8*, 838. [CrossRef]
59. Valentina, U.; Jerneja, F.; Franci, Š. Sugars, organic acids, phenolic composition and antioxidant activity of sweet cherry (*Prunus avium* L.). *Food Chem.* **2008**, *107*, 185–192.
60. Ali, A.H.; Nurullah, D. Physicochemical characteristics, antioxidant activity, organic acid and sugar contents of 12 sweet cherry (*Prunus avium* L.) cultivars grown in Turkey. *J. Food Sci.* **2015**, *80*, 564–570.
61. Girard, B.; Kopp, T.G. Physicochemical Characteristics of Selected Sweet Cherry Cultivars. *J. Agric. Food Chem.* **1998**, *46*, 471–476. [CrossRef]
62. Gloria, D.R.; Daniel, R.V.; Patricia, M.; Merichel, P.; Maria, L.M. Composition of nonextractable polyphenols from sweet cherry pomace determined by dart-orbitrap-hrms and their in vitro and in vivo potential antioxidant, antiaging, and neuroprotective activities. *J. Agric. Food Chem.* **2002**, *70*, 7993–8009.
63. Nuria, A.; Ana, G.; Marta, B.; Antonia, G.; Dolores, M.M. Comparison of phenolic compounds profile and antioxidant properties of different sweet cherry (*Prunus avium* L.) varieties. *Food Chem.* **2019**, *279*, 260–271.
64. Dominika, Ś.T.; Alicja, P.; Ewelina, H.; Agnieszka, G.; Elżbieta, R. The profile and content of polyphenols and carotenoids in local and commercial sweet cherry fruits (*Prunus avium* L.) and their antioxidant activity in vitro. *Antioxidants* **2019**, *8*, 534.
65. Chaovanalikit, A.; Wrolstad, R.E. Anthocyanin and phenolic composition of fresh and processed cherries. *J. Food Sci.* **2004**, *69*, 173–183.
66. Tahir, M.; Farooq, A.; Ijaz, A.B.; Tahira, I. Effect of maturity on proximate composition, phenolics and antioxidant attributes of cherry fruit. *Pak. J. Bot.* **2013**, *45*, 909–914.
67. Cao, J.P.; Jiang, Q.; Lin, J.Y. Physicochemical characterization of four cherry species (*Prunus* spp.) grown in China. *Food Chem.* **2015**, *173*, 855–863. [CrossRef]
68. Aneta, W.; Paulina, N.; Piotr, L.; Jan, O. Evaluation of Sour Cherry (*Prunus cerasus* L.) Fruits for Their Polyphenol Content, Antioxidant Properties, and Nutritional Components. *J. Agric. Food Chem.* **2014**, *62*, 12332–12345.
69. Sílvia, A.; Ivo, V.O.; Anne, S.M.; Alfredo, A.; Maria, J.S.; Berta, G. Phenolic Profile and Bioactive Potential of Stems and Seed Kernels of Sweet Cherry Fruit. *Antioxidants* **2020**, *9*, 1295.
70. Taki, D. Determination of carotenoid, organic acid and sugar content in some sweet cherry cultivars grown in Sakarva, Turkey. *J. Food. Agric. Environ.* **2013**, *11*, 73–75.
71. Mauro, C.; Martino, B.; Flavia, D.C.; Stefania, P.; Alessandra, B.; Francesca, M.; Stefano, N.; Matteo, S.; Stefania, C.; Linda, A.; et al. Multi-approach metabolomics analysis and artificial simplified phytocomplexes reveal cultivar-dependent synergy between polyphenols and ascorbic acid in fruits of the sweet cherry (*Prunus avium* L.). *PLoS ONE* **2017**, *12*, e0180889. [CrossRef]
72. Zan, S.Y.; Wang, R.; Zhang, F.; Zhang, D.Y.; Liu, B.J.; Meng, X.H. Composition analysis of rootstock cherry (*Prunus mahaleb* L.), a potential source of human nutrition and dietary supplements. *Eur. Food Res. Technol.* **2022**, *248*, 1421–1435. [CrossRef]
73. Ali, K.; Mahsa, M.; Kimiya, A. Morphological and pomological characterizations of sweet cherry (*Prunus avium* L.), sour cherry (*Prunus cerasus* L.) and duke cherry (*Prunus ×gondouinii* Rehd.) to choose the promising selections. *Sci. Hortic.* **2019**, *257*, 108719.
74. Mirko, S.; Grafe, C.; Wolfram, B.; Schmidt, H. Cultivars resulting from cherry breeding in Germany. *Erwerbs-obstbau* **2014**, *56*, 67–72.
75. Julia, P.; Craig, H.; Cai, L.C.; Zhao, Y.Y.; Amy, I.; Cameron, P. Genomic heritability estimates in sweet cherry reveal non-additive genetic variance is relevant for industry-prioritized traits. *BMC Genet.* **2018**, *19*, 23.
76. Erdal, A.; Onur, S.; Orhan, K.; Burhan, O.; Sefa, G. The relationship between fruit color and fruit quality of sweet cherry (*Prunus avium* L. cv. '0900 Ziraat'). *Turk. J. Food Agric. Sci.* **2019**, *1*, 1–5.
77. Chen, C.Q.; Chen, H.X.; Yang, W.L.; Li, J.; Tang, W.J.; Gong, R.G. Transcriptomic and metabolomic analysis of quality changes during sweet cherry fruit development and mining of related genes. *Int. J. Mol. Sci.* **2022**, *23*, 7402. [CrossRef]
78. Verena, O.; Michaela, S.; Michael, B. Non-destructive sensor-based prediction of maturity and optimum harvest date of sweet cherry fruit. *Sensors* **2017**, *17*, 277.
79. Tahir, M.; Farooq, A.; Mateen37 Mary, C.B.; Nazamid, S. Compositional variation in sugars and organic acids at different maturity stages in selected small fruits from Pakistan. *Int. J. Mol. Sci.* **2012**, *13*, 1380–1392.
80. María, S.; Fabián, G.; Domingo, M.R.; Salvador, C.; Daniel, V. Chemical constituents and antioxidant activity of sweet cherry at different ripening stages. *J. Agric. Food Chem.* **2005**, *53*, 2741–2745.
81. Marco, B.; Ariel, M. Impact of potassium pre-harvest applications on fruit quality and condition of sweet cherry (*Prunus avium* L.) cultivated under plastic covers in southern Chile orchards. *Plants* **2021**, *10*, 2778.
82. Özgür, A.; Volkan, A.; Ece, T.; Gülser, Y. Effects of potassium fertilization on sweet cherry fruit (*Prunus avium* L.) quality and mineral content. *Commun. Soil Sci. Plant Anal.* **2022**, *53*, 1777–1782.
83. Kadir, U.; Mesut, A. Effects of increasing doses of nitrogen, phosphorus, and potassium on the uptake of other nutrients in sweet cherry trees. *Commun. Soil Sci. Plant Anal.* **2021**, *52*, 1248–1255.
84. Deniz, E. Effect of Preharvest Calcium Treatments on Sweet Cherry Fruit Quality. *Not. Bot. Horti Agrobot.* **2014**, *42*, 150–153.

85. Andreas, W.; Bennet, F.; Moritz, K. Calcium physiology of sweet cherry fruits. *Trees* **2020**, *34*, 1157–1167.
86. Felicia, C.; Sina, N.C. Morphological and biochemical characteristics of fruits of different cornelian cherry (*Cornus mas* L.) genotypes from spontaneous flora. *Not. Sci. Biol.* **2017**, *9*, 577–581.
87. James, W.O.; Amy, F.I.; Matthew, D.W. Genotypic differences in sweet cherry fruit size are primarily a function of cell number. *J. Am. Soc. Hortic. Sci.* **2007**, *132*, 697–703.
88. Sansavini, S.; Lugli, S. Sweet cherry breeding programs in Europe and Asia. *Acta Hortic.* **2008**, *795*, 41–58. [CrossRef]
89. Chen, L. A new middle ripening sweet cherry cultivar "Linglongcui". *Acta Hortic. Sin.* **2018**, *45*, 1419–1420.
90. Diana, N.; Rim, M.; Walid, E.K. Quality and phytochemical composition of sweet cherry cultivars can be influenced by altitude. *Plants* **2023**, *12*, 2254.
91. Danilo, C.; Francesca, A.; Carolina, T.; Corrado, C.; Emilia, C.; Roberto, C. Can environment and genotype influence sweet cherry qualitative parameters? *Plant Biosyst.* **2022**, *156*, 581–589.
92. Iryna, I.; Maryna, S.; Vira, M.; Iryna, B.; Ihor, K.; Tetiana, T.; Oksana, T.; Oleksandr, T.; Mikhailo, M.; Alina, O. The study of soluble solids content accumulation dynamics under the influence of weather factors in the fruits of cherries. *Slovak J. Food Sci.* **2021**, *15*, 350–359.
93. Legua1, P.; Domenech, A.; Martínez, J.J.; Sánchez-Rodríguez, L.; Hernández, F.; Carbonell-Barrachina, A.A.; Melgarejo, P. Bioactive and volatile compounds in sweet cherry cultivars. *J. Food Nutr. Res.* **2017**, *5*, 844–851. [CrossRef]
94. Zhang, C.X.; Matthew, W. The occurrence of protruding pistil in sweet cherry (*Prunus avium* L.) and its consequence on fertilization. *Sci. Hortic.* **2012**, *140*, 149–156. [CrossRef]
95. Cristóbal, P.P.; Marjorie, R.D.; Jorge, G.V.; Alejandra, R.F. The potential roles of the N and P supplies on the internal browning incidence in sweet cherries in the Southern Chile. *Horticulturae* **2022**, *8*, 1209.
96. Valentinuzzi, F.; Mason, M.; Scampicchio, M.; Andreotti, C.; Cesco, S.; Mimmo, T. Enhancement of the bioactive compound content in strawberry fruits grown under iron and phosphorus deficiency. *J. Sci. Food Agric.* **2015**, *95*, 2088–2094. [CrossRef]
97. William, E.C. Organic matter in the agricultural soils of Tasmania, Australia—A review. *Geoderma* **2018**, *312*, 170–182.
98. Tirhas, G.; Paige, M.; Tom, A.F.; Melanie, D.J.; Louise, M.N. Mulching improved soil fertility, plant growth and productivity, and postharvest deficit irrigation reduced water use in sweet cherry orchards in a semi-arid region. *Arch. Agron. Soil Sci.* **2022**, *69*, 1419–1436.
99. Cao, F.; Guan, C.; Dai, H.; Li, X.; Zhang, Z. Soluble solids content is positively correlated with phosphorus content in ripening strawberry fruits. *Sci. Hortic.* **2015**, *195*, 183–187. [CrossRef]
100. Erdal, A.; Onur, S. Role of the foliar fertilization treatments on quality attributes of sweet cherry fruits (*Prunus avium*). *Akad. Ziraat Derg.* **2018**, *7*, 131–136.
101. Andreas, W.; Moritz, K. Calcium and the physiology of sweet cherries: A review. *Sci. Hortic.* **2019**, *245*, 107–115.
102. Tingzhen, M.; Renaud, R.; Franz, D.O.; Robert, T.; Sylvain, C. A nature-based negative emissions technology able to remove atmospheric methane and other greenhouse gases. *Atmos. Pollut. Res.* **2021**, *12*, 101035.
103. Zhou, B.B.; Zhang, Q.; Sun, J.; Li, X.L.; Wei, Q.P. Study and application of partial least squares regression on relationship between soil nutrient and fruit quality. *J. Agric. Sci. Technol.* **2016**, *17*, 362–366.
104. Liu, Z.B.; Huang, Y.; Tan, F.J.; Chen, W.C.; Ou, L.J. Effect of soil type on trace element absorption and fruit quality of pepper. *Front. Plant Sci.* **2021**, *12*, 698796. [CrossRef]
105. Xu, M.G.; Zeng, X.B.; Li, J.M. Effect of pH on adsorption and desorption of Cu^{2+} in latersol and yellow brown earth. *Chin. J. Soil Sci.* **2005**, *36*, 349–351.
106. Cassiano, N.L.; Geovani, S.L.; Evandro, M.S.; Reginaldo, G.N.; Hans, R.G.; Lauriane, A.A.S. Fruit quality of West Indian cherry under saline water irrigation and nitrogen-potassium fertilization. *Braz. J. Agric. Environ. Eng.* **2021**, *25*, 741–749.
107. Yogaratnam, N.; Sharples, R.O. Supplementing the nutrition of bramley's seedling apple with phosphorus sprays ii. effects on fruit composition and storage quality. *J. Hortic. Sci. Biotech.* **1982**, *57*, 53–59. [CrossRef]
108. Johnson, C.R. Phosphorus nutrition on mycorrhizal colonization, photosynthesis, growth and nutrient composition of *Citrus aurantium*. *Plant Soil* **1984**, *80*, 35–42. [CrossRef]
109. Fredeen, A.L.; Rao, I.M.; Terry, N. Influence of phosphorus nutrition on growth and carbon partitioning in Glycine max. *Plant Physiol.* **1989**, *89*, 225–230. [CrossRef]
110. López-Arredondo, D.L.; Leyva-González, M.A.; González-Morales, S.I.; López-Bucio, J.; Herrera-Estrella, L. Phosphate nutrition: Improving low-phosphate tolerance in crops. *Annu. Rev. Plant Biol.* **2014**, *65*, 95–123. [CrossRef] [PubMed]
111. Guo, K.; Peng, L.; Hong, Y.; Qiao, G. Optimizing Nitrogen, Phosphorus, and Potassium Fertilization Rates for Fruit Performance of Chinese Cherry (*Prunus pseudocerasus* Lindl.). *Int. J. Fruit Sci.* **2022**, *22*, 769–778. [CrossRef]
112. Olivos, A.; Johnson, S.; Xiaoqiong, Q.; Crisosto, C.H. Fruit phosphorous and nitrogen deficiencies affect 'Grand Pearl' nectarine flesh browning. *Horstscience* **2012**, *47*, 391–394. [CrossRef]
113. Hina, M.; Vandana Sandeep, D.; Renu, P. Phosphorus nutrition: Plant growth in response to deficiency and excess. In *Plant Nutrients and Abiotic Stress Tolerance*; Hasanuzzaman, M., Fujita, M., Oku, H., Nahar, K., Hawrylak-Nowak, B., Eds.; Springer: Singapore, 2018. [CrossRef]
114. Karl, K.; Hanne, L.P. Effects of potassium, phosphorus and nitrogen fertilization on endogenous ethylene and quality characteristics of apples (*Malus domestica* L.). *J. Plant Nutr.* **2014**, *37*, 1148–1155.

115. Changwei, S.; Ding, Y.; Lei, X.; Zhao, P.; Wang, S.; Xu, Y.; Dong, C. Effects of foliar potassium fertilization on fruit growth rate, potassium accumulation, yield, and quality of 'kousui' Japanese pear. *J. Am. Soc. Hortic. Sci.* **2016**, *26*, 270–277.
116. Santos, M.; Ferreira, H.; Egea-Cortines, M.; Sousa, J.R.; Raimundo, F.; Matos, M.; Gonçalves, B. Evaluation of fruit quality, chromatic parameters and anthocyanin's content under foliar application of magnesium and potassium on sweet cherry (*Prunusavium* L.) cv. Burlat. *Biol. Life Sci. Forum* **2021**, *3*, 47.
117. Pant, R.C.; Singh, B. Potassium Level and Physiological Response and Fruit Quality in Hydroponically Grown Tomato. *Int. J. Veg. Sci.* **2009**, *16*, 85–99.
118. Liu, J.; Hu, T.; Feng, P.; Yao, D.; Gao, F.; Hong, X. Effect of potassium fertilization during fruit development on tomato quality, potassium uptake, water and potassium use efficiency under deficit irrigation regime. *Agric. Water Manag.* **2021**, *250*, 106831. [CrossRef]
119. Perkins-Veazie, P.; Roberts, W. Can potassium application affect the mineral and antioxidant content of horticultural crops? In Proceedings of the Symposium on Fertilizing Crops for Functional Food, Indianapolis, IN, USA, 11 November 2002.
120. José, G.V.; Sonia, O. *Postharvest Physiology and Biochemistry of Fruits and Vegetables*; Elsevier: Amsterdam, The Netherlands, 2019; pp. 207–244.
121. Ghasemzadeh, A.; Jaafar, H.Z.E.; Rahmat, A. Elevated Carbon Dioxide Increases Contents of flavonoids and Phenolic Compounds, and Antioxidant Activities in Malaysian Young Ginger (*Zingiber officinale* Roscoe.) Varieties. *Molecules* **2010**, *15*, 7907–7922. [CrossRef]
122. Wei, H.; Tye, L.; Bresnick, E.; Birt, D.F. Inhibitory effect of epigenin, a plant flavonoid, on epidermal ornithine decarboxylase and skin tumor promotion in mice. *Cancer Res.* **1990**, *50*, 499–502.

horticulturae

MDPI

Article

Chemical Composition and Polyphenol Compounds of *Vaccinium floribundum* Kunth (Ericaceae) from the Volcano Chimborazo Paramo (Ecuador)

Jorge M. Caranqui-Aldaz [1], Raquel Muelas-Domingo [2], Francisca Hernández [3] and Rafael Martínez [3,*]

1 Herbarium, Escuela Superior Politécnica del Chimborazo, Miguel Hernández University, Riobamba 060106, Ecuador

2 Centro de Investigación e Innovación Agroalimentaria y Agroambiental (CIAGRO-UMH), Miguel Hernández University, Carretera de Beniel, km 3.2, 03312 Orihuela, Spain

3 Grupo de Investigación en Fruticultura y Técnicas de Producción, Centro de Investigación e Innovación Agroalimentaria y Agroambiental (CIAGRO-UMH), Miguel Hernández University, Carretera de Beniel, km 3.2, 03312 Orihuela, Spain

* Correspondence: rafa.font@umh.es; Tel.: +34-96-6749-644

Abstract: Mortiño (*Vaccinium floribundum* Kunth) is considered a "superfruit" due to its antioxidant capacity and possible health benefits. To date, there is no known study that addresses the biochemical characterization of mortiño berries from the paramo of the Chimborazo volcano (Ecuador). So, the aim of this research was to evaluate for the first time the effect of the stage of development of the mortiño berries (two stages) and environment of origin (three sampling areas) on fruit quality. Polyphenol compounds were identified by high-performance liquid chromatography (HPLC) coupled to electrospray ionization mass spectrometric (ESI-MS^n) and quantified by high-performance liquid chromatography with a diode array detector (HPLC-DAD). Moreover, antioxidant properties ($ABTS^{\bullet+}$, and DPPH), sugar and organic acids, and minerals were examined. The main organic acids were quinic and citric acid, while glucose, fructose, sucrose, mannose, and sorbitol were the main sugars determined in the mortiño fruits. The main constituents of the mortiño berries included hydroxycinnamic acids (5-*O*-caffeoylquinic acid), flavonols (quercetin 3-hexoside, quercetin 5-hexoside, quercetin 3-pentoside, and quercetin-3-*O*-rhamnoside) and anthocyanins. Seven anthocyanins were identified: glycosides of cyanidin, delphinidin, petunidin, peonidin, and pelargonidin. The research confirms that the mortiño berries produced in the Ecuadorian paramo area are a valuable source of polyphenolics, rich in sugars and organic acids, and can be classified as a good source of microelements.

Keywords: antioxidant activity; polyphenolic; anthocyanins; mortiño; minerals

Citation: Caranqui-Aldaz, J.M.; Muelas-Domingo, R.; Hernández, F.; Martínez, R. Chemical Composition and Polyphenol Compounds of *Vaccinium floribundum* Kunth (Ericaceae) from the Volcano Chimborazo Paramo (Ecuador). *Horticulturae* **2022**, *8*, 956. https://doi.org/10.3390/horticulturae8100956

Academic Editor: Alessandra Durazzo

Received: 21 September 2022
Accepted: 13 October 2022
Published: 16 October 2022

1. Introduction

Mortiño (*Vaccinium floribundum* Kunth)—also known as Andean blueberry—is a deciduous perennial shrub endemic to the high Andes of South America and it is considered a "superfruit" due to its antioxidant capacity and possible health benefits [1,2]. This species is particularly common in the Andes from Venezuela to Bolivia and has been used since immemorial times. The berry is consumed fresh, dried, in sausages, jellies, jams, desserts, and in a special beverage called "colada morada" [3,4]. Mortiño berries are used by local communities in Ecuador for medical uses (allegedly as an aliment for rheumatism, fevers, colics, common colds, hangovers, and liver and kidney problems), as well as for ornamental and other uses as dye, fodder, or firewood [5]. One of the current limitations of *V. floribundum* fruit production is that the plant has not been domesticated and the technical difficulties to cultivate it, in addition to the continuous fragmentation experienced by mortiño populations due to anthropogenic processes such as deforestation, productive land reconversion, and overexploitation [6]. Mortiño berries can be found at local markets in the northern Andes (Colombia and Ecuador), where these fruits are usually collected from wild plants. Actually, with the advancement in food processing technologies, commercial mortiño products are available in the markets in different presentations such as capsules, powder, and wines, known as Wine of the Andes [7].

Mortiño, which can be found in Ecuador in the paramos at high altitudes ranging from 3000 to 4500 m above sea level (masl) [3], is a source of anthocyanins, proanthocyanidins, and polyphenolic compounds, which have been shown to possess antioxidant and anti-inflammatory properties, as well as lipid accumulation inhibition activity in adipocytes [8]. Freire [9] and Coba et al. [10] reported that berries of mortiño have relatively high concentrations of sugar, antioxidants such as vitamin C and the vitamin B complex, and minerals such as potassium, calcium, and phosphorus. The concentration of phytonutrients is influenced by many factors, such as variety, state of maturity, location, environmental conditions, agricultural practices, and pre-/postharvest handling [11–13]. During fruit ripening several biochemical and physiological processes take place, producing changes in fruit quality parameters [14]. Thus, due to oxidative stress in advanced stages of development, an increase in the content of phenol levels and a high content of antioxidant compounds have been observed [10]; however, during berry storage, the content of anthocyanins is reduced, as temperature is the main factor for destabilization of molecular structure [15]. Studies carried out show that the lyophilized extract of *Vaccinium floribundum* does not present toxicity and could be safely included as an ingredient in food. Because of its functional properties, the mortiño extract can also be handled as an antioxidant or natural colorant in the food industry, or even for the development of nutraceuticals in the pharmaceutical industry [16]. Thus, Guijarro-Fuertes et al. [17] developed bread with healthier properties by adding mortiño pulp. Moreover, it has even been used for the synthesis of nanoparticles and solar cells [18]. Additionally, the mortiño bagasse also contains a high amount of gallic acid, chlorogenic acid, caffeic acid, epicatechin, and coumaric acid, which makes it a very interesting source of antioxidant compounds [19]; the berries present as well antimicrobial activity, which makes the blueberry a potential source of bioproducts that can be used to develop new antimicrobials [1]. In addition, medicinal properties have been attributed to *V. floribundum*, such as potential applications in managing the symptoms of diabetes [20] and protection against oxidative stress [21].

Biochemical, nutraceutical and functional evaluation of *mortiño* plant material is essential to acknowledge its potential as a health-promoting aliment. To our knowledge, there are no studies that have addressed the biochemical characterization of *Vaccinium floribundum* from the paramo of the Chimborazo volcano (Ecuador).

For all the above-mentioned reasons, the objective of this research was the novel evaluation of the effect of the stage of development of mortiño berries (two stages) and location of origin (three sampling areas) on: (i) antioxidant activity, (ii) mineral composition, (iii) profile of sugars and organic acids, (iv) total phenolic content, and (v) anthocyanin profile and non-anthocyanin phenolic profile of mortiño (*Vaccinium floribundum* Kunth) produced in the volcano Chimborazo paramo (Ecuador). This information can be used to improve the market for mortiño, which can provide sustainable economic opportunities for farmers, and can be useful in promoting the conservation and sustainable use of this natural resource.

2. Materials and Methods

2.1. Plant Material

Three different local habitats with presence of *V. floribundum* were sampled in the paramo of the Chimborazo volcano in the central Andes of Ecuador. The sampling areas were to observe "mortiño" individuals selected based on verbal information from park rangers. In each study area, different adult plants were randomly selected.

Fresh berries of *V. floribundum* were harvested in three sampling areas: Culebrillas, Polylepis, and Cubillín, in the paramo of the Chimborazo volcano (Ecuador), the native habitat of the species. All growing environments showed loamy-sandy texture soils, with fairly poor percentages of organic matter (0.90% in Polylepis and Culebrillas, and 0.80% in Cubillín), acidic pH values of 5.27 (Culebrillas and Polylepis) and 5.63 (Cubillín). The predominant vegetation in the three monitored areas was the herbaceous paramo, and only Cubillín also showed an Alpine steppe (alpine grassland).

Berries were randomly picked from different parts of wild bushes on mountain slopes at an altitude between 3500 and 4100 masl. More information about each environmental area is shown in Table 1 and Figure 1. The berries were classified according to their maturity state into: (i) Stage 7: Fruit development; the berries began to develop anthocyanins, which was identified by their reddish coloration from the apical to the basal part of the fruit, and (ii) Stage 8: Ripening or fruit maturity; 100% of cluster berries shows a purple epicarp [22].

Table 1. Sampling areas where the mortiño berries were harvested in the paramo of the Chimborazo volcano, Ecuador.

Locality Name	Province	Coordinates	Altitude (m)	Vegetation Type	Mean Temperature(°C)	Mean Precipitation(mm)
Culebrillas	Bolívar	01°34.20′ S 78.55.5′ W	4000	Herbaceous paramo	3.1	967
Polylepis	Chimborazo	01°32.41′ S 78°53.5′ W	4076	Herbaceous paramo	3.1	967
Cubillín	Chimborazo	01°45′ S 78°31′ W	3500	High mountain forest	7.0	1000

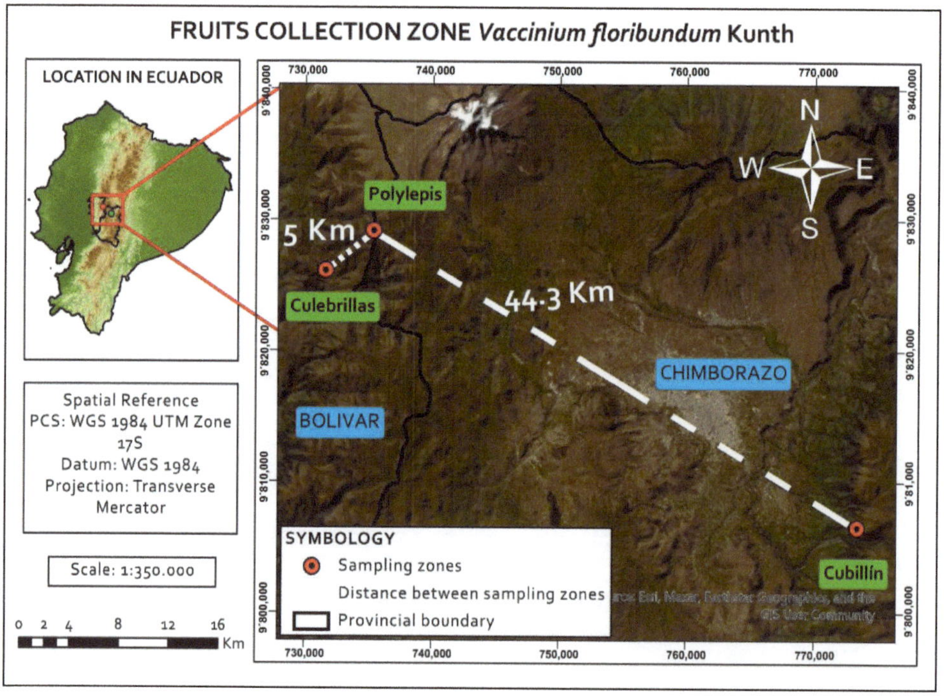

Figure 1. The three paramo zones where the mortiño berries were harvested.

2.2. Sample Preparation

After sorting, the berries were cleaned by removing leaves, stems, and damaged berries and were washed with drinking water to reduce the microbial load, dirt, and organic matter. Then, the berries were immediately frozen with liquid nitrogen and later freeze-dried in an Alpha 2–4 freeze drier (Alpha 2–4; Christ, Osterode am Harz, Germany) for 24 h at a pressure reduction of 0.220 mbar. The temperature in the drying chamber was −25 °C, while the heating plate reached 15 °C. At the end of freeze-drying, the samples were powdered, and vacuum packed at −20 °C until analyzed. Moisture content in freeze-dried mortiño was 5 g 100 g^{-1}. For conversion from DM to FW, the moisture contents in fresh and freeze-dried mortiño were used.

2.3. Extraction Procedure for Total Polyphenols Content (TPC) and Antioxidant Activity (AA)

The extraction procedure for TPC and AA quantification was prepared as described by Wojdyło et al. [23]. The extractions were performed in triplicate.

2.3.1. Quantification of Total Polyphenols Content (TPC)

The TPC was determined using the Folin–Ciocalteu colorimetric method described by Singleton et al. [24], with some modifications. The absorbance of the blue complex formed was read at 765 nm using a UV–visible spectrophotometer (Termospectromic Helios Gamma UVG 1002 E, Cambridge, UK). Calibration curves, with a concentration range between 0 and 0.25 g GAE L^{-1}, were used for the quantification of TPC and showed good linearity ($r^2 \geq 0.996$). All determinations were performed in triplicate, and results were expressed as milligrams of gallic acid equivalent per 100 g of sample dry matter (mg of GAE 100 g^{-1} of DM).

2.3.2. Determination of Antioxidant Activity by Two Different Methods

ABTS Method

The ABTS [2,2-azinobis-(3-ethylbenzothiazoline-6-sulfonic acid) assay was performed according to Re et al. [25] with some modifications. The absorbance was measured by UV–visible spectrophotometer (Termospectromic Helios Gamma UVG 1002 E, Cambridge, UK). All determinations were performed by triplicate, and the results were expressed in millimoles of Trolox per kilogram of sample dry matter (mmol Trolox kg^{-1} of DM).

DPPH Method

The radical scavenging activity was evaluated using the DPPH$^{\bullet}$ radical (2,2-diphenyl-1-picrylhydrazyl) method, as described by Brand-Williams et al. [26], with a modification in the reaction time. The decrease in absorbance was measured at 515 nm using a UV–visible spectrophotometer (Termospectromic Helios Gamma UVG 1002 E, Cambridge, UK). All determinations were performed by triplicate, and the results were expressed in millimoles of Trolox per kilogram of sample dry matter (mmol Trolox kg^{-1} of DM).

Calibration curves in the range 0.01–5.00 mmol Trolox L^{-1} were used for the quantification of the two methods of antioxidant activity, both showing good linearity ($r^2 \geq 0.998$).

2.4. Determination of Sugars and Organic Acids Profile

Organic acids and sugars profile were identified and quantified according to Hernández et al. [27], with some modifications. Briefly, half a gram of freeze-dried mortiño berry was mixed with 5 mL of phosphate buffer (50 mmol L^{-1}) pH 7.8; the mixture was homogenized, centrifuged, and filtered. Then, 10 μL of the supernatant was injected into a Hewlett Packard (Wilmington, DE, USA) series 1100 high-performance liquid chromatography equipped with a refractive index detector for sugars detection, and UV–Vis detector for organic acids analysis. A Supelcogel TM C-610H column (30 cm × 7.8 mm) with a pre-column (Supelguard 5 cm × 4.6 mm; Supelco, Bellefonte, PA, USA) was used for the analyses of both organic acids and sugars. Absorbance of organic acids was measured at 210 nm. Analyses were run in triplicate and the results expressed as g kg^{-1} dry matter (DM).

2.5. Minerals Analysis

To determine mineral content in the mortiño berries weighed accurately to a weight of 0.1 g of freeze-dried powdered into 75 mL Teflon (TFM) vessels. Then, 4 mL HNO_3 (69 vol.%) and 2 mL of ultra-high-purity deionized water were added and left to stand for 15 min to pre-digest the samples. Next, samples were microwave (CEM Mars One 240/50, Matthews, NC, USA) digested. The quantification of macro-elements [calcium (Ca), sodium (Na), potassium (K), and magnesium (Mg)] and micro-elements [iron (Fe), copper (Cu), manganese (Mn), and zinc (Zn)] was carried using an Inductively Coupled Plasma Mass Spectrometer (ICPMS-2030, Shimazu, Kyoto, Japan). Calibration curves were used for the quantification of minerals and showed good linearity ($R^2 \geq 0.998$). The analyses were run in triplicate and results expressed as g kg^{-1} and mg kg^{-1} for macro and micro-elements, respectively.

2.6. Identification and Quantification of Phenolic Compounds by HPLC-DAD-ESI-MS^n

2.6.1. Extraction and Determination of Phenolic Compounds Non-Anthocyanin

Extraction method: Samples (50 mg) were mixed with 1 mL of extractant (methanol/water (80:20, *v/v*) + 1% formic acid), and this mix was stirred during 5 min. After that time, the samples were centrifuged for 10 min at 12,000 rpm and 4 °C. The supernatant was filtered through a 0.45 µm PTFE filter (Millipore Billerica, MA, USA) and then was stored at −20 °C until further use. All extractions were carried out in triplicate.

Phenolic compounds non-anthocyanin: For the analysis of the different samples of mortiño berries, HPLC-ESI-DAD-MS^n Ion Trap (Agilent 1100 series System) was used, which allows us to make successive breaks of the precursor ion for its identification of unknowns. Chromatographic separation was carried out on a C18 column (Poroshell 120, 100 mm × 3 mm i.d., 2.7 µm particle size). The mobile phases consisted of two solvents: water/formic acid (95:1, *v/v*) as solvent A and acetonitrile as solvent B at a flow rate of 1 mL min^{-1}. For the determination of polyphenols, the gradient started with 5% B to reach 60% B at 37 min, at 40 min the percentage of B increased to 98% and was maintained for 2 min before returning to the initial conditions. The injection volume was 20 µL. Relative quantification of the phenolic compounds present in the samples was performed by chromatographic comparison with pure standards (caffeic acid, rutinoside quercetin, pelargonidin and cyanidin), and their absorbance spectra at four wavelengths emitted at 280, 320, 360, and 520 nm through a diode array UV detector (DAD) integrated in the HPLC and connected on-line to the mass spectrometer.

2.6.2. Identification and Quantification of Anthocyanins

Extraction method: Anthocyanins extraction and analysis were determined using the method proposed by Hong et al. [28] with some modifications. Briefly, half a gram of freeze-dried samples of mortiño berries was mixed by shaking in an orbital bath, with 4 mL cold extractant [methanol/water/formic acid (80:19.9:0.1, *v/v/v*)] for 10 min. The mixture was sonicated by ultrasonic bath for 10 min. Next, the samples were centrifuged for 10 min, 4000 rpm at 4 °C, and the supernatants were collected; the pellets residue was re-extracted twice using the same steps, being the supernatants definitely combined. Following, two milliliters of the supernatant were filtered through a 0.45 µm nylon Millipore membrane filter, and then stored at −20 °C until further use. All extractions were carried out in triplicate.

Analytical method: Anthocyanins profile was determined by high-performance liquid chromatography triple quadrupole mass spectrometer (LC-MS/MS 8050; Shimadzu, Kyoto, Japan). The molecules were ionized using Atmospheric Pressure Ionization (Electrospray Ion-ization-ESI). Chromatographic separations were performed with a C18 column (Mediterranean SEA 18, 10 mm × 0.21 mm i.d., 2.2 µm particle size) from Teknokroma (Barcelona, Spain). The mobile phase A consisted of 0.1% (*v/v*) formic acid (FA) in water (Milli-Q), and the mobile phase B consisted of 0.1% (*v/v*) formic acid in acetonitrile (ACN) at a flow rate of 0.4 mL min-1, an injection volume of 10 µL and oven temperature of 50 ° C. The gradient condition was 0–2 min 5% B, 2–10 min 95% B, 10–11 min 95% B, 11–12 min 5%B, and 12–16 min 5% B. For the quantification of the anthocyanins (Delphinidin 3-*O*-glucoside; Cyanidin-3,5-di-*O*-glucoside; Cyanidin 3-*O*-glucoside; Cyanidin-3-*O*-arabinoside; Petunidin-3-*O*-glucoside; Peonidin-3-*O*-glucoside; Pelargonidin-3-*O*-rutinoside), a stock of 100 ppm was made, and from it, a calibration plot of concentrations 0.1, 0.3, 0.5, 0.8 and 1 ppm was obtained. All analyses were performed in triplicate.

2.7. Statistical Analyses

The data were subjected to one-way analysis of variance (ANOVA) and later to Tukey's multiple-range test to compare the means. The confidence interval was 95%, and the significant difference was defined as $p < 0.05$. Principal component analysis (PCA) was used to reduce this complex dataset to a lower dimension and reveal simplified structures and relations between localities and states of maturity. To perform these statistical analyses, the software XLSTAT Premium 2016 (Version 9-Addinsoft, New York, NY, USA) was used.

3. Results and Discussion

Since the processed experimental data are so diverse, the results are differently presented into main subheadings.

3.1. Antioxidant Activity (AA) and Total Polyphenol Content (TPC)

The results for AA and TPC obtained in mortiño berries are summarized in Table 2. These parameters are important from a functional point of view because oxidative stress is reported to be the key factor for many diseases such as cardiovascular, hypertension, atherosclerosis, neurodegenerative or cancer, mainly caused by an imbalance between reactive oxygen species (ROS) and the antioxidative defense system [29]. The AA of mortiño berries was evaluated using two spectrophotometric assays: ABTS and DPPH, as each antioxidant compound has different mechanism of action. Significant differences were observed for "sampling area", "stage" and the interaction "sampling area x stage" factors. The highest value of AA by ABTS method was found in mortiño berries collected in the zone of Culebrillas (65.74 mmol Trolox kg^{-1} DM) while for DPPH method the highest value was obtained in mortiño berries collected in Polylepis (77.22 mmol Trolox kg^{-1} DM). These results were slightly higher than those reported by Vasco et al. [3] in mortiño berries and in Andean blackberry, and that those reported by Garzon et al. [4] for Colombian bilberries (*V. meridionale*). Furthermore, our AA values were higher compared to the values reported by Sellappan et al. [30] for other *Vaccinium* species as *V. corymbosum* L. hybrids and *V. ashei*. On the other hand, and with respect to the stage of fruit development, the AA for both methods decreased during fruit development, ranging from 63.34 (stage 7) to 39.74 (stage 8) mmol Trolox kg^{-1} DM for ABTS method, and from 72.16 (stage 7) to 55.37 (stage 8) mmol Trolox kg^{-1} DM for DPPH method. The same trend was reported by Esquivel-Alvarado et al. [31] for berries of *V. consanguineum, V. poasanum and V. floribundum* collected between 3000 and 3200 masl. Analyzing the interaction "sampling area x stage" factors (Table 2), it can be seen that the highest AA content, measured by both methods, was found in the mortiño fruits collected in the Polylepis and Culebrillas areas (>4000 masl). The results indicate that mortiño berries have a great inhibitory activity against free radicals, as can be seen when contrasted with other reports in the literature. Therefore, it can be concluded that mortiño berries are an accessible source of antioxidants.

Table 2. Antioxidant activity (mmol Trolox kg^{-1} dry matter, DM) and total polyphenol content [mg gallic acid equivalent (GAE) 100 g^{-1} DM] in Mortiño berries as affected by sampling area and stage of fruit development.

Factor	ABTS	DPPH	TPC
	(mmol Trolox kg^{-1} DM)		(mg GAE 100 g^{-1} DM)
	ANOVA Test †		
Zone	***	***	***
Stage	***	***	***
Sampling zone x Stage	***	***	***
	Tukey's multiple range test ‡		
	Zone		
Polylepis	47.76 b	77.22 a	2231.45 b
Cubillín	41.12 c	44.85 c	1649.69 c
Culebrillas	65.74 a	69.22 b	2651.57 a
	Stage ¥		
Stage 7	63.34 a	72.16 a	2559.12 a
Stage 8	39.74 b	55.37 b	1796.02 b
	Study zone × Stage		
Polylepis * Stage 7	79.53 a	86.64 a	3089.38 a
Polylepis * Stage 8	16.00 e	67.80 b	1373.52 e
Cubillín * Stage 7	41.43 d	50.97 c	1717.46 d
Cubillín * Stage 8	40.81 d	38.72 d	1581.91 d
Culebrillas* Stage 7	69.07 b	78.86 a	2870.53 b
Culebrillas* Stage 8	62.42 c	59.57 bc	2432.62 c

† NS: not significant at $p < 0.05$; * and ***, significant at $p < 0.05$ and 0.001, respectively. ‡ Values (mean of 3 replications) followed by the same letter, within the same column was not significantly different ($p < 0.05$), Tukey's least significant difference test. ¥ Stage 7: Fruit development; the berries began to develop anthocyanins; Stage 8: Ripening or maturity of fruit; 100% of cluster berries epicarp is purple.

With respect to the TPC, significant differences were also observed for the "sampling zone", "stage" and the interaction "sampling zone × stage" factors for $p < 0.001$. For the "sampling area" factor, the highest values were found for mortiño berries collected in Polylepis and Culebrillas, ranging those from 2231.45 (Polylepis) to 2651.57 (Culebrillas) mg GAE 100 g^{-1} DM; while the lowest values were obtained in the location of Cubillín at lower altitude (values 0.67- fold lower). For the "stage" factor, mortiño berries presented the highest levels of TPC in stage 7 (2559.12 mg GAE 100 g^{-1} DM); and for the interaction "sampling zone × stage" factors, the highest levels were found for the "Polylepis × Stage 7" (3089.38 mg GAE 100 g^{-1} DM) and the "Culebrillas × Stage 7" (2870.53 mg GAE 100 g^{-1} DM). TPC decreased during fruit development by 0.70-fold for stage 8 (Table 2). A similar trend was reported by Esquivel-Alvarado et al. [31] for *V. floribundum* collected at an altitude of 3200 m, who reported values from 1787 mg GAE 100 g^{-1} DM for early fruit development stage to 1441 mg GAE 100 g^{-1} DM for the late fruit development stage. This trend can be attributed to changes such as the hydrolysis of glycosides, the oxidation of phenols by polyphenol oxidases, and the polymerization of free phenols [32]. In the current study, the content of total polyphenol was higher than those reported by Prior et al. [11], who reported mean TPC values for *V. corymbosum.*, *V. angustifolia*, and *V. myrtillus* of 347, 398, and 525 mg GAE 100 g^{-1} FW, respectively, and were also higher than those found in other fruits as pomegranate (777 to 1660 g GAE kg^{-1} DM), plum (440 mg GAE 100 g^{-1} FW) and strawberry (238 mg GAE 100 g^{-1} FW) [3,33], whereas were lower than those reported by Llerena et al. [34] for blackberry (6352.28 mg GAE 100 g^{-1} DM).

In this study, total polyphenols content and antioxidant activity were clearly influenced by altitude. The higher the altitude, the higher the contents of TPC and AA presented by mortiño berries. These results demonstrate that altitude is an important factor affecting the antioxidant activity and total polyphenols content in *V. floribundum*. Similarly, previous studies have indicated that AA and TPC levels can vary significantly depending on the geographical location of the mortiño plants, as well as the growing conditions such as altitude, radiation, and temperature [1]. The high content of TPC in mortiño berries can significantly contribute to the use of this material as a source of natural antioxidants.

3.2. Sugars and Organic Acids Profile

Table 3 summarizes the effects of environmental origin and stage of fruit development on organic acids (OA) and sugars profile and content. The main organic acids found in *V. floribundum* were quinic acid, followed by citric and malic acids. The same main organic acids were reported by Wang et al. [35] in 10 populations of *Vaccinium uliginosum* from the Changbai Mountains of China. The "Sampling zone" factor significantly ($p < 0.001$) affected the three acids content.

Table 3. Organic acids and sugar content (g kg^{-1} dry matter, DM) in Mortiño berries as affected by location and stage of fruit development.

Factor	Organic Acids			Sugars				
	Citric	Malic	Quinic	Sucrose	Glucose	Fructose	Mannose	Sorbitol
ANOVA Test †								
Zone	***	***	***	***	***	***	***	***
Stage	***	***	***	***	NS	***	***	***
Zonex Stage	***	***	***	***	***	***	***	***
Tukey's multiple range test ‡								
Zone								
Polylepis	41.98 b	22.30 c	108.39 c	9.96 b	110.05 a	60.80 b	31.80 b	106.46 c
Cubillín	65.05 a	28.35 b	199.48 a	61.42 a	71.75 b	50.61 b	35.82 b	190.55 a
Culebrillas	58.06 a	36.71 a	155.99 b	13.05 b	106.94 a	79.19 a	134.04 a	157.76 b
Stage ¥								
Stage 7	63.10 a	26.22 b	177.32 a	41.56 a	91.91	39.08 b	79.30 a	178.97 a
Stage 8	46.96 b	32.02 a	131.92 b	14.73 b	100.58	87.99 a	55.14 b	124.20 b
Zone × Stage								
Polylepis * Stage 7	50.76 b	10.21 e	153.20 b	12.90 bc	116.43 a	64.35 b	34.14 cd	145.12 b
Polylepis * Stage 8	33.20 c	34.96 b	63.57 c	7.02 c	103.66 ab	57.25 b	29.45 cd	67.80 c
Cubillín * Stage 7	75.98 a	25.23 cd	216.84 a	104.74 a	60.04 c	24.19 c	21.00 d	217.02 a
Cubillín * Stage 8	54.11 b	31.47 bc	182.12 ab	18.10 b	83.46 bc	77.03 b	50.64 c	164.07 b
Culebrillas * Stage 7	62.56 ab	19.04 de	161.91 b	7.04 c	99.27 ab	28.69 c	182.76 a	174.77 ab
Culebrillas * Stage 8	53.56 b	54.37 a	150.07 b	19.07 b	114.61 a	129.69 a	85.32 b	140.75 b

† NS: not significant at $p < 0.05$; * and ***, significant at $p < 0.05$ and 0.001, respectively. ‡ Values (mean of 3 replications) followed by the same letter, within the same column were not significantly different ($p < 0.05$), Tukey's least significant difference test. ¥ Stage 7: Fruit development; the berries began to develop anthocyanins; Stage 8: Ripening or maturity of fruit; 100% of the epicarp of the cluster berries is purple.

Quinic and citric acids predominated over malic acid in all environmental areas. This result agreed with that obtained by Mikulic-Petkovsek et al. [36], who indicated that fruit of the Ericaceae family generally contained very little malic acid. Mikulic-Petkovsek et al. [37] reported that wild bilberry fruits (*V. myrtillus*) from high altitude (up to 636 masl) had more organic acids content, compared with bilberry fruit from lower altitudes (up to 217 m). While mortiño berries collected in Cubillín (at an altitude of 3500 m and at a mean temperature of 7 °C) showed the highest content of total OA (287.88 g kg^{-1}), Culebrillas and Polylepis (at more than 4000 m altitude and at a mean temperature of 3.1 °C) showed total OA contents of 250.76 g kg^{-1} and 172.67 g kg^{-1}, respectively. It is important to note that changes in organic acids in response to temperature also depend on other factors as plant age or fruit type. The temperature at which fruits are grown affects both their titratable acidity and the content of stored organic acids [38]. Our results agree with those found by Mikulic-Petkovsek et al. [37]. However, the results obtained in this study suggest that above 3500 m of altitude the content of organic acids decreases strongly; from 3500 m of altitude (Cubillín) to 4076 (Polylepis), the total AO content decreased to 60%.

Different studies indicate that the organic acid content of the flesh of fruits is affected by environmental factors and cultivation practices such as temperature, light intensity, cultivar, rootstock, mineral nutrition, water availability, and fruit load/pruning. Nonetheless, how these factors alter metabolism to bring about changes in organic acid content is in most cases uncertain [39]. Thus, Wang et al. [35] reported that in *V. uliginosum* the organic acid content was not related to altitude. These differences may be due to a combination of different growing environments (e.g., microclimate), genetics, or other environmental factors. The "stage" and the interaction "zone x stage" factors also affected significantly ($p < 0.001$) the organic acid content (Table 3). Organic acids accumulate in the flesh of many types of fruits at certain stages of their development [38]. Organic acids are related to maturation, in particular malic acid which confers a bitter taste. Thus, the results obtained in this study showed that malic acid increases with maturation, ranging from 26.22 g kg^{-1} DM (Stage 7) to 32.02 g kg^{-1} DM (Stage 8), while both quinic and citric acids decreased. Ayaz et al. [40] also reported in *V. artostaphylos* and *V. myrtillus* that the level of malic acid increased gradually during the maturation of fruits. Compared to previous results reported by Kafkas et al. [41] for blackberries and Correia et al. [42] for Highbush blueberries (*V. corymbosum*), our study showed much lower contents of malic acid for *V. floribundum*.

Regarding the sugar content, glucose, fructose, sucrose, mannose, and sorbitol were the main sugars determined in the fruits of mortiño (Table 3). The sugar content was significantly ($p < 0.001$) affected by the "location", "stage" and the interaction "location × stage" factors. The higher values of glucose and fructose were obtained in the areas of higher altitudes (>4000 m), Polylepis, and Culebrillas. Additionally, Wang et al. [35] reported that the higher the location altitude (>1200 m considered high altitude), the higher the contents of glucose and fructose in bog bilberry (*V. uliginosum*). On the other hand, during fruit development of mortiño, fructose and glucose increased while sucrose, mannose, and sorbitol decreased; the low sucrose content may be due to enzymatic hydrolysis or its transformation into other sugars during the ripening process. Our values agreed with the results reported by Kalt and McDonal [43] for lowbush blueberry (*V. angustifolium*), Correia et al. [42] for highbush blueberry (*V. corymbosum*), and Ayaz et al. [40] for *V. arcostaphylos* and *V. myrtillus*. However, Mikulic-Petkovsek et al. [37] indicated that bilberries grown at low altitude sites (217 m considered low altitude) contained higher levels of total sugars compared to bilberries grown at higher altitudes (636 m considered high altitude). Additionally, since the sucrose content of mortiño was quite low, this fruit should be recommended for low-carbohydrate diets.

Sorbitol is a sugar alcohol characteristic of higher plants. It is a major final product of photosynthesis and, together with sucrose, represents the main forma of carbon translocated in some fruit species [36]. While the highest sorbitol content (190.55 g kg^{-1} DM) was obtained in mortiño fruits grown at low altitudes (Cubillín), the lowest content was found in those fruits grown at higher altitudes (>4000 m). Though Mikulic-Petkovsek [36] also detected sorbitol in chokeberry, rowanberry, and eastern shadbush, this sugar was not detected at all in other berry species [44]. Our results suggest that climate factors such as altitude and temperature play an important role in the sugar content of mortiño berries. Likewise, Cobo et al. [45] also observed this variability in the chemical composition of *V. floribundum* fruits, as a result of climatic and geographic influences.

3.3. Mineral Content

The content of the macronutrients and micronutrients was significantly affected by the "growing environment" and the interaction "growing environment × stage" factors.

Only the macronutrients potassium (K), sodium (Na), and magnesium (Mg) and the micronutrient iron (Fe) were significantly affected by the "stage" factor (Table 4). While the highest quantity of macronutrients (K > Ca > Mg > Na) was found in the mortiño fruits grown in Polylepis, the highest content of micronutrients (Fe > Mn) was observed in fruits grown in Cubillín. Unfortunately, to date there are no studies that provide complete information on the content of minerals in mortiño fruits, and neither how these contents can be influenced by the factors previously outlined such as altitude, stage of development of the fruit, temperature, etc. Vasco et al. [3] reported that mortiño berries are rich in potassium; a serving of 100 g could provide 13% of the recommended adequate intake (AI) of 4.7 g/day for adults. Our results revealed that the macronutrient (K, Na, and Mg) contents decrease with fruit development, while calcium (Ca) and iron (Fe) are the only macro and micronutrients that increase with fruit development. Several studies showed that Ca is an effective pressure-lowering agent [46]; thus, a high Ca content can be beneficial for health. Likewise, it was observed that mortiño immature berries (development stage 7) grown at high altitudes (>4000 m) had a higher potassium content (10.13 and 8.27 g kg^{-1} DM for Polylepis and Culebrillas, respectively). Our results indicated that the mineral content in mortiño fruits is clearly influenced by the growing conditions, the state of fruit development, and the altitude. Karlsons et al. [47] studied the mineral composition of four species of *Vaccinium* (*V. corymbosum*, *V. myrtillus*, *V. macrocarpon*, and *V. oxycoccos*), and reported that the berries of these species were characterized by having a high content of Fe, Ca, Mg, and Mn. In our study, mortiño berries showed levels of macro and micronutrients comparable to those obtained by Karlsons et al. [47] and by Miljkovič et al. [48] in Serbia for *V. myrtillus*. The mineral composition shown by mortiño fruits indicates that these berries are an excellent source of K, Ca, and Fe. In addition, due to their low levels of sodium, mortiño fruits could be properly recommended for low-sodium diets.

3.4. Identification and Quantification of Phenolic Compounds Non-Anthocyanin and Anthocyanins

A total of sixteen different compounds, nine non-anthocyanin (Table 5) and seven anthocyanins (Table 6) have been identified in mortiño berries. To make the discussion easy to follow, phenolic compounds non-anthocyanin and anthocyanins were discussed separately. Quantification of each identified compound is shown in Table 7.

Table 4. Minerals content (g or mg kg^{-1} dry matter, DM) in Mortiño berries as affected by environmental zone and stage of fruit development.

Factor	K	Na	Ca	Mg	Cu	Mn	Fe	Zn
	Macro-Elements (g kg^{-1})				Micro-Elements (mg kg^{-1})			
ANOVA Test †								
Zone	**	***	***	***	**	***	***	***
Stage	**	***	NS	***	NS	NS	***	NS
Zonex Stage	**	***	***	***	**	***	***	***
Tukey's multiple range test ‡								
Zone								
Polylepis	8.85 a	0.30 a	4.88 a	1.24 a	5.63 a	47.63 b	80.95 b	29.61 a
Cubillín	6.83 b	0.24 b	1.80 c	0.41 c	4.74 b	93.05 a	126.89 a	14.40 c
Culebrillas	8.10 a	0.22 b	3.47 b	0.93 b	4.84 ab	22.96 c	70.90 b	17.79 b
Stage ¥								
Stage 7	8.52 a	0.28 a	3.18	0.92 a	5.11	56.76	70.84 b	21.60
Stage 8	7.33 b	0.23 b	3.58	0.79 b	5.03	52.34	114.99 a	19.60

Table 4. *Cont.*

Factor	K	Na	Ca	Mg	Cu	Mn	Fe	Zn
		Macro-Elements (g kg^{-1})				Micro-Elements (mg kg^{-1})		
Zone × Stage								
Polylepis * Stage 7	10.13 a	0.36 a	3.92 b	1.28 a	5.06 ab	48.97 b	69.81 bc	32.42 a
Polylepis * Stage 8	7.57 b	0.25 bc	5.84 a	1.20 a	6.20 a	46.30 b	92.10 b	26.80 a
Cubillín * Stage 7	7.15 b	0.27 b	1.87 c	0.43 c	5.02 ab	93.04 a	61.29 c	15.82 bc
Cubillín * Stage 8	6.50 b	0.21 bc	1.72 c	0.39 c	4.42 b	93.07 a	192.49 a	12.99 c
Culebrillas * Stage 7	8.27 ab	0.19 a	3.74 b	1.06 a	5.22 ab	28.26 c	81.41 bc	16.55 bc
Culebrillas * Stage 8	7.93 b	0.24 ab	3.20 bc	0.81 b	4.46 b	17.67 c	60.38 c	19.02 b

† NS: not significant at $p < 0.05$; *, **, and ***. significant at $p < 0.05$, 0.01, and 0.001, respectively. ‡ Values (mean of 3 replications) followed by the same letter, within the same column, were not significantly different ($p < 0.05$), Tukey's least significant difference test. ¥ Stage 7: Fruit development; the berries began to develop anthocyanins; Stage 8: Ripening or fruit maturity; 100% of cluster berries epicarp is purple.

Table 5. Phenolic compounds (non-anthocyanin) identified by HPLC-DAD-ESI-MS^n in Mortiño berries.

Peak No.	[a] Rt (min)	[b]MS/MS (*m/z*)	Name of Compounds [c]	Chemical Family
P1	7.3	353,191	3-*O*-Caffeoylquinic acid	Hydroxycinnamic Acid
P2	9.1	337,163	3-Coumaroylquinic acid	Hydroxycinnamic Acid
P3	9.9	707,353,191	5-*O*-Caffeoylquinic acid	Hydroxycinnamic Acid
P4	12.6	335,179	Caffeoylshikimic acid	Hydroxycinnamic Acid
P5	15.9	433,323	Caffeic acid derivate	Hydroxycinnamic Acid
P6	16	463,301	Quercetin 3-hexoside	Flavonols
P7	17.8	463,301	Quercetin 5-hexoside	Flavonols
P8	18	433,301	Quercetin 3-pentoside	Flavonols
P9	18.1	447,301	Quercetin-3-O-rhamnoside	Flavonols

[a] Rt = retention time; [b] MS/MS = tandem mass spectrometry; [c] Compounds were numberer by their elution time.

Table 6. Anthocyanins identified by HPLC-ESI-MS^n in Mortiño berries.

Peak No.	[a] Rt (min)	Molecular Ion [M + H] (*m/z*)	[b]MS/MS (*m/z*)	Name of Compounds [c]
An1	2.3	627	303,465	Delphinidin 3-*O*-glucoside
An2	4.3	611	287,449	Cyanidin-3,5-di-*O*-glucoside
An3	4.8	449	287,213,137	Cyanidin 3-*O*-glucoside
An4	4.9	419	287,137,213	Cyanidin-3-*O*-arabinoside
An5	4.9	479	317,302,274	Petunidin-3-*O*-glucoside
An6	5.1	463	301,286,201	Peonidin-3-*O*-glucoside
An7	5.0	579	271,433	Pelargonidin-3-*O*-rutinoside

[a] Rt = retention time; [b] MS/MS = tandem mass spectrometry; [c] Compounds were numberer by their elution time.

Table 7. Phenolic compounds quantified in Mortiño berries as affected by sampling zone and stage of fruit development (mg 100 g^{-1} DM).

Factor	Hydroxycinnamic Acid				Flavonols					Anthocyanins							∑Total Polyphenols
	P1	P2	P3	P4	P5	P6	P7	P8	P9	An1	An2	An3	An4	An5	An6	An7	
ANOVA Test †																	
Zone	***	***	***	***	***	***	***	***	***	***	***	***	***	***	***	***	***
Stage	***	***	***	***	***	***	***	***	***	***	***	***	***	***	***	***	***
Zone × Stage	***	***	***	***	***	***	***	***	***	***	***	***	***	***	***	***	***
Tukey's multiple range test ‡																	
Zone																	
Polylepis	1.56 c	1.56 c	291.68 c	140.89 c	69.73 c	61.40 c	75.54 c	35.08 b	24.34 c	0.07 a	0.04 a	6.48 b	12.30 b	0.00 b	0.35 b	0.03 a	721.14 c
Cubillín	247.37 a	95.05 b	595.05 b	343.98 a	221.44 a	165.64 b	224.88 b	149.97 a	367.79 b	0.00 b	0.001 c	10.71 a	18.69 a	0.91 a	0.67 a	0.006 c	2442.21 b
Culebrillas	192.67 b	142.15 a	2032.71 a	196.16 b	138.23 b	196.03 a	281.66 a	1.29 c	509.90 a	0.00 b	0.02 b	6.08 b	11.47 b	0.01 b	0.12 c	0.01 b	3708.58 a
Stage ¥																	
Stage 7	179.22 a	93.33 a	577.24 b	271.86 a	183.31 a	174.01 a	228.07 a	122.94 a	339.93 a	0.00 b	0.001 b	4.46 b	9.74 b	0.06 b	0.42 a	0.008 b	2184.64 a
Stage 8	115.18 b	65.85 b	1369.06 a	182.17 b	102.96 b	108.03 b	159.98 b	1.29 b	261.43 b	0.05 a	0.04 a	11.06 a	18.57 a	0.54 a	0.34 b	0.02 a	2396.65 a
Zone × Stage																	
Polylepis * Stage 7	1.56 d	1.56 d	581.80 b	280.22 bc	137.89 cd	105.71 c	124.17 c	68.88 b	47.40 c	0.00 b	0.00 c	5.55 c	10.44 c	0.00 c	0.70 b	0.02 b	1365.96 c
Polylepis * Stage 8	1.56 d	1.56 d	1.56 c	1.56 e	1.56 e	17.08 d	26.91 d	1.29 c	1.29 c	0.15 a	0.09 a	7.42 c	14.17 bc	0.00 c	0.00 e	0.04 a	76.32 d
Cubillín * Stage 7	302,36 a	110.06 b	667.27 b	363.20 a	235.87 a	172.4 b	226.14 b	298.64 a	370,36 b	0,00 b	0.00 c	6.77 c	12.97 bc	0.19 b	0.47 c	0.00 d	2766.40 b
Cubillín * Stage 8	192,39 bc	80.05 c	522.84 b	324.75 ab	207.01 ab	159.23 b	223.62 b	1.29 c	365.23 b	0.00 b	0.002 c	14.65 a	24.41 a	1.63 a	0.87 a	0.01 c	2118.03 b
Culebrillas * Stage 7	233,75 b	168.36 a	482.64 b	172.14 d	176.16 bc	244.28 a	333.90 a	1.29 c	602.03 a	0.00 b	0.004 c	1.06 d	5.81 d	0.00 c	0.08 de	0.00 d	2421.57 b
Culebrillas * Stage 8	151,59 c	115.95 b	3582.78 a	220.19 cd	100.31 d	147.78 bc	229.42 b	1.29 c	417.78 b	0.00 b	0.05 b	11.10 b	17.12 b	0.01 c	0.15 d	0.02 b	4995.60 a

† NS: not significant at $p < 0.05$; * and ***, significant at $p < 0.05$ and 0.001, respectively. ‡ Values (mean of 3 replications) followed by the same letter, within the same column, were not significantly different ($p < 0.05$), Tukey's least significant difference test. ¥ Stage 7: Fruit development; berries began to develop anthocyanins; Stage 8: Ripening or fruit maturity; 100% of the epicarp of cluster berries is purple.

The nine non-anthocyanin phenolic compounds were classified into two chemical families: (i) hydroxycinnamic acid (5 compounds), and (ii) flavonols (4 compounds). The content of the phenolic compounds non-anthocyanins was significantly affected by the "growing zone", "stage" and the interaction "growing zone x stage" factors (Table 7). Five hydroxycinnamic acids were detected, four caffeoyl acid derivates, and one coumaroylquinic acid. Baenas et al. [16] also identified in mortiño berries purchased at a local market in Machaci, Ecuador, the same four caffeoyl acid derivates, but at very low concentrations compared to our values; yet they could not identify the presence of coumaroylquinic acid. Furthermore, the highest content of hydroxycinnamic acids was noted for mortiño berries grown in Culebrillas (2701.92 mg 100 g^{-1} DM), followed by Cubillín (1502.89 mg 100 g^{-1} DM), while the lowest content was presented by mortiño fruits grown in Polylepis (505,42 mg 100 g^{-1} DM). From the hydroxycinnamic acids, the isomer of chlorogenic acid, 5-*O*-caffeoylquinic acid, was the most representative in mortiño berries. Our results agree with previous studies carried out by Vasco et al. [3] in *V. floribundum*, by Garzón et al. [4] in *V. meridionale*, by Prencipe et al. [49] in *Vaccinium* berries and by Baenas et al. [16] in *V. floribundum*. Moreover, Wojdyło et al. [50] reported that hydroxycinnamic acids such as 5-caffeoylquinic and caffeoylquinic acid are good sources of antioxidants in vitro that protect low-density lipoprotein (LDL) from oxidation and therefore, supposedly prevent various age-related diseases. Therefore, the consumption of mortiño berries would reduce the risk of cardiovascular disease since its antioxidants lower low-density lipoprotein (LDL) cholesterol levels.

The flavanol glycosides were the second group after the hydroxycinnamic acid derivatives that contributed to the final concentration of polyphenols in mortiño berries. Additionally, those fruits grown in Culebrillas presented the highest levels of flavonols, the quercetin 3-hexoside, quercetin 5-hexoside, and quercetin-3-*O*-rhamnoside were significantly high (Table 7). Other studies also reported these flavonols as the predominant ones in mortiño berries [3,16]. Garzón et al. [4] reported in *V. meridionale* that 93% of the total flavonoids are represented by quercetin derivatives. High content of flavonols may well reflect plant responses to biotic and abiotic stresses or just acclimation to environmental stressors such as heat, cold, UV radiation, drought, salinity, or an attack of herbivores or pathogens [51]. In this study, the highest content of flavonols and hydroxycinnamic acids was obtained for mortiño fruits at the early stages of fruit development (stage 7); the only exception was the 5-*O*-Caffeoylquinic acid, which showed a higher content in the late stage of development. According to Garzón et al. [4], a fairly common feature in the *Vaccinium* family is the presence of quercetin glycosides and hydroxycinnamic acids. Our results confirm that fruit maturity stage and altitude definitely influence the content of non-anthocyanin phenolic compounds; in such a way that above 4000 m of altitude there is a strong reduction in the content of these phenolic compounds. Jaakola and Hohtola [52] reported that flavonol accumulation in fruit skin, as a result of sunlight exposition, is well documented and is the most important environmental factor inducing flavonol biosynthesis, just like that; fruits with sun-exposed peel have higher levels of anthocyanins and flavonols than those grown in the shade. In the literature, there is a great variability regarding the non-anthocyanin phenolic compounds in mortiño berries and in other several species of the *Vaccinium* genus. This variability is due to several factors such as stage of maturity, agronomic factors, cultivars and varieties, geographic region, storage conditions, ripeness, and climate, among others [3,16,31].

Anthocyanins are coloring pigments that give a wide range of colors such as orange, red, purple, and blue in flowers, seeds, fruits, and vegetative tissues [53]. Blueberry and bilberry (*Vaccinium* spp.) are one of the richest sources of anthocyanins [3,16]. In this study, seven anthocyanins have been identified in mortiño berries: glycosides of cyanidin (peaks An_2, An_3, and An_4), delphinidin (peak An_1), petunidin (peak An_5), peonidin (peak An_6) and pelargonidin (peak An_7) (Table 6). Baenas et al. [16] and Esquivel-Alvarado et al. [31] reported the presence of six and five anthocyanins in mortiño berries, respectively, namely derivatives of delphinidin and cyanidin. Garzón et al. [4] and Vasco et al. [3] reported that Colombian bilberry and Andean blueberry contained only cyanidin and delphinidin glycosides. To our knowledge, the current study identifies the presence of petunidin, peonidin, and pelargonidin in *V. floribundum*. In addition, the anthocyanins contents were significantly affected by the "growing environment", "stage" and the interaction "growing environment x stage" factors (Table 7). Analyzing the environmental area factor, it was observed that the predominant anthocyanins were cyanidin-3-*O*-arabinoside (ranging from 18.69 mg 100 g^{-1} DM of Cubillín to 11.47 mg 100 g^{-1} DM of Culebrillas), followed by cyanidin 3-*O*-glucoside (ranging from 10.71 mg 100 g-1 DM of Cubillín to 6.08 mg 100 g^{-1} DM of Culebrillas). Likewise, higher contents of anthocyanins were shown when mortiño berries reached the late developmental stage (stage 8), being the cyanidin derivatives, followed by petunidin and peonidin, the main anthocyanins. The mortiño fruits showing the highest anthocyanin contents were those grown at an altitude of 3500 m and in a more advanced developmental stage (stage 8). Furthermore, our results indicate that the anthocyanin content slightly decreases above 3500 m of altitude.

It is fairly known that temperature plays a vital role in the synthesis of anthocyanins, and these are more prone to oxidation and relatively unstable. However, the mechanisms are not well understood [54]. Low temperature induces anthocyanin synthesis in various species [55]. However, the accumulation of anthocyanins in cold temperatures is light dependent; in the absence of light, low temperatures prevent anthocyanin biosynthesis. The regulation of cold induction of anthocyanins and the role of light are not well understood yet [52]. The higher the solar radiation at high altitudes, the greater the influence on the secondary metabolite profiles [52]. Li et al. [56] reported that warm weather was related to low levels of anthocyanins, and cool weather was associated with the rapid accumulation of anthocyanins in fruit skin. Maier and Hoecker [57] suggested that high light intensity stimulates anthocyanins production in most plants. In view of the results obtained in this research, well-designed long-term studies are necessary to better understand the plant–environment interaction regarding anthocyanin biosynthesis.

3.5. Principal Component Analysis (PCA)

PCA was used because it is one of the beneficial statistical tools for analyzing several samples and variables in order to establish their differences and similarities. Figure 2 shows that 64.19% of the total variance in the data are represented by PC1 and PC2. Of these two top principal components, PC1 described 42.29% of the total variation and PC2 explained 21.90% of the variation. It is important to note that the higher the distance between two parameters, the lower their correlation. Considering F1 as the dimension that explained the main differences among growing areas and fruit developmental stage Culebrillas-red and Culebrillas-green were positively linked with hydroxycinnamic acids (3-coumaroylquinic acid, and 5-caffeoylquinic acid), flavonols (quercetin derivatives), (antioxidant activity (ABTS), TPC, and organic acids (quinic and malic acids). Instead, Cubillín-red and Cubillín-green were positively linked with anthocyanins (cyanidin 3-*O*-glucoside, cyanidin-3-*O*-arabinoside, pelargonidin-3-*O*-rutinoside, and peonidin-3-*O*-glucoside), hydroxycinnamic acids (caffeoylshikimic acid), flavonols (quercetin-3-pentoside) and minerals (Mn and Fe).

On the other hand, whereas Polylepis-red was negatively linked with anthocyanins (cyanidin-3,5-di-*O*-glucoside, petunidin-3-*O*-glucoside, and delphinidin 3-*O*-glucoside), minerals (Cu) and sugar (fructose), Polylepis-green was positively linked with minerals (Ca, Na, Zn, K, and Mg), antioxidant activity (DPPH) and sugars (glucose).

The biplot (Figure 2) showed that Culebrillas-red and Cubillín-green were laid relatively close to each other along the *X*-axis (PC1). Polylepis-red had large negative scores on the PC2, and it was quite separated from the other locations across PC1. Polylepis-green had large positive scores on the PC2 axis, and it was opposed to Polylepis-red.

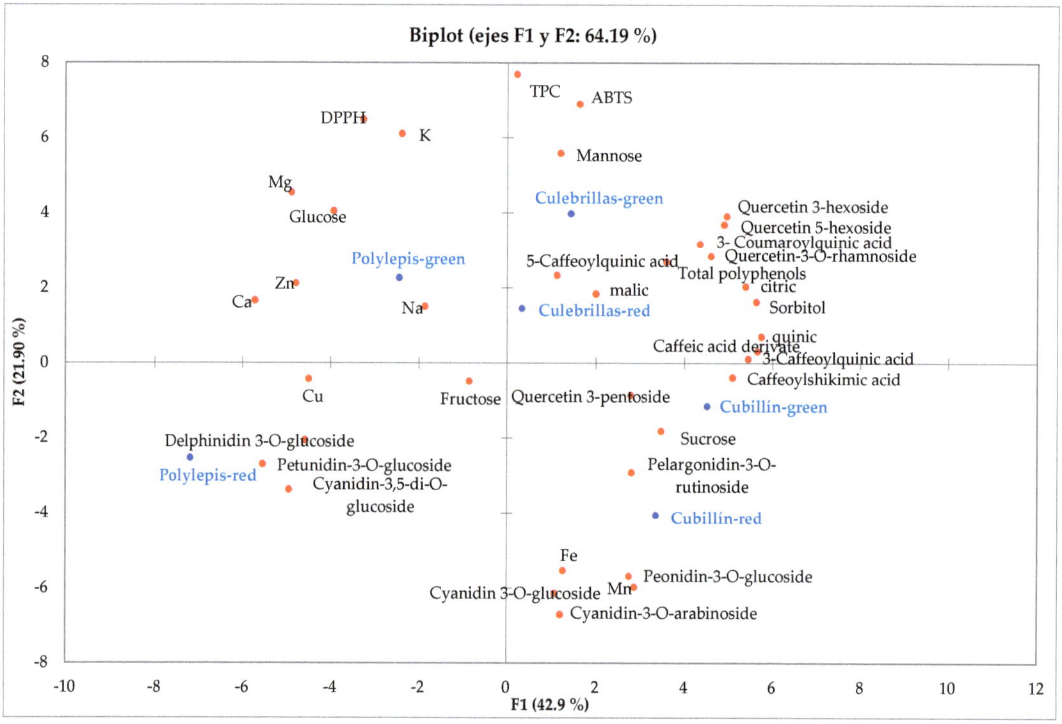

Figure 2. Biplot of principal components analysis (PCA) means showing the relationship among phytochemical parameters and effects of environmental zones (Polylepis, Cubillín and Culebrillas) and stage of development of the fruit of mortiño berries (stage 7 green and stage 8 red).

The results clearly indicated that the sampling zone and fruit developmental stage factors had a crucial effect on the chemical composition and polyphenol compounds of *Vaccinium floribundum*. Therefore, more research considering both the growing environmental factor and the stage of development of the fruit is a must to obtain mortiño berries with high contents of bioactive compounds, becoming an exceptional ingredient to be used in both the cosmetic and pharmacological industries, as well as in the agri-food one.

4. Conclusions

The present study investigated the chemical composition and polyphenol compounds of *Vaccinium floribundum* produced in the volcano Chimborazo paramo, Ecuador. The study was carried out in three growing areas (Polylepis, Culebrillas, and Cubillín) located above 3500 m of altitude and with mortiño berries showing two different stages of fruit development. Despite the fact that the altitude at which this mountain fruit species is found is the main limitation for observation and monitoring, the research confirms that the mortiño berries produced in the Ecuadorian paramo area are a valuable source of polyphenols, rich in sugars and organic acids and can be classified as a good source of microelements, an excellent source of K, Ca and Fe. In addition, due to their low sodium levels, mortiño berries could be recommended for low-sodium diets. The main constituents of mortiño berries include hydroxycinnamic acids (5-*O*-caffeoylquinic acid), flavonols (quercetin derivatives), and anthocyanins. Three anthocyanins (petunidin, peonidin, and pelargonidin) were reported for the first time in mortiño berries, which have never been before identified and quantified in *V. floribundum*. Overall, our data indicate that altitude and stage of fruit development significantly affect mortiño berries quality. This research may successfully contribute to improving market sales for the mortiño and, at the same time, can provide sustainable economic opportunities for farmers. Finally, our contribution to improving knowledge about this wild fruit species will help improve the sustainability and preservation of this rich natural resource.

Author Contributions: Conceptualization, F.H. and R.M.; investigation, J.M.C.-A.; R.M.-D. and F.H.; methodology, R.M.-D., R.M. and F.H.; supervision, R.M. and F.H.; writing—original draft, J.M.C.-A., R.M.-D. and F.H.; writing—review and editing, R.M. All authors have read and agreed to the published version of the manuscript.

Funding: The ICPMS and LC-MS/MS have been acquired thanks to Grant EQC2018-004170-P funded by MCIN/AEI/10.13039/501100011033 and by ERDF A way of making Europe.

Institutional Review Board Statement: Not applicable.

Informed Consent Statement: Not applicable.

Data Availability Statement: Not applicable.

Conflicts of Interest: The authors declare no conflict of interest.

References

1. Llivisaca, S.; Manzano, P.; Ruales, J.; Flores, J.; Mendoza, J.; Peralta, E.; Cevallos-Cevallos, J.M. Chemical, Antimicrobial, and Molecular Characterization of Mortiño (*Vaccinium floribundum* Kunth) Fruits and Leaves. *Food Sci. Nutr.* **2018**, *6*, 934–942. [CrossRef]
2. Chang, S.; Alasalvar, C.; Shahidi, F. Superfruits: Phytochemicals, Antioxidant Efficacies, and Health Effects—A Comprehensive Review. *Crit. Rev. Food Sci. Nutr.* **2019**, *59*, 1580–1604. [CrossRef]
3. Vasco, C.; Riihinen, K.; Ruales, J.; Kamal-Eldin, A. Chemical Composition and Phenolic Compound Profile of Mortiño (*Vaccinium floribundum* Kunth). *J. Agric. Food Chem.* **2009**, *57*, 8274–8281. [CrossRef]
4. Garzón, G.A.; Narváez, C.E.; Riedl, K.M.; Schwartz, S.J. Chemical composition, anthocyanins, non-anthocyanin phenolics and antioxidant activity of wild bilberry (*Vaccinium meridionale* Swartz) from Colombia. *Food Chem.* **2010**, *122*, 980–986. [CrossRef]
5. Muñoz, V.; Caviedes, M. *Determinación de Métodos Para Producción de Mortiño (Vaccinium floribundum Kunth), Con Fines de Propagación y Producción Comercial*; Universidad San Francisco de Quito—USFQ: Quito, Ecuador, 2011.
6. Torres, M.; Trujillo, D.; Arahana, V. Cultivo In Vitro del Mortiño (*Vaccinium floribundum* Kunth). *ACI AV. Ciencias Ing.* **2010**, *2*, B9–B15. [CrossRef]
7. Ortiz, J.; Marín-Arroyo, M.-R.; Noriega-Domínguez, M.-J.; Navarro, M.; Arozarena, I. Color, Phenolics, and Antioxidant Activity of Blackberry (*Rubus Glaucus* Benth.), Blueberry (*Vaccinium floribundum* Kunth.), and Apple Wines from Ecuador. *J. Food Sci.* **2013**, *78*, C985–C993. [CrossRef]
8. Schreckinger, M.; Wang, J.; Yousef, G.; Lila, M.; Gonzalez, E. Antioxidant Capacity and in Vitro Inhibition of Adipogenesis and Inflammation by Phenolic Extracts of Vaccinium floribundum and Aristotelia Chilensis. *J. Agric. Food Chem.* **2010**, *58*, 8966–8976. [CrossRef]
9. Freire, A. *Botánica Sistemática Ecuatoriana*; Missouri Botanical Garden: St Louis, MO, USA, 2004; pp. 1–209.
10. Coba, P.; Coronel, D.; Verdugo, K.; Paredes, M.F.; Yugsi, E.; Huachi, L. Estudio etnobotánico del mortiño (*Vaccinium floribundum*) como alimento ancestral y potencial alimento funcional. *Granja Rev. Cienc.* **2012**, *16*, 5–13. [CrossRef]
11. Prior, R.L.; Cao, G.; Martin, A.; Sofic, E.; McEwen, J.; O'Brien, C.; Lischner, N.; Ehlenfeldt, M.K.; Kalt, W.; Krewer, G.; et al. Antioxidant capacity as influenced by total phenolic and anthocyanin content, maturity, and variety of Vaccinium species. *J. Agric. Food Chem.* **1998**, *46*, 2686–2693. [CrossRef]
12. Cho, M.J.; Howard, L.R.; Prior, R.L.; Clark, J.R. Flavonol glycosides and antioxidant capacity of various blackberry and blueberry genotypes determined by high-performance liquid chromato graphy/mass spectrometry. *J. Sci. Food Agric.* **2005**, *85*, 2149–2158. [CrossRef]
13. Gu, C.; Howell, K.; Dunshea, F.R.; Suleria, H.A.R. LC-ESI-QTOF/MS Characterisation of Phenolic Acids and Flavonoids in Polyphenol-Rich Fruits and Vegetables and Their Potential Antioxidant Activities. *Antioxidants* **2019**, *8*, 405. [CrossRef]
14. Acosta-Montoya, O.; Vaillant, F.; Cozzano, S.; Mertz, C.; Pérez, A.M.; Castro, M.V. Phenolic content and antioxidant capacity of tropical highland blackberry (*Rubus adenotrichus* Schltdl.) during three edible maturity stages. *Food Chem.* **2010**, *119*, 1497–1501. [CrossRef]
15. Chaovanalikit, A.; Wrolstad, R.E. Total anthocyanins and total phenolics of fresh and processed cherries and their antioxidant properties. *J. Food Sci.* **2004**, *69*, FCT67–FCT72. [CrossRef]
16. Baenas, N.; Ruales, J.; Moreno, D.A.; Barrio, D.A.; Stinco, C.M.; Martínez-Cifuentes, G.; Meléndez-Martínez, A.J.; García-Ruiz, A. Characterization of Andean Blueberry in bioactive compounds, evaluation of biological properties, and in vitro bioaccessibility. *Foods* **2020**, *9*, 1483. [CrossRef]
17. Guijarro-Fuertes, M.; Andrade-Cuvi, M.; Bravo-Vásquez, J.; Ramos-Guerrero, L.; Vernaza, M. Andean blueberry (Vaccinium floribundum) bread: Physicochemical properties and bioaccessibility of antioxidants. *Food Sci. Technol.* **2019**, *39*, 56–62. [CrossRef]

18. Vizuete, K.S.; Kumar, B.; Vaca, A.V.; Debut, A.; Cumbal, L. Mortiño (*Vaccinium floribundum* Kunth) Berry Assisted Green Synthesis and Photocatalytic Performance of Silver–Graphene Nanocomposite. *J. Photochem. Photobiol. A Chem.* **2016**, *329*, 273–279. [CrossRef]
19. Rojas-Ocampo, E.; Torrejón-Valqui, L.; Muñóz-Astecker, L.D.; Medina-Mendoza, M.; Mori-Mestanza, D.; Castro-Alayo, E.M. Antioxidant capacity, total phenolic content and phenolic compounds of pulp and bagasse of four Peruvian berries. *Heliyon* **2021**, *7*, e07787. [CrossRef] [PubMed]
20. Schreckinger, M.; Lila, M.; Yousef, G.; Gonzalez, E. Inhibition of -Glucosidase and -Amylase by *Vaccinium floribundum* and *Aristotelia chilensis* Proanthocyanidins. *ACS Symp. Ser.* **2012**, *1109*, 71–82. [CrossRef]
21. Alarcón-Barrera, K.S.; Armijos-Montesinos, D.S.; García-Tenesaca, M.; Iturralde, G.; Jaramilo-Vivanco, T.; Granda-Albuja, M.G.; Giampieri, F.; Alvarez-Suarez, J.M. Wild Andean Blackberry (*Rubus glaucus* Benth) and Andean Blueberry (*Vaccinium floribundum* Kunth) from the Highlands of Ecuador: Nutritional Composition and Protective Effect on Human Dermal Fibroblasts Against Cytotoxic Oxidative Damage. *J. Berry Res.* **2018**, *8*, 223–236. [CrossRef]
22. Caranqui-Aldaz, J.M.; Romero-Saltos, H.; Hernández, F.; Martínez, R. Reproductive phenology of *Vaccinium floribundum* Kunth (Ericaceae) and codification according to the BBCH scale based on evidence from the volcano Chimborazo paramo (Ecuador). *Sci. Hortic.* **2022**, *303*, 111207. [CrossRef]
23. Wojdyło, A.; Oszmianski, J.; Laskowski, P. The polyphenolic compounds and antioxidant activity of new and old apple varieties. *J. Agric. Food Chem.* **2008**, *56*, 520–6530. [CrossRef]
24. Singleton, V.L.; Orthofer, R.; Lamuela-Reventos, R.M. Analysis of total phenols and other oxidation substrates and antioxidants by means of Folin-Ciocalteu reagent. *Methods Enzymol.* **1999**, *299*, 152–178. [CrossRef]
25. Re, R.; Pellegrini, N.; Proteggente, A.; Pannala, A.; Yang, M.; Rice-Evans, C. Antioxidant activity applying an improved ABTS radical cation decolorization assay. *Free Radic. BiolMed.* **1999**, *26*, 1231–1237. [CrossRef]
26. Brand-Williams, W.; Cuvelier, M.E.; Berset, C. Use of a free radical method to evaluate antioxidant activity. *LWT Food Sci. Technol.* **1995**, *28*, 25–30. [CrossRef]
27. Hernández, F.; Noguera-Artiaga, L.; Burló, F.; Wojdyło, A.; Carbonell-Barrachina, Á.A.; Legua, P. Physico-chemical, nutritional, and volatile composition and sensory profile of Spanish jujube (*Ziziphus jujuba* Mill.) fruits. *J. Sci. Food Agric.* **2016**, *96*, 2682–2691. [CrossRef] [PubMed]
28. Hong, H.T.; Netzel, M.E.; O'Hare, T.J. Optimisation of extraction procedure and development of LC-DAD-MS methodology for anthocyanin analysis in anthocyanin-pigmented corn kernels. *Food Chem.* **2020**, *319*, 126515. [CrossRef] [PubMed]
29. Shrivastava, A.; Mishra, S.P.; Pradhan, S.; Choudhary, S.; Singla, S.; Zahra, K.; Aggarwal, L.M. An Assessment of Serum Oxidative Stress and Antioxidant Parameters in Patients Undergoing Treatment for Cervical Cancer. *Free Radic. Biol. Med.* **2021**, *167*, 29–35. [CrossRef] [PubMed]
30. Sellappan, S.; Akoh, C.C.; Krewer, G. Phenolic compounds and antioxidant capacity of Georgia-grown blueberries and blackberries. *J. Agric. Food Chem.* **2002**, *50*, 2432–2438. [CrossRef]
31. Esquivel-Alvarado, D.; Muñoz-Arrieta, R.; Alfaro-Viquez, E.; Madrigal-Carballo, S.; Krueger, C.; Reed, J.D. Composition of Anthocyanins and proanthocyanins in three tropical *Vaccinium* species from Costa Rica. *J. Agric. Food Chem.* **2020**, *68*, 2872–2879. [CrossRef] [PubMed]
32. Remorini, D.; Tavarini, S.; Degl, E.; Loreti, F.; Massai, R.; Guidi, L. Effect of root stock and harvesting time on the nutritional quality of peel and flesh of nectarine fruits. *Food Chem.* **2008**, *110*, 361–367. [CrossRef]
33. Nuncio-Jáuregui, N.; Munera-Picazo, S.; Calín-Sánchez, A.; Wojdyło, A.; Hernández, F.; Carbonell-Barrachina, A.A. Bioactive compound composition of pomegranate fruits removed during thinning. *J. Food Compos. Anal.* **2015**, *37*, 11–19. [CrossRef]
34. Llerena, W.; Samaniego, I.; Angós, I.; Brito, B.; Ortiz, B.; Carrillo, W. Biocompounds Content Prediction in Ecuadorian Fruits Using a Mathematical Model. *Foods* **2019**, *8*, 284. [CrossRef] [PubMed]
35. Wang, Y.; Yang, H.; Zhong, S.; Liu, X.; Li, T.; Zong, C. Variation in sugar and organic acid content of fruit harvested from different *Vaccinium uliginosum* populations in the Changbai Mountains of China. *J. Am. Soc. Hort. Sci.* **2019**, *144*, 420–428. [CrossRef]
36. Mikulic-Petkovsek, M.; Schmitzer, V.; Slatnar, A.; Stampar, F.; Veberic, R. Composition of sugars, organic acids, and total phenolics in 25 wild or cultivated berry species. *J. Food Sci.* **2012**, *77*, C1064–C1070. [CrossRef]
37. Mikulic-Petkovsek, M.; Schmitzer, V.; Slatnar, A.; Stampar, F.; Veberic, R. A comparison of fruit quality parameters of wild bilberry (*Vaccinium myrtillus* L.) growing at different locations. *J. Sci. Food. Agric.* **2015**, *95*, 776–785. [CrossRef]
38. Famiani, F.; Battistelli, A.; Moscatello, S.; Cruz-Castillo, J.G.; Walker, R.P. The organic acids that are accumulated in the flesh of fruits: Occurrence, metabolism and factors affecting their contents—A review. *Rev. Chapingo Ser. Hortic.* **2015**, *21*, 97–128. [CrossRef]
39. Etienne, A.; Génard, M.; Lobit, P.; Mbeguié-A-Mbéguié, D.; Bugaud, C. What controls fleshy fruit acidity? A review of malate and citrate accumulation in fruit cells. *J. Exp. Bot.* **2013**, *64*, 1451–1469. [CrossRef] [PubMed]
40. Ayaz, F.A.; Kadioglu, A.; Bertoft, E.; Acar, C.; Turna, I. Effect of fruit maturation on sugar and organic acid composition in two blueberries (*Vaccinium arctostaphylos* and *V. myrtillus*) native to Turkey. *N. Z. J. Crop Hortic. Sci.* **2001**, *29*, 137–141. [CrossRef]
41. Kafkas, E.; Kosar, M.; Türemis, N.; Baser, K.H.C. Analysis of sugars, organic acids and vitamin C contents of blackberry genotypes from Turkey. *Food Chem.* **2006**, *97*, 732–736. [CrossRef]

42. Correira, S.; Gonçalves, B.; Aires, A.; Silva, A.; Ferreira, L.; Carvalho, R.; Fernandes, H.; Freitas, C.; Carnide, V.; Silva, A. Effect of harvest year and altitude on nutritional and biometric characteristics of blueberry cultivars. *J. Chem.* **2016**, *2016*, 8648609. [CrossRef]
43. Kalt, W.; McDonald, J.E. Chemical Composition of Lowbush Blueberry Cultivars. *J. Am. Soc. Hort. Sci.* **1996**, *121*, 142–146. [CrossRef]
44. Milivojevic, J.; Maksimović, V.; Nikolić, M.; Bogdanović, J.; Maletić, R.; Milatović, D. Chemical and antioxidant properties of cultivated and wild Fragaria and Rubus berries. *J. Food Qual.* **2011**, *34*, 1–9. [CrossRef]
45. Cobo, M.; Gutiérrez, B.; Torres, A.; Torres, M. Preliminary Analysis of the Genetic Diversity and Population Structure of Mortiño (*Vaccinium floribundum* Kunth). *Biochem. Syst. Ecol.* **2016**, *64*, 14–21. [CrossRef]
46. Osborne, C.A.; Lulich, J.; Thumchai, R.P.; Ulrich, L.K.; Koehler, L.A.; Bird, K.A.; Bartges, J.W. Feline Urolithiasis: Etiology and Pathophysiology. *Vet. Clin. North Am. Small Anim.* **1996**, *26*, 217–232. [CrossRef]
47. Karlsons, A.; Osvalde, A.; Čekstere, G.; Pormale, J. Research on the mineral composition of cultivated and wild blueberries and cranberries. *Agron. Res.* **2018**, *16*, 454–463. [CrossRef]
48. Miljković, V.M.; Nikolić, G.S.; Zvezdanović, J.; Mihajlov-Krstev, T.; Arsić, B.B.; Miljković, M.N. Phenolic profile, mineral content and antibacterial activity of the methanol extract of *Vaccinium myrtillus* L. *Not. Bot. Horti Agrobot.* **2018**, *46*, 122–127. [CrossRef]
49. Prencipe, F.P.; Bruni, R.; Guerrini, A.; Rossi, D.; Benvenuti, S.; Pellati, F. Metabolite Profiling of Polyphenols in Vaccinium Berries and Determination of Their Chemopreventive Properties. *J. Pharm. Biomed. Anal.* **2014**, *89*, 257–267. [CrossRef]
50. Wojdyło, A.; Nowicka, P.; Grimalt, M.; Legua, P.; Almansa, M.S.; Amorós, A.; Carbonell-Barrachina, A.A.; Hernández, F. Polyphenol Compounds and Biological Activity of Caper (*Capparis spinosa* L.) Flowers Buds. *Plants* **2019**, *8*, 539. [CrossRef]
51. Wang, S.Y.; Zheng, W. Effect of plant growth temperature on antioxidant capacity in strawberry. *J. Agric. Food Chem.* **2001**, *49*, 4977–4982. [CrossRef]
52. Jaakola, L.; Hohtola, A. Effect of latitude on flavonoid biosynthesis in plants. *Plant Cell Environ.* **2010**, *33*, 1239–1247. [CrossRef]
53. Bueno, J.M.; Sáez-Plaza, P.; Ramos-Escudero, F.; Jiménez, A.M.; Fett, R.; Asuero, A.G. Analysis and antioxidant capacity of anthocyanin pigments. Part II: Chemical structure, color, and intake of anthocyanins. *Crit. Rev. Anal. Chem.* **2012**, *42*, 126–151. [CrossRef]
54. Sunil, L.; Shetty, N.P. Biosynthesis and regulation of anthocyanin pathway genes. *Appl. Microbiol. Biotechnol.* **2022**, *106*, 1783–1798. [CrossRef] [PubMed]
55. Chalker-Scott, L. Environmental Significance of Anthocyanins in Plant Stress Responses. *Photochem. Photobiol.* **1999**, *70*, 1–9. [CrossRef]
56. Li, X.-J.; Hou, J.-H.; Zhang, G.-L.; Liu, R.-S.; Yang, Y.-G.; Hu, Y.-X.; Lin, J.-X. Comparison of anthocyanin accumulation and morphoanatomical features in apple skin during color formation at two habitats. *Sci. Hortic.* **2004**, *99*, 41–53. [CrossRef]
57. Maier, A.; Hoecker, U. COP1/SPA ubiquitin ligase complexes repress anthocyanin accumulation under low light and high light conditions. *Plant Siganl. Behav.* **2015**, *10*, e970440. [CrossRef] [PubMed]

horticulturae

MDPI

Article

Physiological and Metabolic Traits Linked to Kiwifruit Quality

Vaia Styliani Titeli [1], Michail Michailidis [1,*], Georgia Tanou [2,3] and Athanassios Molassiotis [1]

[1] Laboratory of Pomology, Department of Horticulture, Aristotle University of Thessaloniki, 57001 Thermi, Greece; titelivg@agro.auth.gr (V.S.T.); amolasio@agro.auth.gr (A.M.)
[2] Institute of Soil and Water Resources, Hellenic Agricultural Organisation-DIMITRA (ELGO-DIMITRA), 57001 Thermi, Greece; gtanou@swri.gr
[3] Joint Laboratory of Horticulture, Hellenic Agricultural Organisation-DIMITRA (ELGO-DIMITRA), 57001 Thermi, Greece
* Correspondence: msmichai@agro.auth.gr

Abstract: The assessment of fruit quality traits is a key factor in increasing consumer acceptance of kiwifruit. Here, an experiment was performed to evaluate the relationship between dry matter (DM) and soluble solids concentration (SSC), evaluated by both destructive (D) and non-destructive (ND) approaches, with acidity content and sensory evaluation, particularly taste, in fully ripened 'Hayward' kiwifruits from 20 orchards. Nutrient content and metabolomic analysis were also performed in ripened kiwifruit tissues (pericarp, placenta, and seeds) from four selected orchards of kiwifruits of high taste scores (HTS) and four orchards of low taste scores (LTS). The results suggest that ND-DM measurement positively correlated with D-DM and may serve as an indicator of kiwifruit taste. Moreover, the taste of kiwifruit was affected by both SSC and acidity. Based on the nutrient content of the pericarp and the primary metabolites of the pericarp and placenta, a clear separation was observed between kiwifruits with HTS and those with LTS, while no differences were found in seed samples. Metabolites such as fructose, maltose, mannobiose, tagatose, and citrate were accumulated in kiwifruits with a strong taste in the pericarp, whereas others, such as serine in the pericarp and placenta, have a negative impact on taste. The current study contributes to a greater understanding of the influence of dry matter, ripening characteristics, primary metabolites, and nutrient content on the taste of kiwifruits.

Keywords: dry matter; metabolomics; sensory evaluation; taste; elements

Citation: Titeli, V.S.; Michailidis, M.; Tanou, G.; Molassiotis, A. Physiological and Metabolic Traits Linked to Kiwifruit Quality. *Horticulturae* **2023**, *9*, 915. https://doi.org/10.3390/horticulturae9080915

Academic Editors: Paola Sánchez-Bravo and Luis Noguera-Artiaga

Received: 10 July 2023
Revised: 8 August 2023
Accepted: 9 August 2023
Published: 10 August 2023

1. Introduction

The dietary and nutritional value of kiwifruit is highly appreciated by consumers [1,2]. Fruit ripening, as a process, is signaled by changes in the sensory components, including taste (e.g., increased sweetness), visual appearance (internal color change), texture (softening of the flesh), and aroma (formation of volatile organic compounds, VOCs) [3]. Aroma and taste primarily develop during the final stages of ripening in climacteric fruit such as kiwifruit, where they are associated with autocatalytic ethylene production [4]. Consequently, flavor, texture, aroma, nutritional elements, and expiration life are the primary characteristics that influence consumers' preferences and determine a product's quality. Kiwifruit flavor is the most important quality indicator and characterized by balance of sweetness and acidity with a fine blend of volatile aromatic compounds [5], illustrating how these specific characteristics are affected by the ratio between sugars and acids [6,7].

The soluble solid concentration (SSC) is recognized as the primary ripening criterion for kiwifruits, which are harvested when the SSC content exceeds 6.25% Brix. SSC content depends on the accumulation of sugars in the form of starch/dry matter during fruit ripening [7], and there is a direct correlation between SSC content and sweetness in flavor [8]. Dry matter is also another crucial quality indicator for kiwifruit; it refers to the non-water components of the fruit and, whether there are photosynthesizing leaves, it increases as the fruit ripens [9]. According to Feng [9], the anticipated concentration of SSC in ripe fruit corresponds to the percentage of dried weight at harvest, reduced by three percent. Dry weight estimation is an accepted indicator of fruit ripening and a predictor of SSC in New Zealand [10], allowing producers to obtain higher prices [11]. Crisosto [8] noted the reliability of dry matter as an indicator of quality, but not ripening, of kiwifruit, which confirms the increased consumer acceptability of high dry weight kiwifruits [5]. Liao [12] found that cultivation practices such as kiwifruit canopy formation and summer pruning influence the fruit's dry matter content. In addition, the final estimation of all these fruit quality traits that may affect flavor is being studied through sensory evaluation. Sensory analysis is a technique that provides valuable information that food processing industries use to evaluate the quality of their products [13]. To be more representative, comparisons and correlations between outcomes of sensory analysis must be quantified and rated [14]. Consequently, the correlation between the sensory evaluation's results and objective measurements provides the highest degree of reliability when calculating fruit quality index [15].

The purpose of this study is to (i) establish a relationship between destructive and non-destructive methods for evaluating the quality of fully ripened kiwifruits, and (ii) reveal possible kiwifruit quality indicators, utilizing sensory evaluation and particularly kiwifruits' taste from different orchards and from separated fruit tissues (pericarp, placenta, and seeds) and their association with nutrient content and metabolites abundance.

2. Materials and Methods

2.1. Experiment and Fruit Sampling

Kiwifruits (*Actinidia chinensis* var. *deliciosa* (A. Chev.) A. Chev. 'Hayward') were harvested from 20 productive (all above 5 years old) commercial orchards (codes 1–20) that are located in the Nestos area (East Macedonia, North Greece; Figure S1). All orchards were subjected to standard cultural practices for sustainable kiwifruit production. During the commercial harvest period, 200 fruits per orchard, with an average soluble solid concentration of SSC = 6.25% Brix and fruit weight of approximately 100 g, were collected. Fruit were transferred to the laboratory of Pomology at Aristotle University and, following postharvest storage at 0 °C (RH = 95%) for three weeks, were then maintained at 20 °C (RH = 70%) until ripened (pericarp firmness lower than 10 Newtons). In the fully ripe stage of kiwifruits, both non-destructive and destructive approaches were used to determine dry matter and SSC; additionally, acidity as well as sensory evaluation were performed. In parallel, the kiwifruits were divided into three distinct tissues: pericarp, placenta, and seeds, followed by sampling with liquid nitrogen and storage at −80 °C for further analysis.

2.2. Fruit Quality Traits

2.2.1. Non-Destructive Estimation of the Dry Matter and Soluble Solid Concentration

The non-destructive determination of kiwifruit dry matter (%) and SSC (%) were carried out in 30 kiwifruits per orchard at the fully ripe stage using an F-750 device (NIR spectroscopy-based technology, Produce Quality Meter, Felix Instruments, Camas, WA, USA) with the software of F-751 Kiwi Quality Meter. Initially, the F-750 device was calibrated with 20 fruits at 20 °C (RH = 70%). The fruit was placed in the special holder and the reading was taken according to the manufacturer's instructions. Thereafter, non-destructive determination of dry matter (%) and SSC (%) were conducted in the kiwifruits from 20 orchards.

2.2.2. Destructive Determination of Dry Matter, Soluble Solid Concentration, and Acidity

Dry matter content was determined in a transverse slice (1 cm) from the equatorial region of each fruit, when dried in an oven (65 °C, 48 h), and expressed as a percentage (%) of dry weight against fresh slice weight. Soluble solid concentration (SSC, %) and titratable acidity (TA) were determined in juice from three biological replicates of 10 kiwifruits per orchard. SSC measurement was performed using a digital refractometer (Atago PR-101,

Atago Co., Ltd., Tokyo, Japan) while TA was performed by potentiometric titration with 0.01 NaOH up to pH 8.2 and was expressed in citric acid (%), as previously described [16].

2.3. Sensory Evaluation Analysis

Panelists evaluated the ripe kiwifruits of cultivar 'Hayward' under normal light, temperature, and relative humidity conditions, as previously described [17]. The sensory descriptors that were determined based on intensity scale were appearance (1 = abnormal fruit with defects and 9 = oval shape without any defect), internal color (pericarp green intensity; 1 = white; R, G, B = 255, 255, 255; 5 = light green; R, G, B = 0, 180, 0; 9 = dark green; R, G, B = 0, 120, 0), texture/firmness (1 (4 N) to 9 (10 N)) from low to high firmness based on chewing of pericarp and placenta) [18], aftertaste intensity (1 = absent and 9 = very strong), taste intensity (1 = without taste; taste of mineral water and 9 = high; very tasteful), as previously described with some modifications [19]. Each sample was labeled with a random four-digit code, containing five whole kiwifruits of each orchard, and then transverse slices (1 cm) of each fruit were presented on white polystyrene plates balanced in a randomized order across the nine judges/panelists. A training session was organized with the purpose of acquainting the panelists with the sensory evaluation approach, attribute definitions, and reference standards. Additional information about the tested parameters is provided in Table S1.

2.4. Mineral Element Analysis

Determination of nutrients (K, P, Ca, Mg, Na, Zn, Fe, Mn, Cu) was performed in the pericarp of ripe kiwifruits (dry samples) by the inductively coupled plasma optical emission spectrometry (ICP–OES) system (Perkin Elmer Optima 2100DV, PerkinElmer Inc., Waltham, MA, USA). A batch of five kiwifruits in three biological replicates per orchard was used, and extraction was carried out on ash dissolution in 6 N HCl, after sample incineration at 550 °C for 6 h. The Kjeldahl method was used to perform the analysis of nitrogen (N) with a Vapodest 50 s system (Gerhardt, Königswinter, Germany) [20].

2.5. Primary Polar Metabolites Analysis

Primary polar metabolites extraction of ripe kiwifruits in each tissue (pericarp, placenta, seeds) and derivatization processes were employed, as previously described [16], with slight modifications. Frozen grinding tissue (0.5 g) from eight selected orchards (four orchards that received a high score based on taste and four orchards that received a low score based on taste) were used. Samples were extracted with 1.4 mL of methanol (100%, pre-cooled at −20 °C), and 0.1 mL adonitol (1 mg mL^{-1}) was added as an internal quantitative standard, and then incubated for 10 min (70 °C). In the supernatant, 0.75 mL of chloroform (100%, pre-cooled at −20 °C) and 1.5 mL dH_2O (100%, pre-cooled at 4 °C) was added and then centrifugated (2200× *g*, 4 °C, 10 min). An aliquot of 0.15 mL of the supernatant was transferred into a vial glass and placed to dry in a desiccator under vacuum. The residues were redissolved in 0.04 mL methoxyamine hydrochloride (20 mg mL^{-1}) and then in 0.07 mL N–methyl–N–(trimethylsilyl) trifluoroacetamide reagent (MSTFA) for 120 min and 30 min at 37 °C, respectively. The GC–MS analysis was carried out with a Perkin Elmer Clarus™ SQ 8 (Waltham, MA, USA), as previously described in detail [16]. Compounds were determined using standards or NIST11 database or GOLM metabolome database [21]. The metabolites were expressed as the relative abundance of adonitol and are provided in Table S2.

2.6. Statistical Processing and Analysis

The statistical analysis of all quality traits, sensory attributes, and mineral elements was conducted using SPSS (SPSS v27.0., Chicago, IL, USA) by one-way ANOVA or by multivariate analysis of variance (MANOVA) [22]. Metabolomic analysis was performed by analysis of variance (ANOVA) between high- and low-taste-score kiwifruits. Mean values were compared based on the least significant difference (LSD) or Student's *t*-test; $p \leq 0.05$.

Pearson correlation analysis was conducted using SPSS whereas principal components analysis (PCA) and hierarchical clustering along with a heatmap were employed using ClustVis software 2.0 [23].

3. Results

3.1. Non-Destructive Application to Define Fruit Quality

To examine the internal kiwifruit quality at the fully ripe stage from the 20 orchards, both non-destructive and destructive approaches were used to determine dry matter (DM) and soluble solid concentration (SSC), as well as the acidity content and the sensory attributes (Figure 1).

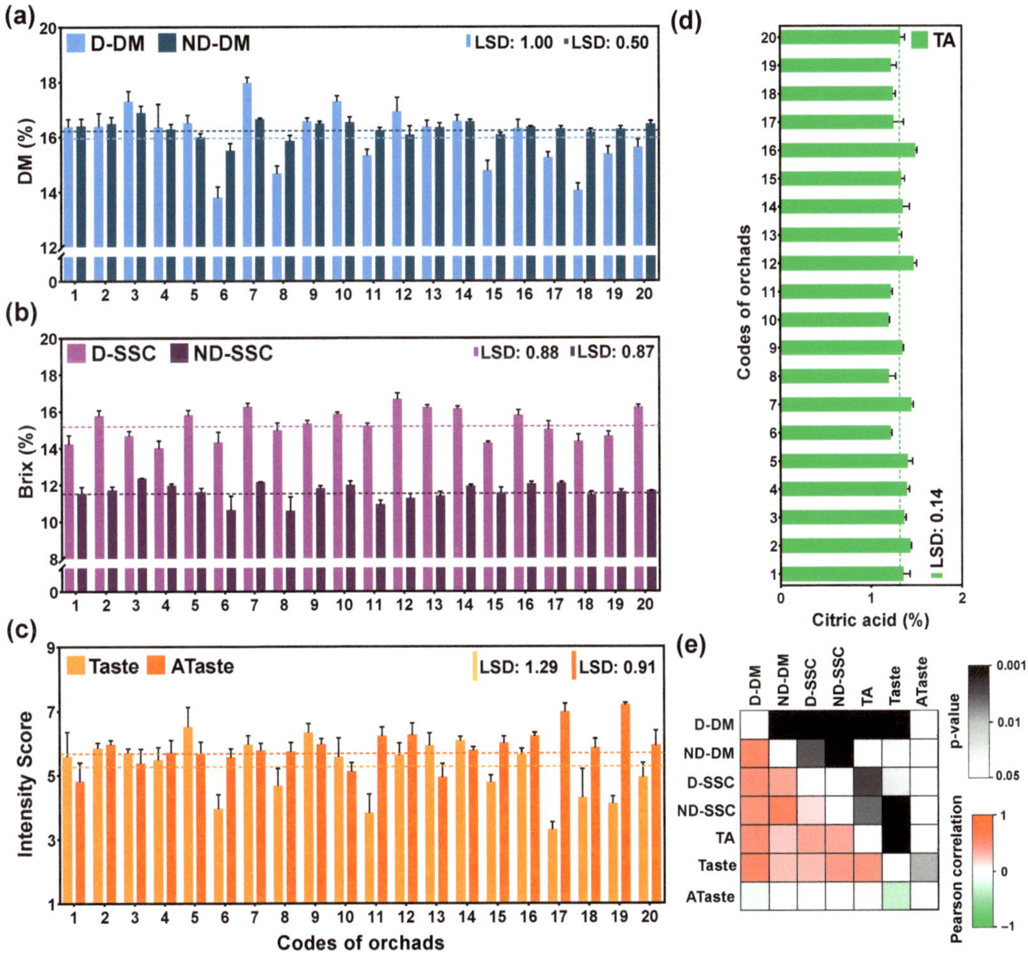

Figure 1. Quality traits of fully ripe fruits from 20 kiwifruit orchards: (**a**) dry matter (%) of fruit using destructive (D-DM) and non-destructive (ND-DM) approach, (**b**) soluble solids concentration (% Brix) of fruit using destructive (D-SSC) and non-destructive (ND-SSC) methods, (**c**) sensory evaluation of kiwifruit base on intensity score ranged from 1 (without taste) to 9 (very strong taste/aftertaste) of taste and aftertaste (ATaste), (**d**) titratable acidity (TA), and (**e**) Pearson correlation of seven variables depicted with a heatmap (red indicates positive whereas green indicates negative correlation). The dotted line indicates the grand mean of its variable. The means of its quality and sensory trait were compared based on least significant difference (LSD) $p \leq 0.05$.

The average dry matter (DM) content at the 20 orchards was 16% by the destructive (D) and 16.3% by the non-destructive (ND) approach. The higher DM was recorded at orchard codes 3 and 7 and the lower DM was recorded at 6 and 8 orchard codes both in D and ND approaches (Figure 1a). The average soluble solid concentration (SSC) at the 20 orchards was 15.3% and 11.6% by D and ND approaches, respectively. It is worth noting that ND-SSC was significantly underestimated with respect to D-SSC. Nevertheless, a higher SSC was recorded in both D and ND approaches at orchard code 7. Additionally, high SSC was observed at orchard codes 12 for D and 3 for ND, whereas the lower SSC was at orchard codes 4 and 1 for D and orchard codes 6 and 8 for ND (Figure 1b). We should also highlight that orchard codes 3 and 7 received high values of DM and SSC, while orchards 6 and 8 received low values of DM and SSC.

Sensory evaluation revealed an acceptable intensity score (≥5) for most of the orchards analyzed for descriptors such as appearance, internal color, and texture/firmness intensity in fully ripe kiwifruits (Table S1). We assumed that taste and aftertaste are two crucial descriptors in kiwifruit sensory evaluation. Thus, we were focused on these two descriptors; the average taste score was 5.2 and the average aftertaste score was 5.9 among the 20 orchards. The four higher scores in taste were observed at 7, 14, 9, and 5 codes of orchards whereas the four lower scores were observed at 17, 11, 6, and 19 codes of orchards (Figure 1c). In contrast, aftertaste (Ataste) results were recorded at a higher intensity in low-taste-score orchards, such as 17 and 19, indicating a negative relationship between taste and aftertaste in kiwifruits. Finally, the average value of acidity (TA) content at the ripe kiwifruits of 20 orchards was 1.3%, with the higher TA being recorded at orchard codes 12 and 16 and the lower TA at orchard codes 10 and 8 (Figure 1d).

To uncover possible associations among the seven variables tested, a Pearson correlation analysis was performed. This examination revealed positive correlations between destructive and non-destructive approaches among DM, SSC, TA, and taste, whereas a negative correlation was observed between taste and Ataste intensity (Figure 1e). Hence, considering the above-mentioned correlations and to evaluate the factors affecting the taste in ripe kiwifruit, we chose the four orchards with the highest scores (7, 14, 9, and 5 codes of orchards) and four orchards with the lowest scores (17, 11, 6, and 19 codes of orchards) according to their taste assessment for further analysis.

3.2. *Taste-Dependent Quality Traits and Mineral Content*

To estimate the impact of taste in ripe kiwifruit, the four orchards that received a high taste score (HTS) were compared with the four orchards that received a low taste score (LTS). Physiological analysis revealed that the HTS orchards exhibited higher values of dry matter, SSC, and acidity than the LTS orchards, whereas LTS orchards had higher Ataste intensity than HTS (Figure 2a). In contrast to the physiological data, the mineral content in the pericarp of ripe kiwifruit displayed a decrease in HTS compared with LTS orchards, concerning the content of nitrogen, phosphorus, potassium, iron, and zinc (Figure 2b). However, the content of calcium, magnesium, sodium, copper, and manganese was similar between the ripe kiwifruits that received high and low taste scores (Figure 2b).

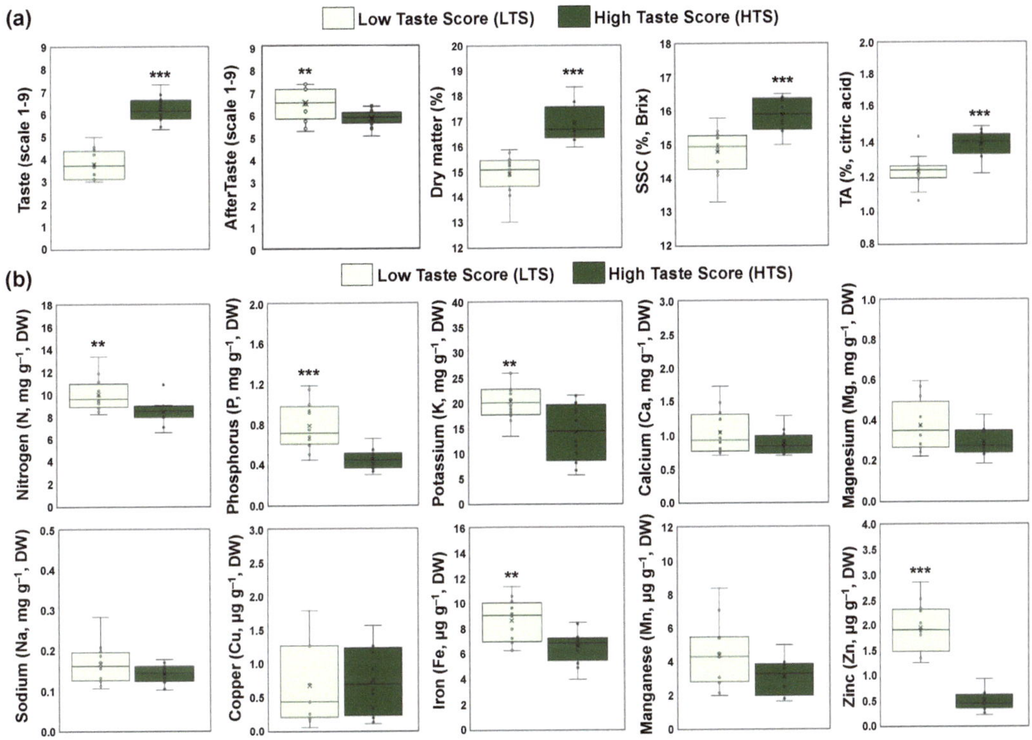

Figure 2. (**a**) Sensory evaluation (taste and aftertaste intensity) and physiological traits (dry matter, SSC, and TA) of fully ripe kiwifruits with high taste scores (HTS, dark green) and low taste scores (LTS, light green), and (**b**) mineral content in the pericarp of ripe kiwifruits with high taste scores (HTS, dark green) and low taste scores (LTS, light green). Each box plot was constructed by 12 replicates (four codes of orchards × three biological replicates). Means were compared based on Student's *t*-test; * $p \leq 0.05$, ** $p \leq 0.01$, *** $p \leq 0.001$.

3.3. Taste-Associated Primary Metabolites in the Various Kiwifruit Tissues

To identify the contribution of primary metabolites in the taste of kiwifruit at the fully ripe stage, a primary polar metabolomic analysis in three kiwifruit tissues (pericarp, placenta, and seeds) was performed. The orchards that received a high taste score (HTS) were compared with the orchards of LTS in each tissue separately. In pericarp tissue, 42 metabolites were identified belonging to five classes, namely sugars (16), acids (11), alcohols (6), amino acids (7), and other compounds (3). A clear separation was observed between HTS and LTS in the pericarp of ripe kiwifruits based on hierarchical cluster analysis, as depicted in Figure 3a. Pericarp metabolomic analysis exhibited a decrease of two amino acids (valine and serine), one alcohol (inositol), and one other compound (phosphoric acid) in the HTS kiwifruits (Figure 3a, Table S2). On the contrary, an increase in nine sugars (tagatose, fructose, glucose, talose, sucrose, turanose, lactulose, mannobiose, and maltose), three acids (oxalic acid, citric acid, and quininic acid), one alcohol (galactinol), and one other compound (aucubin) was detected in the HTS kiwifruits (Figure 3a, Table S2). Furthermore, we found an increase in sugars and acids as well as a reduction in amino acids and other compounds in the HTS kiwifruits (Figure 3b).

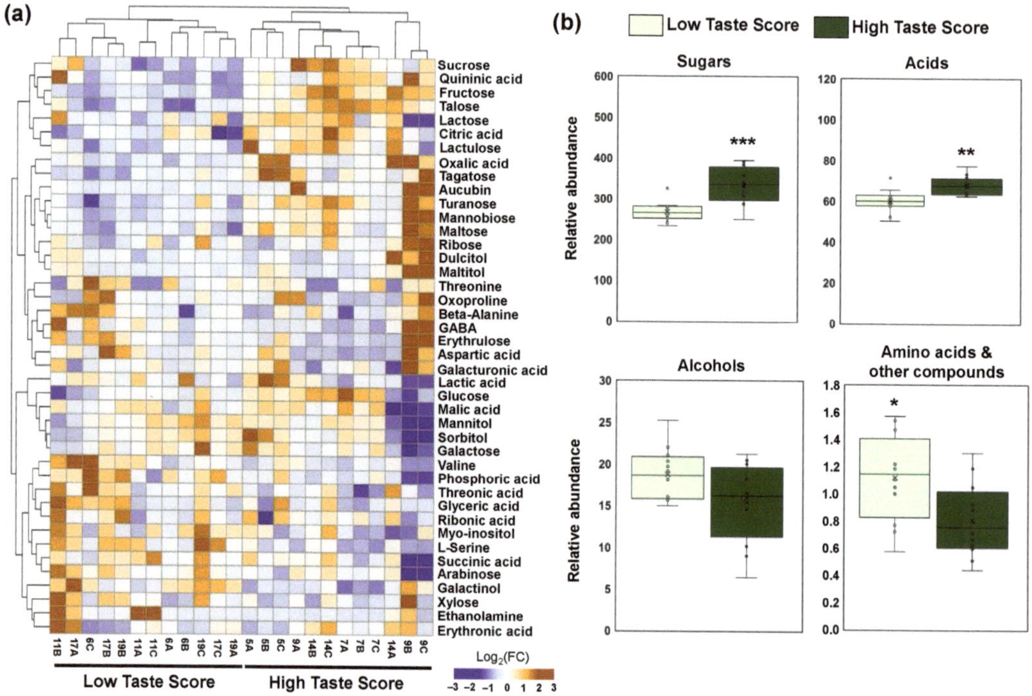

Figure 3. (**a**) Heatmap and hierarchical cluster analysis of the primary metabolites in the pericarp of the fully ripe kiwifruits with high taste scores (HTS; codes of orchards 5, 7, 9, 14) and low taste scores (LTS; codes of orchards 6, 11, 17, 19) using ClustVis software. The orange color indicates an increase in abundance; the purple color indicates a decrease in the abundance of each metabolite. (**b**) Metabolite classes in the pericarp of fully ripe kiwifruits with high taste scores (HTS, dark green) and low taste scores (LTS, light green). Each box plot was constructed by 12 replicates (four codes of orchards × three biological replicates). Means were compared based on Student's *t*-test; * $p \leq 0.05$, ** $p \leq 0.01$, *** $p \leq 0.001$. Metabolite data are provided in Table S2.

The current metabolomic analysis also identified 42 metabolites in the placenta tissue. In addition to this, 32 metabolites were detected in seeds tissue that correspond to five classes, namely sugars (16), acids (6), alcohols (6), amino acids (3), and other compounds (1). Based on hierarchical cluster analysis, a distinct separation was also noticed between HTS and LTS in the placenta of ripe kiwifruits but not in seeds, as illustrated in Figure 4. Metabolomic analysis in the placenta tissue disclosed a decrease in HTS of two sugars (xylose and galactose), two alcohols (sorbitol and galactinol), one acid (lactic acid), and three amino acids (serine, GABA, and oxoproline) (Figure 4a, Table S2).

An increase in metabolic abundance in the placenta of HTS kiwifruits was detected regarding six sugars (tagatose, turanose, lactulose, mannobiose, maltose, and lactose), one acid (erythronic acid), one alcohol (maltitol), and one other compound (aucubin) (Figure 4a, Table S2). In addition to this, seed metabolomic analysis showed a decrease in oxalic acid (acid) in HTS kiwifruits (Figure 4b, Table S2). In the seed of HTS kiwifruits, we also noticed an accumulation of three sugars (glucose, trehalose, and turanose), citric acid (acid), and aucubin (other compounds) (Figure 4b, Table S2).

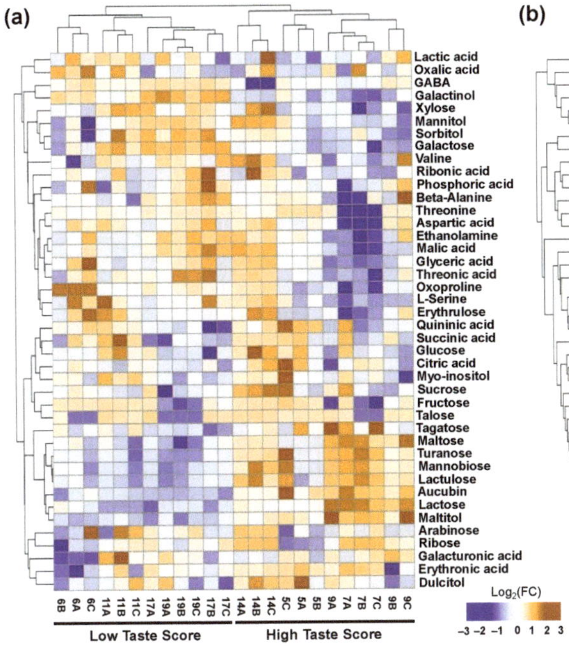

Figure 4. Heatmap and hierarchical cluster analysis of primary metabolites in the (**a**) placenta and (**b**) seeds of fully ripe kiwifruits with high taste scores (HTS; codes of orchards 5, 7, 9, 14) and low taste scores (LTS; codes of orchards 6, 11, 17, 19) using ClustVis software 2.0. The orange color indicates an increase in abundance; the purple color indicates a decrease in abundance. Metabolite data are provided in Table S2.

3.4. Taste-Based Discrimination of Fully Ripe Kiwifruits

To understand the effect of nutrients and metabolites on taste, a principal component analysis (PCA) was performed (Figure 5). Based on PCA of nutrients in the pericarp, we detected a clear separation between kiwifruits with high taste scores (HTS) and low taste scores (LTS) (Figure 5a). Similarly, a distinct separation was observed when PCA was performed in the metabolites of the pericarp and placenta but not in seeds between HTS kiwifruits and LTS ones (Figure 5b). It was also obvious that the discrimination of kiwifruits based on taste in the pericarp and placenta was mainly due to PC1, which explained the 20.7% variability of metabolites (Figure 5b). Focusing on metabolites that mainly participated in the construction of PC1, we noticed that six sugars (fructose, turanose, mannobiose, maltose, talose, and tagatose) and citric acid (the main acid in kiwifruits) received a positive score, above 0.6, indicating an increase in their abundance in the pericarp of HTS (Figure 5c). On the contrary, two amino acids (serine and valine) and phosphoric acid scored below −0.6, indicating an increase in their abundance in the pericarp of LTS (Figure 5c). In parallel, six sugars (lactose, turanose, maltose, lactulose, tagatose, and mannobiose), maltitol, and aucubin received scores above 0.6, implying an increase in their abundance in the placenta of HTS, whereas three amino acids (aspartic acid, GABA, and serine), two sugars (xylose, galactose), three acids (glyceric acid, threonic acid, and malic acid), galactinol, and ethanolamine scored below −0.6, indicating an increase in their abundance in the placenta of LTS kiwifruit (Figure 5c).

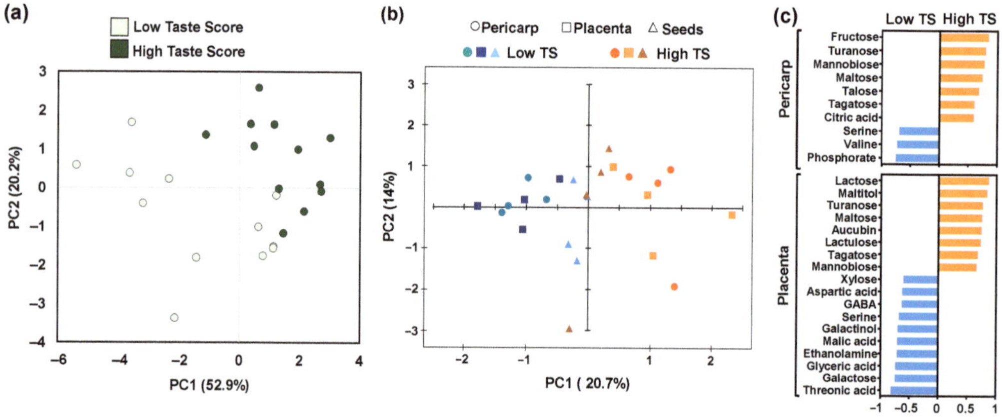

Figure 5. Principal component analysis (PCA) of the (**a**) minerals in fully ripe kiwifruits with high taste scores (HTS) and low taste scores (LTS), and (**b**) primary metabolites in the pericarp (cycle), placenta (square), and seeds (triangle) of fruit with high taste scores (HTS) and low taste scores (LTS). (**c**) Discrimination of LTS and HTS in the pericarp and placenta samples based on the metabolites using PCA models.

4. Discussion

It is well known that, during kiwifruit ripening, an increase in levels of SSC, ethylene, and volatiles, as well as a decrease in acids, starch content, and firmness, occur [5]. Flavor generation compounds are linked to both aroma (volatile compounds) and taste (sweet and acidic) in kiwifruit undergoing ripening. It is also well documented that soft kiwifruits, due to the activation of cell-wall-related enzymes such as polygalacturonase [24–26], are sweeter, whereas firmer kiwifruits are more acidic. Generally, values of firmness of approximately 8–10 Newtons make the kiwifruit ready to eat [27], while fruits with a slightly softer texture are more acceptable [18]. Therefore, our study focused on ripe kiwifruits, below 8 Newtons in the pericarp firmness, to determine quality traits such as SSC, DM, acidity, sensory attributes such as taste, etc. (Figure 1).

In the present study, considerable variation in fruit DM and SSC were evident among orchard samples of kiwifruit (Figure 1). Such variation in kiwifruit DM and SSC observed in this study was similar to that reported elsewhere [6]. For kiwifruit, the application of NIR-spectroscopy has been proposed as a tool for a quality metric and a harvest maturity metric [28]. It was also interesting that the non-destructive determination of DM and SSC using NIR-spectroscopy could develop strategies for appropriate postharvest handling of kiwifruits [2]. Especially, the SSC in kiwifruits should be higher than 6.2% Brix at harvest [29], while high DM (above 16%) was proposed as a quality indicator in kiwifruits [30]. It has been suggested that SSC should range between 10–14 °Brix in ready-to-eat kiwifruits [6]. In the current study, SSC ranged above these values from 14 to 16.7% Brix (Figure 1).

Since the DM of kiwifruit shows only minor change over the postharvest period [30], the measure of fruit DM at harvest may be used to predict the kiwifruit sensory potential following cold storage. In addition, the high level of dry matter in kiwifruits is linked to the acceptance of consumers with respect to the taste, since their sweetness increased [6,10,29,31–33]. In the present study, we also noticed that quality characteristics of kiwifruits (DM, SSC, and acidity) were positively associated with their taste scores (Figures 1 and 2a). As a result, DM content, SSC, and acidity can be considered as indicators related to other intrinsic quality attributes to improve the precision of the sorting process.

The nutrient content of the fruit may be contributed to the proportion of dry matter [5]. Nevertheless, there is still an extensive gap in our understanding concerning the connection between nutrient homeostasis and dry matter of kiwifruit. In this regard, an interesting finding that emerged from this work is the fact that nutrient elements tend to elevate their content in the pericarp tissue of LTS kiwifruits (Figure 2b), suggesting that it can be possibly used for discrimination between HTS and LTS kiwifruits (Figure 5a). This connection between fruit nutrient concentration and DM needs further attention, particularly considering the fact that mineral elements have been previously analyzed to achieve fruit-part discrimination [34].

Central carbon metabolism, which in fruits involves the pathways of sucrose, starch, major organic acids, and respiration, provides energy and biosynthetic precursors to support fruit growth and, eventually, ripening [35]. It is also essential for fruit quality, as sweetness and sourness are conditioned by sugars and organic acids, respectively, which are major components in most fruits [35]. Focusing on taste via discrimination in high- and low-taste-score kiwifruits, we examined the effect of the primary metabolites in the different parts of kiwifruits, including pericarp, placenta, and seeds (Figures 3 and 4). The metabolic profile in kiwifruit pericarp displayed an enhancement of sugars and acids in kiwifruit that received a high taste score (Figure 3b). The interplay between SSC and acidity has been previously pointed out in kiwifruits for achieving high rates of consumer acceptance [5,6]. Moreover, no strong changes in the primary metabolites of seed samples were observed between the HTS and LTS groups of kiwifruit (Figure 4b). Recently, it was proposed that seeds' presence (depending on the pollen donor) in kiwifruits did not affect taste balance [36], reinforcing our observation.

Based on the high taste score (HTS), a clear discrimination in primary metabolites in both pericarp and placenta tissues was recorded (Figures 3–5), probably due to hydrolysis of starch to glucose during the postharvest period resulting in sugar accumulation in HTS kiwifruit [37]. It was documented that metabolites along with sensory properties influence taste in a variety of foods [38,39]. The discrimination between HTS and LTS pericarp samples may be related to the levels of nine metabolites (Figure 5c), among them, citric acid which is the main acid in cv. 'Hayward' [40,41] and fructose which is one of the main sugars in kiwifruits [16,42]. Hence, the elevated abundance of fructose and citric acid in HTS could be associated with the taste as well as with the observed increase in acids and sugars in HTS kiwifruits (Figure 3). It is also worth noting that other sugars such as turanose, maltose, tagatose, and mannobiose (these sugars in kiwifruits were found in low abundance [16,42]) were also responsible for the distinct patterns between HTS and LTS in the pericarp and placenta tissues (Figure 5c). Moreover, an amino acid, serine, increased in the placenta and pericarp of LTS fruit (Figure 5c), indicating that this metabolite may be associated with the low-taste feature in kiwifruit. Recently, serine has been positively related to quality features in guava fruits [43].

5. Conclusions

This study indicates that non-destructive methods for determining dried matter can be used as an internal indicator of kiwifruit taste. In addition, the taste of kiwifruit was influenced by both SSC and acidity and it was the most important sensory characteristic in terms of overall acceptability. There was also evidence of discrimination between kiwifruits with high and low taste scores based on the nutrient content and primary metabolites of the pericarp and placenta. Several metabolites (fructose, maltose, mannobiose, tagatose, and citrate) were elevated in kiwifruits with a strong taste, whereas others, such as serine, may have a negative impact on taste. This study revealed the significance of certain minerals and metabolites in taste discrimination and set a basis for future research on the quality of kiwifruit.

Supplementary Materials: The following are available online at https://www.mdpi.com/article/10.3390/horticulturae9080915/s1, Figure S1. Sampling area and location of kiwifruit orchards; Table S1. Sensory analysis of ripe kiwifruits; Table S2. Quantitative results of metabolites in ripe kiwifruit tissues.

Author Contributions: Conceptualization, M.M. and A.M.; data curation, M.M., V.S.T., G.T. and A.M.; formal analysis, M.M. and V.S.T.; investigation, V.S.T.; methodology, V.S.T.; project administration, A.M.; software, M.M.; supervision, A.M.; validation, M.M., G.T. and A.M.; visualization, M.M. and V.S.T.; writing—original draft, M.M. and V.S.T.; writing—review and editing, G.T. and A.M. All authors have read and agreed to the published version of the manuscript.

Funding: This work was funded by the Action "Investment Plans of Innovation" of the Operational Program "Eastern Macedonia—Thrace 2014-2020", which is co-funded by the European Regional Development Fund and Greece (Project code: AMTHR4-0040267).

Data Availability Statement: The data presented in this study are available in Supplementary Material.

Acknowledgments: We would like to thank Chrysanthi Polychroniadou and candidate/students Elpida Nasiopoulou, Christina Skodra, and Iasonas Zacharis for helping us during the sampling process and determination of kiwifruits' physiological quality traits.

Conflicts of Interest: The authors declare no conflict of interest.

References

1. Nieuwenhuizen, N.J.; Maddumage, R.; Tsang, G.K.; Fraser, L.G.; Cooney, J.M.; Nihal de Silva, H.; Green, S.; Richardson, K.A.; Atkinson, R.G. Mapping, complementation, and targets of the cysteine protease actinidin in kiwifruit. *Plant Physiol.* **2012**, *158*, 376–388. [CrossRef]
2. Kaur, L.; Boland, M. *Influence of Kiwifruit on Protein Digestion*, 1st ed.; Elsevier Inc.: Amsterdam, The Netherlands, 2013; Volume 68, ISBN 9780123942944.
3. Rodríguez, A.; Alquézar, B.; Peña, L. Fruit aromas in mature fleshy fruits as signals of readiness for predation and seed dispersal. *New Phytol.* **2013**, *197*, 36–48. [CrossRef]
4. Atkinson, R.G.; Gunaseelan, K.; Wang, M.Y.; Luo, L.; Wang, T.; Norling, C.L.; Johnston, S.L.; Maddumage, R.; Schröder, R.; Schaffer, R.J. Dissecting the role of climacteric ethylene in kiwifruit (*Actinidia chinensis*) ripening using a 1-aminocyclopropane-1-carboxylic acid oxidase knockdown line. *J. Exp. Bot.* **2011**, *62*, 3821–3835. [CrossRef]
5. Garcia, C.V.; Quek, S.Y.; Stevenson, R.J.; Winz, R.A. Kiwifruit flavour: A review. *Trends Food Sci. Technol.* **2012**, *24*, 82–91. [CrossRef]
6. Harker, F.R.; Carr, B.T.; Lenjo, M.; MacRae, E.A.; Wismer, W.V.; Marsh, K.B.; Williams, M.; White, A.; Lund, C.M.; Walker, S.B.; et al. Consumer liking for kiwifruit flavour: A meta-analysis of five studies on fruit quality. *Food Qual. Prefer.* **2009**, *20*, 30–41. [CrossRef]
7. Nardozza, S.; Gamble, J.; Axten, L.G.; Wohlers, M.W.; Clearwater, M.J.; Feng, J.; Harker, F.R. Dry matter content and fruit size affect flavour and texture of novel *Actinidia deliciosa* genotypes. *J. Sci. Food Agric.* **2011**, *91*, 742–748. [CrossRef]
8. Crisosto, C.H.; Zegbe, J.; Hasey, J.; Crisosto, G.M. Is dry matter a reliable quality index for "Hayward" kiwifruit? *Acta Hortic.* **2011**, *913*, 531–534. [CrossRef]
9. Feng, J. Segregation of "Hayward" Kiwifruit for Storage Potential. Ph.D. Thesis, Massey University, Palmerston North, New Zealand, 2003.
10. Burdon, J.; McLeod, D.; Lallu, N.; Gamble, J.; Petley, M.; Gunson, A. Consumer evaluation of "Hayward" kiwifruit of different at-harvest dry matter contents. *Postharvest Biol. Technol.* **2004**, *34*, 245–255. [CrossRef]
11. Woodward, T.J.; Clearwater, M.J. Relationships between "Hayward" kiwifruit weight and dry matter content. *Postharvest Biol. Technol.* **2008**, *48*, 378–382. [CrossRef]
12. Liao, G.L.; Xu, X.B.; Liu, Q.; Zhong, M.; Huang, C.H.; Jia, D.F.; Qu, X.Y. A special summer pruning method significantly increases fruit weight, ascorbic acid, and dry matter of Kiwifruit ('Jinyan', Actinidia eriantha 3 A. chinensis). *HortScience* **2020**, *55*, 1698–1702. [CrossRef]
13. Vivek, K.; Subbarao, K.V.; Routray, W.; Kamini, N.R.; Dash, K.K. Application of Fuzzy Logic in Sensory Evaluation of Food Products: A Comprehensive Study. *Food Bioprocess Technol.* **2020**, *13*, 1–29. [CrossRef]
14. Chakraborty, D. Structural quantization of vagueness in linguistic expert opinions in an evaluation programme. *Fuzzy Sets Syst.* **2001**, *119*, 171–186. [CrossRef]
15. Vaclavik, V.A.; Christian, E.W. *Food Science Text Series Essentials of Food Science*; Springer: New York, NY, USA, 2014; ISBN 9781461491378.
16. Polychroniadou, C.; Karagiannis, E.; Michailidis, M.; Adamakis, I.D.S.; Ganopoulos, I.; Tanou, G.; Bazakos, C.; Molassiotis, A. Identification of genes and metabolic pathways involved in wounding-induced kiwifruit ripening. *Plant Physiol. Biochem.* **2022**, *179*, 179–190. [CrossRef] [PubMed]
17. Karagiannis, E.; Sarrou, E.; Michailidis, M.; Tanou, G.; Ganopoulos, I.; Bazakos, C.; Kazantzis, K.; Martens, S.; Xanthopoulou, A.; Molassiotis, A. Fruit quality trait discovery and metabolic profiling in sweet cherry genebank collection in Greece. *Food Chem.* **2021**, *342*, 128315. [CrossRef]

18. Stec, M.G.H.; Hodgson, J.A.; Macrae, E.A.; Triggs, C.M. Role of fruit firmness in the sensory evaluation of kiwifruit (*Actinidia deliciosa* cv Hayward). *J. Sci. Food Agric.* **1989**, *47*, 417–433. [CrossRef]
19. Colaric, M.; Veberic, R.; Stampar, F.; Hudina, M. Evaluation of peach and nectarine fruit quality and correlations between sensory and chemical attributes. *J. Sci. Food Agric.* **2005**, *85*, 2611–2616. [CrossRef]
20. Michailidis, M.; Karagiannis, E.; Tanou, G.; Sarrou, E.; Adamakis, I.D.; Karamanoli, K.; Martens, S.; Molassiotis, A. Metabolic mechanisms underpinning vegetative bud dormancy release and shoot development in sweet cherry. *Environ. Exp. Bot.* **2018**, *155*, 1–11. [CrossRef]
21. Hummel, J.; Strehmel, N.; Selbig, J.; Walther, D.; Kopka, J. Decision tree supported substructure prediction of metabolites from GC-MS profiles. *Metabolomics* **2010**, *6*, 322–333. [CrossRef]
22. Karlsen, A.M.; Aaby, K.; Sivertsen, H.; Baardseth, P.; Ellekjær, M.R. Instrumental and sensory analysis of fresh Norwegian and imported apples. *Food Qual. Prefer.* **1999**, *10*, 305–314. [CrossRef]
23. Metsalu, T.; Vilo, J. ClustVis: A web tool for visualizing clustering of multivariate data using Principal Component Analysis and heatmap. *Nucleic Acids Res.* **2015**, *43*, W566–W570. [CrossRef]
24. Tanou, G.; Minas, I.S.; Karagiannis, E.; Tsikou, D.; Audebert, S.; Papadopoulou, K.K.; Molassiotis, A. The impact of sodium nitroprusside and ozone in kiwifruit ripening physiology: A combined gene and protein expression profiling approach. *Ann. Bot.* **2015**, *116*, 649–662. [CrossRef]
25. Minas, I.S.; Tanou, G.; Krokida, A.; Karagiannis, E.; Belghazi, M.; Vasilakakis, M.; Papadopoulou, K.K.; Molassiotis, A. Ozone-induced inhibition of kiwifruit ripening is amplified by 1-methylcyclopropene and reversed by exogenous ethylene. *BMC Plant Biol.* **2018**, *18*, 358. [CrossRef]
26. Minas, I.S.; Vicente, A.R.; Dhanapal, A.P.; Manganaris, G.A.; Goulas, V.; Vasilakakis, M.; Crisosto, C.H.; Molassiotis, A. Ozone-induced kiwifruit ripening delay is mediated by ethylene biosynthesis inhibition and cell wall dismantling regulation. *Plant Sci.* **2014**, *229*, 76–85. [CrossRef] [PubMed]
27. Macrae, E.A.; Bowen, J.H.; Stec, M.G.H. Maturation of kiwifruit (*Actinidia deliciosa* cv Hayward) from two orchards: Differences in composition of the tissue zones. *J. Sci. Food Agric.* **1989**, *47*, 401–416. [CrossRef]
28. Ciccoritti, R.; Paliotta, M.; Amoriello, T.; Carbone, K. FT-NIR spectroscopy and multivariate classification strategies for the postharvest quality of green-fleshed kiwifruit varieties. *Sci. Hortic.* **2019**, *257*, 108622. [CrossRef]
29. Wang, M.Y.; MacRae, E.; Wohlers, M.; Marsh, K. Changes in volatile production and sensory quality of kiwifruit during fruit maturation in *Actinidia deliciosa* "Hayward" and A. chinensis "Hort16A". *Postharvest Biol. Technol.* **2011**, *59*, 16–24. [CrossRef]
30. Crisosto, G.M.; Hasey, J.; Zegbe, J.A.; Crisosto, C.H. New quality index based on dry matter and acidity proposed for Hayward kiwifruit. *Calif. Agric.* **2012**, *66*, 70–75. [CrossRef]
31. Marsh, K.; Attanayake, S.; Walker, S.; Gunson, A.; Boldingh, H.; MacRae, E. Acidity and taste in kiwifruit. *Postharvest Biol. Technol.* **2004**, *32*, 159–168. [CrossRef]
32. Harker, F.R.; Hunter, D.; White, A.; Richards, K.; Hall, M.; Fullerton, C. Measuring changes in consumer satisfaction associated with kiwifruit ripening: A new approach to understand human-product interactions with fruit. *Postharvest Biol. Technol.* **2019**, *153*, 118–124. [CrossRef]
33. de Freitas, S.T.; Guimarães, Í.T.; Vilvert, J.C.; do Amaral, M.H.P.; Brecht, J.K.; Marques, A.T.B. Mango dry matter content at harvest to achieve high consumer quality of different cultivars in different growing seasons. *Postharvest Biol. Technol.* **2022**, *189*, 111917. [CrossRef]
34. Michailidis, M.; Karagiannis, E.; Nasiopoulou, E.; Skodra, C.; Molassiotis, A.; Tanou, G. Peach, apple, and pear fruit quality: To peel or not to peel? *Horticulturae* **2021**, *7*, 85. [CrossRef]
35. Beauvoit, B.; Belouah, I.; Bertin, N.; Cakpo, C.B.; Colombié, S.; Dai, Z.; Gautier, H.; Génard, M.; Moing, A.; Roch, L.; et al. Putting primary metabolism into perspective to obtain better fruits. *Ann. Bot.* **2018**, *122*, 1–21. [CrossRef] [PubMed]
36. Chai, Y.; Hong, W.; Liu, H.; Shi, X.; Liu, Y.; Liu, Z. The Pollen Donor Affects Seed Development, Taste, and Flavor Quality in 'Hayward' Kiwifruit. *Int. J. Mol. Sci.* **2023**, *24*, 8876. [CrossRef]
37. Tian, S.; Qu, M.; Xu, H. Establishment of evaluation criterion based on starch dyeing method and implementation of optical and acoustic techniques for postharvest determination of "HongYang" kiwifruit ripeness. *Eur. J. Agron.* **2023**, *142*, 126682. [CrossRef]
38. Fu, H.; Pan, L.; Wang, J.; Zhao, J.; Guo, X.; Chen, J.; Lu, S.; Dong, J.; Wang, Q. Sensory Properties and Main Differential Metabolites Influencing the Taste Quality of Dry-Cured Beef during Processing. *Foods* **2022**, *11*, 531. [CrossRef]
39. Kaleem, M.M.; Nawaz, M.A.; Ding, X.; Wen, S.; Shireen, F.; Cheng, J.; Bie, Z. Comparative analysis of pumpkin rootstocks mediated impact on melon sensory fruit quality through integration of non-targeted metabolomics and sensory evaluation. *Plant Physiol. Biochem.* **2022**, *192*, 320–330. [CrossRef]
40. Dias, M.; Caleja, C.; Pereira, C.; Calhelha, R.C.; Kostic, M.; Sokovic, M.; Tavares, D.; Baraldi, I.J.; Barros, L.; Ferreira, I.C.F.R. Chemical composition and bioactive properties of byproducts from two different kiwi varieties. *Food Res. Int.* **2020**, *127*, 108753. [CrossRef]
41. Nishiyama, I.; Fukuda, T.; Shimohashi, A.; Oota, T. Sugar and organic acid composition in the fruit juice of different *Actinidia* varieties. *Food Sci. Technol. Res.* **2008**, *14*, 67–73. [CrossRef]

42. Polychroniadou, C.; Michailidis, M.; Adamakis, I.D.S.; Karagiannis, E.; Ganopoulos, I.; Tanou, G.; Bazakos, C.; Molassiotis, A. Mechanical stress elicits kiwifruit ripening changes in gene expression and metabolic status. *Postharvest Biol. Technol.* **2022**, *194*, 112102. [CrossRef]
43. Lee, S.; Choi, H.K.; Cho, S.K.; Kim, Y.S. Metabolic analysis of guava (*Psidium guajava* L.) fruits at different ripening stages using different data-processing approaches. *J. Chromatogr. B Anal. Technol. Biomed. Life Sci.* **2010**, *878*, 2983–2988. [CrossRef]

horticulturae

MDPI

Article

Relation between Rind Pigmentation and Internal Quality of Blood Orange 'Sanguinelli': Physicochemical and Sensory Studies

Paula Tarancón, Begoña Cebrián, Paula Fernández-Serrano and Cristina Besada *

Sensory and Consumer Science Research Group, Postharvest Technology Center, Valencian Institute for Agricultural Research (IVIA), Carretera Moncada-Náquera, km. 10, 4.5, 46113 Moncada, Spain; tarancon_pau@gva.es (P.T.); cebrian_beg@gva.es (B.C.); fernandez-serrano_pau@externos.gva.es (P.F.-S.)
* Correspondence: besada_cri@gva.es

Abstract: This study evaluated the relation between rind colour and the internal physicochemical and sensory qualities of 'Sanguinelli' blood oranges, one of the main blood orange cultivars grown in the Mediterranean region. To this end, 400 fruits were harvested in three different orchards and classified according to rind pigmentation intensity (slight, medium, intense, very intense). All fruits were individually evaluated by determining rind and pulp colour, total soluble solids, acidity, maturity index, juice yield, firmness, and size. Moreover, 71 consumers performed a triangle test to evaluate if fruit sensory properties depended on rind pigmentation. Our results revealed (for the first time) that pulp pigmentation and total soluble solid (TSS) content strongly depend on rind colouration. Among the fruit from the same orchard, the redder the pigmented fruit was (externally) the deeper the red pulp, and the higher the TSS became. This pattern was corroborated in the three orchards under study. Other characteristics, such as acidity, juice yield, firmness, and fruit size, did not depend on external pigmentation. Sensory studies showed that the more coloured the fruit, the higher the accumulated sugar content; consumers perceive these fruits as being sweeter than slightly pigmented ones. This information can be very useful for the citrus industry as external colour may become a quality index for blood oranges, as well as for consumers to make purchase decisions.

Keywords: total soluble solids; anthocyanin; colour; triangle test; consumers

Citation: Tarancón, P.; Cebrián, B.; Fernández-Serrano, P.; Besada, C. Relation between Rind Pigmentation and Internal Quality of Blood Orange 'Sanguinelli': Physicochemical and Sensory Studies. *Horticulturae* **2022**, *8*, 448. https://doi.org/10.3390/horticulturae8050448

Academic Editors: Paola Sánchez-Bravo and Luis Noguera-Artiaga

Received: 13 April 2022
Accepted: 14 May 2022
Published: 17 May 2022

1. Introduction

Among the different groups of oranges, blood oranges (*Citrus sinensis* L. Osbeck) stand out for their unique, red-coloured flesh and rinds, due to their anthocyanin content. These soluble pigments belong to the flavonoid compounds family and are responsible for the characteristic colours of other fruits, such as pomegranates [1], grapes [2,3], or different kinds of berries [4,5].

The popularity of these oranges is growing worldwide [6], but they are cultivated mainly in the Mediterranean region. In Italy, blood oranges have a long-standing tradition and are well placed on the markets. The 'Sanguinello', 'Tarocco', and 'Moro' varieties have protected geographical indications (Rossa di Sicilia) [7]. In recent years, consumers in Spain have focused on blood oranges; production is increasing, with a focus mainly on the 'Sanguinelli' cultivar, which comes from spontaneous mutation of the blood orange 'Doble Fina'.

In addition to anthocyanin compounds being used to confer colour to fruit, these compounds are known for their high antioxidant capacity [8]. Awareness about the health benefits of phytonutrients is driving consumer purchases as well as fruit consumption. Thus, anthocyanin levels in the pulps of blood oranges represent important quality indices for fresh and processed products [9].

However, when growers harvest the fruit and when consumers purchase it in stores, they can only evaluate the external fruit appearance. Therefore, it is necessary to investigate if the rind colouring intensity of blood oranges is related to pulp pigmentation and, therefore, to the anthocyanin content of the edible part. Along these lines, two different studies carried out with Tarocco and Moro [9,10] found that exposing fruits to cold storage at 8 °C increased the anthocyanin content of both the pulp and rind. According to these studies, when fruits are exposed to low temperatures after harvest, they undergo an anthocyanin accumulation process and, therefore, colour enhancement, which happens in parallel to the pulp and rind. However, no studies have approached the relation between external and internal colourings at harvest.

Recently, Cebadera-Miranda et al. [11] described varieties with an intense yellow–orange colouring and the reddest pulp, and others with a more reddish peel colouring and a more yellow–orange pulp. However, we herein wished to examine the relation between rind and pulp colouration from a different perspective. Fruits from the same variety may present wide pigmentation variability, even if they come from the same orchard. The causes of such variability are unclear, but Lo Piero [12] suggested that it can be at least partially related to the position of fruits on the canopy. In this study, we wished to investigate if a relation exists between the intensity of rind pigmentation and pulp pigmentation when fruits of the same cultivar are evaluated. That is, if the external fruit colour is related to the internal colouration and if, therefore, the fruit appearance may act as an indicator of flesh anthocyanin content. Moreover, it was necessary to evaluate if internal characteristics, other than the pulp colour, depend on the rind colour. Furthermore, it was necessary to evaluate if sensory fruit properties, other than visual appearance, are linked to rind pigmentation.

In this context, this study aimed to answer two different questions about the relation between the external appearance and the internal quality of 'Sanguinelli' blood oranges: (1) is the physicochemical quality of pulp linked to rind pigmentation intensity? (2) Are sensory properties perceived by consumers linked to rind pigmentation intensity?

2. Materials and Methods

2.1. Plant Material

Blood oranges (*Citrus sinensis* L. Osbeck) cv. Sanguinelli were obtained from a packing house located in Valencia (Valencian Community, Spain). In the second half of February (the 2019 season), three different fruit batches corresponding to the fruit harvested from three commercial orchards were collected at the packing house, the day after harvest (Orchard 1, Orchard 2, and Orchard 3). The mean values of day and night temperatures from 1 January to harvest date for each orchard were as follows (mean ± SD): 12.8 ± 1.9 and 8.8 ± 2.5 for Orchard 1, 12.7 ± 2.4 and 7.8 ± 3.9 Orchard 2, and 12.9 ± 2.2 and 6.2 ± 3.3 for Orchard 3.

Then fruit were transferred to the Postharvest Department of the Valencian Institute for Agricultural Research (IVIA), where they were divided into lots according to rind pigmentation. The fruits from Orchards 1 and 2 were divided into three lots: slight (P1), medium (P2), and intense (P3) rind pigmentation. In Orchard 3, fruits were divided into two lots according to fruit appearance: medium (P2) and very intense (P4) pigmentation (Figure 1).

Three steps were followed to divide fruit into batches according to their pigmentation: (1) upon the arrival of fruit from each orchard to the laboratory, batches of 50 fruits with different rind pigmentations were initially made by one researcher with experience in blood oranges. (2) Then, the fruit previously selected for batches were mixed and another researcher was asked to group the fruit into 50-fruit batches based on the perceived colour. In case of doubts about any specific fruit, it could be removed and substituted for another one that did not cause the researchers to doubt. In general, a maximum of 1 or 2 fruits per lot of 50 fruits generated disagreement between both researchers and were substituted for consensus. (3) Finally, fruit were mixed again, and we asked a consumer to divide the fruit into 50-fruit batches based on the colour he/she perceived. For the three orchards, consumers who made the final batches showed no difference with respect to the lots that were made by researchers in the previous steps. The three consumers who participated in this study were invited to do so based on their consumption of blood oranges (at least once every two weeks during the season) and their interest in participating. They were given a box of chocolates for participating.

a*/b*	+ 0.5	+ 0.7	+ 1.3	+ 1.6
Orchard 1	P1	P2	P3	—
Orchard 2	P1	P2	P3	—
Orchard 3	—	P2	—	P4

Figure 1. 'Sanguinelli' blood oranges grouped according to increased rind pigmentation intensity (P1—slight, P2—medium, P3—intense, P4—very intense).

The 50 fruits per lot were numbered and the following non-destructive measurements were individually taken on each one: rind colour, firmness, diameter, and weight. Then, juice was obtained individually from all 50 fruits to make the following determinations: juice yield (JY), juice colour, total soluble solids (TSS), titratable acidity (TA), and maturity index (MI).

Moreover, in view of the physicochemical results obtained when evaluating the fruit from Orchard 1, sensory studies were planned to be included when analysing the fruit from Orchard 2. Total anthocyanin content was determined in juice samples used for sensory evaluation.

2.2. *Physicochemical Analysis*

Firmness measurements were taken using an Instron Universal Testing Machine (model 3343, Instron, Ltd., Buckinghamshire, UK). The results are expressed as the percentage of millimetres of fruit deformation resulting from 10 N pressure, applied by a 3.5 cm plunger on the longitudinal axis at constant speed. Weight (in grams) was determined using an analytical balance while diameter (in millimetres) was measured using a calliper.

As reported in previous studies [9], among the several CIELAB parameters, the a^*/b^* ratio is the one that best displays changes in rind and pulp pigmentation in blood oranges [9,13,14]. Hence, in this study, the a^*/b^* ratio was also considered as the colour index ($a^*/b^* < -0.5$ correspond to dark green tones, $a^*/b^* \approx 0$ indicates the colour break from green to yellow, $a^*/b^* \approx +0.5$ correspond to orange tones, and $a^*/b^* > +0.5$ correspond to red tones).

Pigmentation of the blood orange rind is usually quite heterogeneous around the fruit surface. Previous studies have described the determination of the rind colour by taking measurements along the equatorial axis [9] or by taking three consecutive measurements in the darkest part of the peel, the clearest part, and on the base of the fruit [11]. These procedures may be valid to evaluate colour evolution during storage [9] or to compare samples from different varieties [11]. However, in order to study in-depth the relation between the external pigmentations and internal characteristics of fruit from the same variety, a method is needed to better reflect the rind pigmentation heterogeneity. Thus, in this study, a new method to measure rind colour was proposed. After visually evaluating fruit, an estimation of the percentage of surface displaying each pigmentation intensity (I1-absence—very light, I2-medium, I3-intense, and I4-very intense) was recorded. To this end, the skin colour of each fruit was evaluated all round its surface; firstly, one 180° fruit face was evaluated before it was rotated to evaluate the opposite 180° fruit face. Two to three different pigmentation areas were generally detected on each fruit (Figure S1). Then measurements of the parameters L*, a*, b* of the CIELAB space were taken in all of these areas using a Minolta colorimeter (model CR-300; Minolta Co., Ltd., Osaka, Japan), and the ratio a*/b* was calculated for each area. Then, a unique a*/b* value was calculated per individual fruit according to the following formula:

a*/b* = [(% of surface with intensity 1 × a*/b*-value in this area) + (% of surface with intensity 2 × a*/b* value in this area) + (% of surface with intensity 3 × a*/b* value in this area) + (% of surface with intensity 4 × a*/b*value in this area)]/100.

Pulp colouration was determined by squeezing each fruit individually via an electric juice extractor with a rotating head (Lomi®, Model 4, Lorenzo Miguel, S.L., Madrid, Spain). The a*/b* value was measured on juice samples, which were also used to determine JY, TA, and TSS. The juice yield was expressed as a percentage, calculated by dividing the volume of juice by the total fruit weight. The TA was determined by titration with a 0.1 N NaOH solution, using phenolphthalein as the indicator, expressed as g of citric acid per 1 L of juice. The TSS in juice was measured by a digital refractometer (Atago PR-1, Atago Co., Ltd., Tokyo, Japan) and data were expressed as %. The maturity index (MI) was calculated as TSS/TA.

2.3. Sensory Evaluation

Sensory studies were performed with the fruit from Orchard 2 to evaluate if rind pigmentation had an effect on the juice sensory properties. The triangle discrimination test technique [15,16] was used to determine whether there were any detectable differences in the sensory properties of the juice samples. A panel of 71 consumers evaluated the juice samples from fruit lots P1 (slight rind pigmentation) and P3 (intense rind pigmentation) (Figure 1). To this end, after separating the volume of juice needed to determine the aforementioned physicochemical characteristics, the remaining juice obtained from fruit lots P1 and P3 was used in the sensory analysis. For each lot, five juice samples were obtained by mixing the juice that remained from 10 fruits in each one. Before the sensory evaluation, colour, TA, TSS, MI, and total anthocyanin content were determined in these five juices of each type. Then consumers compared the juices obtained from fruits with low pigmented rinds (juice P1) versus very intense rind pigmentation (juice P3). Panellists were simultaneously presented with three samples—two from juice P1 and one from juice P3, or vice versa.

Consumers were between the ages of 22 and 61 years; the male/female ratio (%) was between 43/57. All consumers voluntarily agreed to participate in the evaluation session. Panellists were seated in partitioned booths and samples were randomly presented to avoid any positional bias as the middle sample was usually chosen as "odd". The possible combinations of samples from the triad were: AAB, ABA, BAA, BBA, BAB, and ABB. All samples were coded with three-digit random numbers. In order to avoid consumer responses being conditioned by juice colour, samples were served in opaque cardboard glasses with perforated lids and a red-coloured straws (Figure S2). In addition, red lights were used in the tasting booths; 30-mL juice samples were served at room temperature and panellists were provided with glasses of water for palate cleansing, which they used between samples. Then they were asked to taste samples from left to right and to indicate the odd sample. They were also asked to indicate the main reason why they found the odd sample different compared to the other two.

In addition to the physicochemical characterisations (colour, TA, TSS, and MI) of the juice samples used for the sensory evaluation, two 2-mL samples were frozen for the posterior total anthocyanin content (TAC) evaluation. The total anthocyanin content (TAC) was determined by the pH differential method [17]. Anthocyanin pigments undergo reversible structural transformations with changes in pH manifested by strikingly different absorbance spectra. The coloured oxonium form predominates at pH 1.0 and the colourless hemiketal form at pH 4.5. The pH-differential method is based on this reaction, and permits accurate and rapid measurements of the total anthocyanins, expressed as mg/L.

2.4. Statistical Analysis

An ANOVA was applied to the physicochemical data to evaluate if they depended on external pigmentation. A multiple comparison between means was run by Duncan's multiple range test (p = 0.05). To determine if significant differences were perceived by consumers between the evaluated juice samples, the number of correct answers in the triangle test was calculated and the significance of differences according to binomial distribution was established. All statistical analyses were performed with the XL-stat programme (2019 version).

3. Results

3.1. Pulp Pigmentation Is Linked to Rind Pigmentation

In this study, the rind pigmentation of fruit depended on the orchard. The fruit from Orchards 1 and 2 showed similar pigmentation ranges and were divided into three lots according to pigmentation intensity (P1—slight, P2—medium, and P3—intense). However, the fruit from Orchard 3 were generally more coloured and homogenous, and only two lots were obtained (P2—medium and P4—very intense pigmentation). A representation of the different pigmentation intensities is shown in Figure 1.

One of the objectives of this study was to evaluate to what extent of information (i.e., about internal fruit properties) that the 'Sanguinelli' rind colour provides us with. Therefore, we needed to be sure that the colour measurements reflected, as much as possible, the human perception of fruit colour. To this end, the heterogeneous distribution of pigmentation around the rind surface was considered when measuring the external colour. The procedure is explained in detail in the Material and Methods section; it mainly took into account the intensity and area of pigmentation while measuring the rind colour. Our results show a gradual increase in the a*/b* index, parallel to the rind pigmentation intensity. Thus the colour index of samples shown in Figure 1 is as follows: P1= 0.51 ± 0.07; P2= 0.72 ± 0.16; P3 = 1.28 ± 0.3; P4 = 1.59 ± 0.37.

Figure 2 shows the mean values for the different physicochemical characteristics determined per pigmentation group and orchard. For all three orchards, significant differences in the a*/b* values were detected among the different sample groups created based on the visual perception of rind colour (Figure 2A, letters above bars). When the a*/b* values of the different lots (P1, P2, P3, and P4) were compared among orchards, no significant differences were detected among the lots with the same rind pigmentation intensity (Figure 2A, letters inside bars). This reveals that the visual fruit classification was consistent among the orchards.

The mean a*/b* values of the juice obtained from the 50 fruits belonging to each group are shown in Figure 2B. The ANOVA analysis, performed to explore if pulp colour depended on rind pigmentation, revealed a clear pattern that was detected in the three orchards. Thus, for each orchard, the more intense the rind pigmentation was, the more coloured the juice was. In all cases, except for P2 and P3 from Orchard 2, significant differences in juice pigmentations were detected according to the rind colour.

However, when taking into account the fruit from the three orchards to perform the statistical analysis, some differences in the juice colour were detected between fruits with the same rind pigmentation, but from different orchards. For example, the juice of the P1- fruit from Orchard 1 had a significantly lower a*/b* value than that of the juice from the P1-fruit from Orchard 2. Therefore, despite detecting a common pattern in all three orchards (the more intense the rind pigmentation, the more coloured the juice), significant variabilities existed among the orchards.

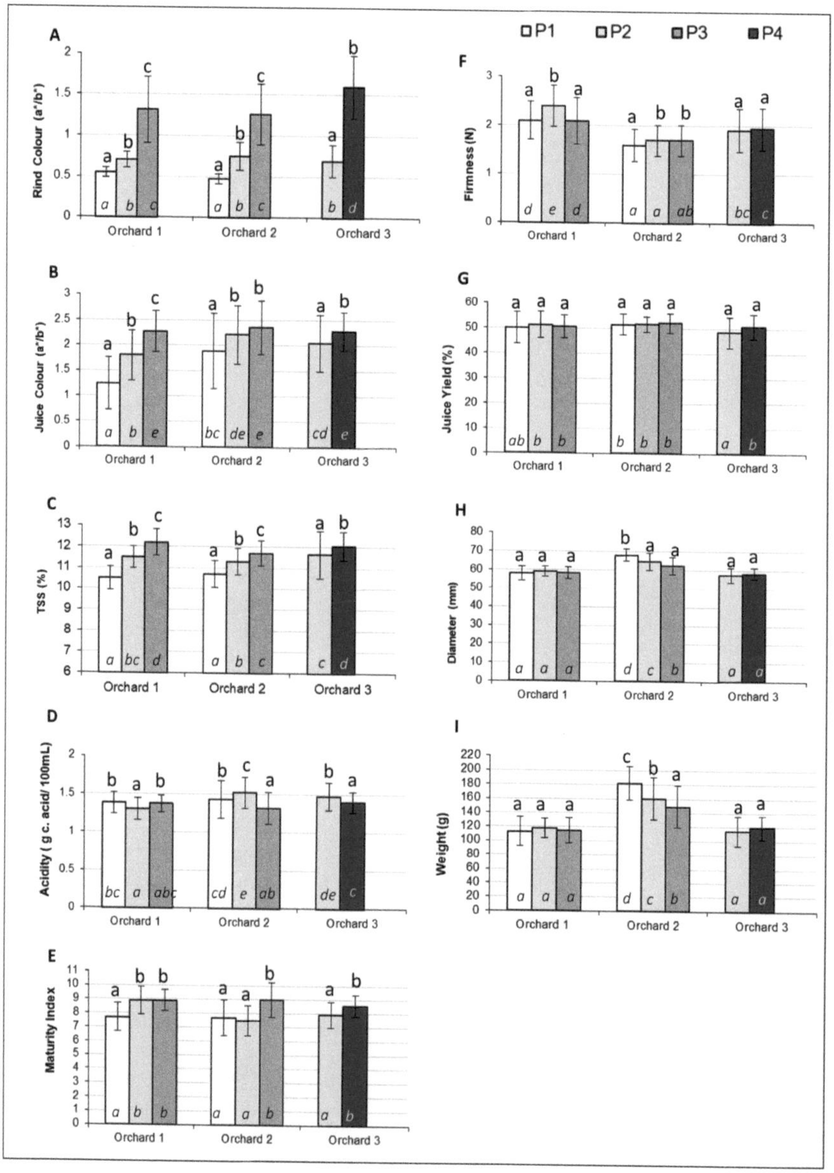

Figure 2. Effect of rind pigmentation intensity (P1—slight, P2—medium, P3—intense, P4—very intense) of 'Sanguinelli' on internal properties ((**A**)—rind colour, (**B**)—juice colour, (**C**)—TSS, (**D**)—titratable acidity, (**E**)—maturity index, (**F**)—firmness, (**G**)—juice yield, (**H**)—diameter, (**I**)—weight). For each orchard, different letters above the bars indicate significant differences among fruit lots. The italics at the bottom of the bars denote differences among fruit lots when comparing all the fruit lots; i.e., when comparing the lots from the three orchards by Duncan's multiple range test ($p = 0.05$).

A Pearson correlation analysis revealed that correlation between a*/b* values of rind and a*/b* values of juice was significant for Orchard 1 but not for Orchard 2 and 3. Despite the coefficient correlation being relatively low, r = 0.29, significance was detected when correlation was performed, taking into account data from the total dataset (Table S1).

To more profoundly study the relation between the rind and pulp colour, the frequency distribution histogram was obtained (Figure 3). To help interpretate these data, a juice colour scale was developed in this study (Figure 4). Bars in the histogram (Figure 3) represent the percentages of juice samples (y-axis) within a certain range of a*/b* values (X-axis). A similar distribution was detected in the three orchards under study. The fruit with slight rind pigmentation (P1) displayed wide pulp pigmentation variability, with a*/b* values ranging between 0.25 and 2.25 (Orchard 1) and between 0.25 and 3.75 (Orchard 2). As observed in Figure 4, these values correspond to the fruit with orange pulp and red-coloured pulp, respectively. In the fruit with rind colour P2, variability was narrower than in P1, as pulp colour was generally more pigmented, and most of the samples had a*/b* values that fell within the range of 1.25–2.50 (Figure 3). In the P3 group, no juice with a*/b* values lower than 1.25 was detected, and most samples concentrated a*/b* values between 2 and 2.75, which corresponded to deep red juice. Finally, the most rind-pigmented fruit (P4) obtained pulp a*/b* values between 1.50 and 3, which were once again related to deep pigmented pulp (Figure 4). Therefore, it can be stated that, although certain intra- and inter-orchard variabilities can be expected, the more pigmented the skin colour is, the higher the probability of finding fruit with a deep red pulp.

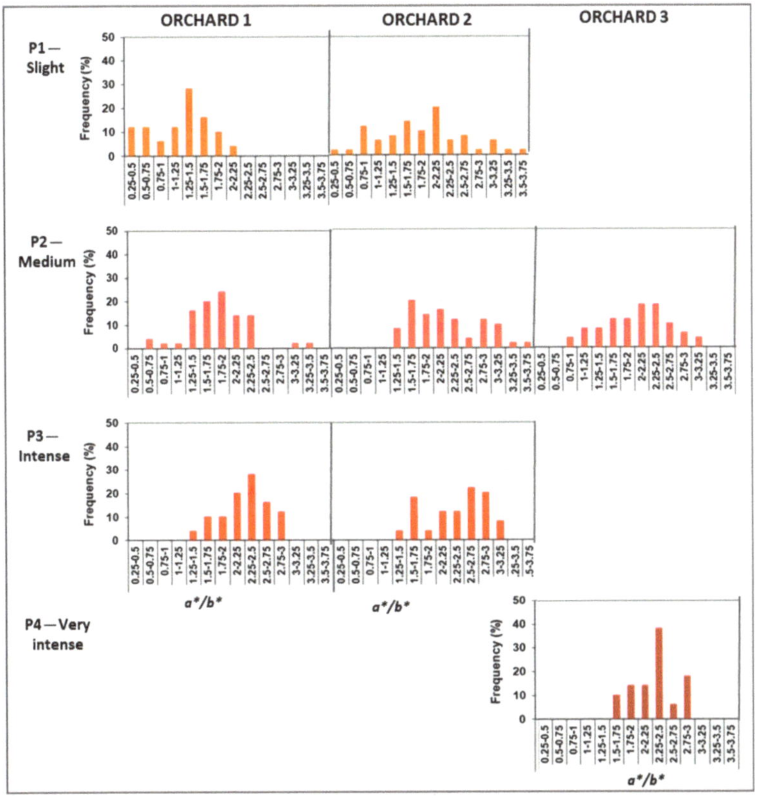

Figure 3. Frequency distribution of the pulp colour (a*/b* values of juice) of 'Sanguinelli' blood oranges grouped according to rind pigmentation (P1—slight, P2—medium, P3—intense, P4—very intense). Each group of intensity per orchard consisted of 50 individually analysed fruits.

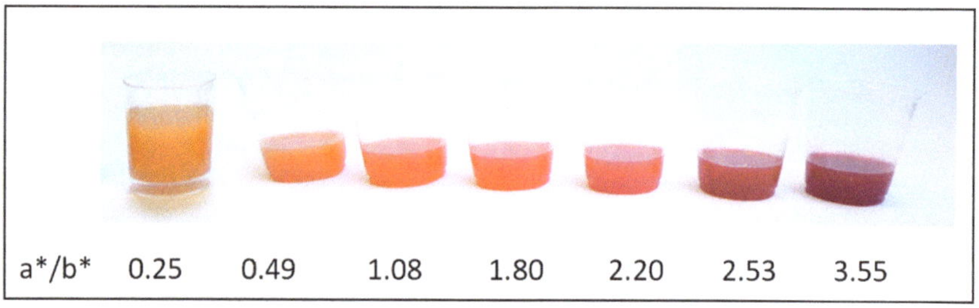

Figure 4. Relation between visual colour of juice and a*/b* index.

3.2. Concentration of Total Soluble Solids Depends on Rind and Pulp Pigmentation

Figure 2C shows the content of total soluble solids depending on external colouration. Interestingly, our results showed for the fruits of each orchard that the more intense the rind pigmentation was, the higher the content of soluble solids in the juice. This effect was corroborated in the fruits from the three orchards under study.

Similar to that reported for juice colouration, certain variabilities among orchards were observed, mainly in the more intensely pigmented fruit.

Once again, we looked closely at the relation between TSS and rind colour by obtaining the frequency distribution histogram, in which bars represent the percentage of juice samples with a certain range of TSS content (Figure 5). TSS ranged between 8.5 (P2-Orchard 3) and 13.5 (P3-Orchard 1). Similar to that described for pulp colouration, TSS showed certain intra- and inter-orchard variability but, in all cases, the more pigmented the rind, the higher the frequency of the samples with high TSS. This was clearly observed in Orchard 2, where the most frequent TSS content was 10.5–11% for P1-fruit, 11–11.5% for P2-fruit, and 11.5–12% in P3-fruit.

Moreover, as a close relation between rind pigmentation and pulp colouration was observed in this study, the relation between pulp colour and total soluble solids was also investigated. To this end, the Pearson correlation between the a*/b* values and the TSS of the juice samples were studied by taking into account the 400 fruits individually evaluated in this study. The correlation coefficient was $r = 0.53$ and the p-value was lower than 0.05, which indicates that TSS content was significantly related to pulp colouration. Significant correlations were also detected when data from each of the orchards were analysed separately (Table S1). As the a*/b values of juice increased, as colouration changed from orange to deep red, a positive correlation between them and TSS indicated that the more pigmented the pulp, the higher the TSS content.

Parallel to the increment in TSS associated with rind pigmentation, the MI was found to be linked to some extent to rind colouration (Figure 2E). Although this relation was not as clear as that of the rind colour and TSS, it was still evident that the most pigmented rind fruit had a higher MI than the less coloured fruit.

3.3. Physicochemical Characteristics Not Linked to Pigmentation

Apart from juice colour, TSS, and MI, the other physicochemical characteristics herein evaluated did not show a relation with rind colour. In the three orchards, firmness values came close to 2% of deformation with juice yield at around 50%, and no differences were found among the sample groups (Figure 2F,G).

Fruit size, determined as diameter and weight, was slightly bigger in the fruit from Orchard 2. In this orchard, an effect of pigmentation was observed, as the most pigmented fruits were smaller. However, this pattern was not detected in Orchards 1 or 3 (Figure 2H,I).

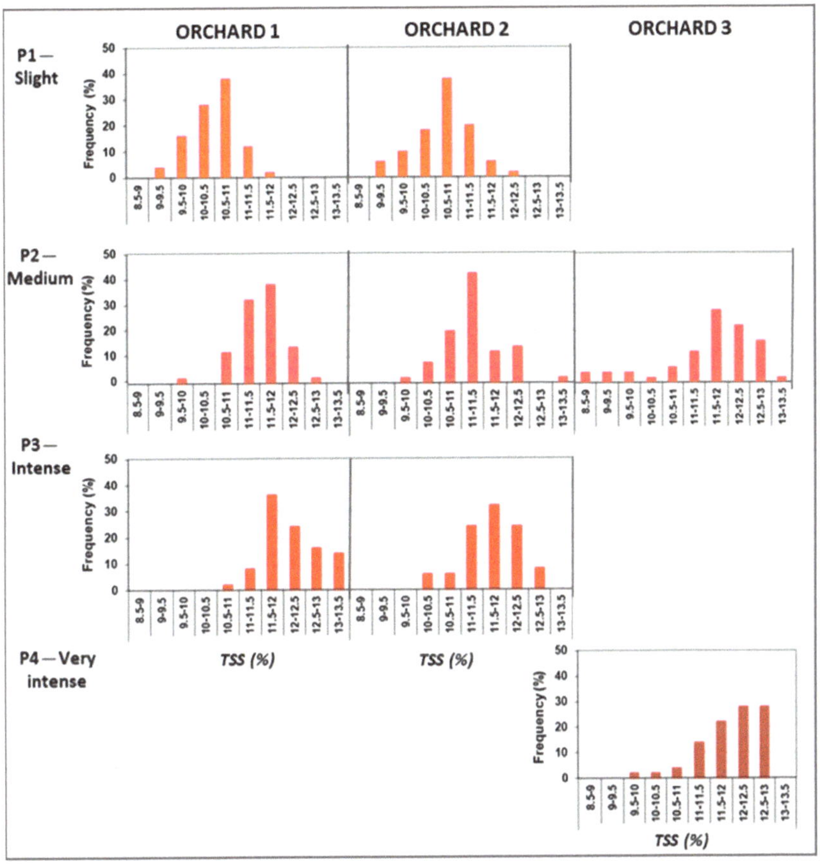

Figure 5. Frequency distribution of the total soluble solid content of 'Sanguinelli' blood oranges grouped according to rind pigmentation (P1—slight, P2—medium, P3—intense, P4—very intense). Each group of intensity for each orchard consisted of 50 individually analysed fruits.

3.4. Effect of Rind Pigmentation on Sensory Properties

The second main objective of this study was to investigate if the sensory properties perceived by consumers when tasting 'Sanguinelli' fruits depended to some extent on rind pigmentation. To this end, a triangle test was carried out.

The test was performed with fruit from Orchard 2, and juice samples were obtained from the fruits with slight ($a^*/b^* = 0.55$) and intense ($a^*/b^* = 1.31$) rind pigmentations (Table 1). The mean values of the TSS content, TA, MI, total anthocyanin content, and the colour of the juice samples obtained from these fruits are shown in Table 1. According to our previous results, the juice obtained from the most pigmented rind fruit (juice P3) was the most coloured and had a higher TSS than the juice obtained from the less pigmented rind fruit (juice P1). As no differences were detected in TA, the higher TSS of juice P3 resulted in a higher MI. In agreement with its more intense pigmentation, juice P3 showed a higher total anthocyanin content than juice P1. The results from the triangle test revealed significant differences in their sensory properties (p-value = 0.04). The main reasons given by consumers (94% of those who gave a correct answer) involved differences in the sweetness/acidity level.

Table 1. Physicochemical characteristics of juice samples with markedly different rind pigmentations (Orchard 2, samples P1—slight rind pigmentation and P3—intense rind pigmentation) and results from the triangular test performed by consumers. TA is expressed as g citric acid/100 mL and TAC as mg/L. For each parameter, * indicates significant differences between the two different juice samples.

Samples			**Physicochemical**			
Juice	a*/b* Rind	a*/b* Juice	TSS (%)	TA	MI	TAC
P1	0.55 *	1.12 *	11.2 *	1.5	7.4 *	40 *
P3	1.31	2.36	12.1	1.4	8.6	59
			Sensory			
Juice	Trials		Correct answers		p-value	
P1 vs. P3	71		51		0.04	

* Significant differences between juices according to the LSD test (p-value < 0.05).

4. Discussion

Rind pigmentation heterogeneity is a characteristic of most blood orange varieties. According to Kafkas et al. [18], the most coloured area usually corresponds to the part of the fruit with a north orientation.

An accurate procedure to measure rind colour, considering the rind pigmentation heterogeneity, was herein stablished. This method allows obtaining colour measurements that better reflect the human eye perception of fruit colouration. Accurate colour data were the basis to evaluate the relation between rind pigmentation and the internal fruit properties. From a practical point of view, when colour is determined by a colorimeter, this method can be slow, but useful, for researchers. Moreover, it could be implemented into colour determination with automatic calibrator machines.

The individual evaluation of 400 fruits revealed a clear relation between juice and rind pigmentation in one of the main varieties cultivated in the Mediterranean region. To date, the relation between rind and pulp colour has been mainly approached by comparing different varieties. Accordingly, Cebadera-Miranda et al. [11] reported that it is possible to find fruit with an intensely coloured rind, but a slight coloured pulp, and vice versa. However, we focused on variability within the same variety, and our results revealed that the more pigmented the rind, the higher the probability of an intensely coloured pulp.

As pulp pigmentation is linked to anthocyanin content in Sanguinelli fruits [11], our results imply that, to some extent, rind pigmentation is an indicator of pulp anthocyanin content. In this sense, as consumer interests in healthy foods increase, one reason to promote blood oranges is based on their high anthocyanin content. Hence, this information can be very useful for the citrus industry and consumers, as intensely pigmented oranges may offer added value due to their high content of antioxidant compounds.

However, it is worth noting that significant inter-orchard variability was detected. Thus the criterion that the more pigmented the rind, the higher probability of obtaining fruit with a high anthocyanin content, seems to apply mainly when comparing fruit from the same lot.

According to Lo Piero [12], anthocyanin accumulation in blood orange cultivars is affected by different factors, such as variety, maturity, cultivation region, cultural practices, and many other environmental factors. Therefore, some of these factors are likely to contribute to the inter-orchard variability herein observed. Among them, it is possible that temperature contributes to the most intense rind pigmentation of fruit from the Orchard 3. Thus, the highest temperatures at day and the lowest at night were recorded for this orchard. According to Butelli et al. [19], all blood orange varieties require strong day–night thermal clines for intense colour formation in fruit flesh, and varieties such as 'Moro', with the potential for high pigmentation, are strongly dependent on the prevailing climatic conditions during fruit ripening for full colour development.

However, we should note that, as herein reported for the first time, under the same conditions (same orchard), anthocyanin accumulation in 'Sanguinelli' blood orange pulp parallels to rind anthocyanin accumulation, to some extent.

Interestingly, our results also revealed that the TSS content was clearly related to rind and pulp pigmentation. Along these lines, in a study in which the development and maturation process of blond and blood oranges were compared, Muccilli et al. [20] found a higher accumulation of enzymes related to sugar in the pulp of blood oranges. These authors linked the higher sugar metabolism required in blood cultivars to the need for carbon skeletons required for anthocyanin biosynthesis [20]. More recently, Carmona et al. [21] carried out a protein analysis while storing 'Moro' at low temperatures and described that when fruits were stored at 9 °C, anthocyanin accumulation took place, which correlated with the promotion, among others, of the protein belonging to the metabolism of sugars. Therefore, some previous information has indirectly reported the accumulation of sugars during processes in which anthocyanins also accumulated, such as fruit development or storage at low temperatures. However, this study demonstrates for the first time that among the fruit of the same lot, the more pigmented the fruit is, the higher the TSS content. If we assume that the MI is an indicator of the fruit maturity stage, it can be stated that the most pigmented fruits were in the most advanced maturity stage. However, it is important to clarify that, at least partially, the accumulation of sugars in blood oranges seems to run parallel to anthocyanin accumulation, and is not directly related to the maturation process. It is well-known that increments in TSS associated with citrus fruit maturation are accompanied by a drop in the acidity level. However, our results showed no differences in the TA depending on the pigmentation for any of the three studied orchards. Similar to TA, other physicochemical characteristics, such as fruit size, firmness, or juice yield, were not related to pigmentation.

Finally, we conducted a sensory test to evaluate if fruit pigmentation could affect the sensory properties perceived by consumers. The initial hypothesis that led us to approach this question was that anthocyanin content was reported to affect the sensory properties of hibiscus drinks [22]. Moreover, in wine, anthocyanin content may affect the sensory profile by a reaction with other compounds [23,24]. Our results showed that consumers found differences between juice samples obtained from slight versus intense pigmented rind oranges. These samples showed significant differences in juice colour, which corroborated the stablished relation between external and internal pigmentation. Moreover, a physicochemical analysis showed that they differed in terms of MI, with an almost 1-point difference. Such a difference in MI was associated with a higher TSS content of the most pigmented fruit. The great majority (94%) of consumers that identified the odd sample, referred the sweetness/acidity perception as the main detected difference. Therefore, the differences in the MI index were identified as the main factors to affect the sensory properties perceived by consumers. Hence, our hypothesis in this regard is that the anthocyanin content itself does not affect sensory properties, but the TSS accumulation that occurs parallel to anthocyanin accumulation may result in sensory differences between highly pigmented fruit and slightly pigmented fruit.

To summarise, our results revealed that the internal properties of 'Sanguinelli' blood oranges are linked to rind pigmentation. For the fruit from the same orchard, the more intense the rind colour, the more pigmented the pulp and, therefore, the higher anthocyanin content. Moreover, TSS content also increases in parallel to anthocyanin accumulation. As the acidity level does not depend on fruit pigmentation, this increment in TSS results in fruit with a higher MI, which leads consumers to perceive that the most pigmented fruit is sweeter than the slightly pigmented fruit. Therefore, the intensity of rind pigmentation may act as a quality parameter linked to nutritional and sensory properties. Sorting fruit by rind colour would allow the industry to commercialize batches of fruit with homogenous properties and to give added value to the most pigmented ones.

Supplementary Materials: The following supporting information can be downloaded at: https://www.mdpi.com/article/10.3390/horticulturae8050448/s1, Figure S1: Visual estimation of rind colour areas where colour measurements were taken per fruit. After evaluating each side of the fruit (one 180° fruit face and the opposite 180° fruit face), the areas with the same intensity (I) were added and then divided by 2; Figure S2: Samples presentation in booths to consumers for the triangle test; Table S1: Pearson correlation between the colour index (a*/b*) of rind and juice, and between the colour index (a*/b*) of juice and total soluble solids.

Author Contributions: Conceptualization, C.B.; formal analysis, P.T., P.F.-S. and B.C.; writing—original draft preparation, C.B.; writing—review and editing, C.B.; supervision, C.B.; funding acquisition, C.B. All authors have read and agreed to the published version of the manuscript.

Funding: This study was co-funded by the European Regional Development Fund (ERDF) of the Generalitat Valenciana (IVIA project number 52201).

Institutional Review Board Statement: Not applicable.

Informed Consent Statement: Not applicable.

Data Availability Statement: Not applicable.

Acknowledgments: Author Paula Fernandez-Serrano thanks the European Social Fund for the financial support of her contract.

Conflicts of Interest: The authors declare no conflict of interest.

References

1. Sharma, A.; Thakur, N.S. Wild pomegranate (*Punica granatum* L.): A review on physical and chemical attributes of himalayan wild pomegranate fruit. *J. Pharmacogn. Phytochem.* **2018**, *7*, 1518–1524.
2. Laitila, J.E.; Suvanto, J.; Salminen, J.P. Liquid chromatography–tandem mass spectrometry reveals detailed chromatographic fingerprints of anthocyanins and anthocyanin adducts in red wine. *Food Chem.* **2019**, *294*, 138–151. [CrossRef] [PubMed]
3. Liu, Y.; Zhang, X.K.; Shi, Y.; Duan, C.Q.; He, F. Reaction kinetics of the acetaldehyde-mediated condensation between (−)-epicatechin and anthocyanins and their effects on the color in model wine solutions. *Food Chem.* **2019**, *283*, 315–323. [CrossRef] [PubMed]
4. Routray, W.; Orsat, V. Blueberries and their anthocyanins: Factors affecting biosynthesis and properties. Comprehensive Rev. Food. Sci. *Food Saf.* **2011**, *10*, 303–320. [CrossRef]
5. Demirbas, A.; Yilmaz, V.; Ildiz, N.; Baldemir, A.; Ocsoy, I. Anthocyanins-rich berry extracts directed formation of Ag NPs with the investigation of their antioxidant and antimicrobial activities. *J. Mol. Liquids.* **2017**, *248*, 1044–1049. [CrossRef]
6. Simons, T.J.; McNeil, C.J.; Pham, V.D.; Suh, J.H.; Wang, Y.; Slupsky, C.M.; Guinard, J.X. Evaluation of California-Grown Blood and Cara Cara Oranges through Consumer Testing, Descriptive Analysis, and Targeted Chemical Profiling. *J. Food Sci.* **2019**, *84*, 3246–3346. [CrossRef] [PubMed]
7. Cebadera-Miranda, L.; Morales, P.; Cámara, M. Bioactive Compounds in Oranges from the Mediterranean Climate Area. In *The Mediterranean Diet. An Evidence-Based Approach*; Preedy, V.R., Ross Watson, R., Eds.; Academic Press: Cambridge, MA, USA, 2020; pp. 293–309. [CrossRef]
8. Pina, F.; Parola, A.J.; Melo, M.J.; Lima, J.C.; de Freitas, V. Chemistry of Anthocyanins. In *Anthocyanins from Natural Sources. Exploiting Targeted Delivery for Improved Health*; Brooks, M.S.L., Celli, G.B., Eds.; RSC, Publishing: London, UK, 2019; pp. 34–76. [CrossRef]
9. Rapisarda, P.; Bellomo, S.E.; Intelisano, S. Storage temperature effects on blood orange fruit quality. *J. Agric. Food Chem.* **2001**, *49*, 3230–3235. [CrossRef] [PubMed]
10. Hamedani, M.; Rabiei, V.; Moradi, H.; Ghanbari, A.; Azimi, M.R. Determination of storage duration and temperature effects on fruit quality parameters of blood orange (*Citrus sinensis* cv. Tarocco). *Biharean Biol.* **2012**, *6*, 10–13.
11. Cebadera-Miranda, L.; Domínguez, L.; Dias, M.I.; Barros, L.; Ferreira, I.C.; Igual, M.; Martinez-Navarrete, N.; Fernández-Ruiz, V.; Morales, P.; Cámara, M. Sanguinello and Tarocco (*Citrus sinensis* L. Osbeck): Bioactive compounds and colour appearance of blood oranges. *Food Chem.* **2019**, *270*, 395–402. [CrossRef] [PubMed]
12. Lo Piero, A.R. The state of the art in biosynthesis of anthocyanins and its regulation in pigmented sweet oranges [(*Citrus sinensis*) L. Osbeck]. *J. Agric. Food Chem.* **2015**, *63*, 4031–4041. [CrossRef] [PubMed]
13. Buccheri, M.; Grassi, M.; Bianchi, G.; Lovati, F.; Vanoli, M.; Rizzolo, A.; Croce, D. Postharvest physiology, nutritional quality and aroma profile of 'Tarocco' blood orange fruit (*Citrus sinensis* L. Osbeck). *Acta Hort.* **2015**, *1079*, 291–298. [CrossRef]
14. Zacarías, L.; Rodrigo, M.J.; Tornero, C.; Zacarías, J.; Rey, F.; Simó, G.; Tornero, L.; Raga, A. Efecto de la Temperatura de Conservación en el Color y la Calidad Interna de Naranjas Sanguinelli. In Proceedings of the 12th Simposio Nacional y X Ibérico de Maduración y Postcosecha, Badajoz, Spain, 4–7 June 2018.
15. Meilgaard, M.C.; Civille, G.V.; Carr, B.T. *Sensory Evaluation Techniques*, 4th ed.; CRC Press: Boca Raton, FL, USA, 2007.

16. Lawless, H.T.; Heymann, H. *Sensory Evaluation of Food: Principles and Practices*; Springer Science & Business Media: New York, NY, USA, 2010.
17. Giusti, M.M.; Wrolstad, R.E. Characterization and measurement of anthocyanins by UV visible spectroscopy. *Curr. Protoc. Food Anal. Chem.* **2001**, *1*, F1–F2. [CrossRef]
18. Kafkas, E.; Ercisli, S.; Kemal, K.N.; Baydar, K.; Yilmaz, H. Chemical composition of blood oranges varieties from Turkey. A comparative study. *Pharmacogn. Mag.* **2009**, *5*, 329–335. [CrossRef]
19. Butelli, E.; Licciardello, C.; Zhang, Y.; Liu, J.; Mackay, S.; Bailey, P.; Reforgiato-Recupero, G.; Martin, C. Retrotransposons control fruit-specific, cold-dependent accumulation of anthocyanins in blood oranges. *Plant Cell* **2012**, *24*, 1242–1255. [CrossRef]
20. Muccilli, V.; Licciardello, C.; Fontanini, D.; Russo, M.P.; Cunsolo, V.; Saletti, R.; Recupero, G.R.; Foti, S. Proteome analysis of *Citrus sinensis* L. (Osbeck) flesh at ripening time. *J. Proteom.* **2009**, *73*, 134–152. [CrossRef] [PubMed]
21. Carmona, L.; Alquézar, B.; Marques, V.V.; Peña, L. Anthocyanin biosynthesis and accumulation in blood oranges during postharvest storage at different low temperatures. *Food Chem.* **2017**, *237*, 7–14. [CrossRef] [PubMed]
22. Bechoff, A.; Cissé, M.; Fliedel, G.; Declemy, A.L.; Ayessou, N.; Akissoe, N.; Touré, C.; Bennett, B.; Pintado, M.; Pallet, D.; et al. Relationships between anthocyanins and other compounds and sensory acceptability of Hibiscus drinks. *Food Chem.* **2014**, *148*, 112–119. [CrossRef] [PubMed]
23. Dufour, C.; Sauvaitre, I. Interactions between anthocyanins and aroma substances in a model system. Effect on the flavor of grape-derived beverages. *J. Agric. Food. Chem.* **2000**, *48*, 1784–1788. [CrossRef] [PubMed]
24. De Freitas, V. Revisiting Wine Polyphenols Chemistry in Relation to Their Sensory Characteristics. *Recent Adv. Polyphenol. Res.* **2019**, *6*, 263–284. [CrossRef]

horticulturae

MDPI

Article

Cultivar and Rootstock Effects on Growth, Yield and Nut Quality of Pistachio under Semi-Arid Conditions of South Mediterranean

Samiha Ouni [1], Luis Noguera-Artiaga [2], Angel Carbonell-Barrachina [2,*], Imen Ouerghui [3], Fadwa Jendoubi [4], Ali Rhouma [5] and Azza Chelli-Chaabouni [3]

1 University Campus, Faculty of Sciences of Tunis, University of Tunis El Manar, El Manar, Tunis 2092, Tunisia; ouni_samiha@yahoo.fr
2 Research Group "Food Quality and Safety", Centro de Investigación e Innovación Agroalimentaria y Agroambiental, Miguel Hernández University of Elche, Carretera de Beniel km 3.2, 03312 Orihuela, Spain; lnoguera@umh.es
3 Laboratory of Horticulture, National Institute of Agronomic Research of Tunisia (INRAT), University of Carthage, Hédi Karray Street, Tunis 1004, Tunisia; a.oimen@hotmail.com (I.O.); azza.chelli@gmail.com (A.C.-C.)
4 National Agronomic Institute of Tunisia, University of Carthage, Charles Nicolle Street, Tunis 1082, Tunisia; fadwa.jendoubi@hotmail.com
5 Laboratory "Integrated Olive Production in the Humid, Sub Humid and Upper Semi-Arid Regions", Olive Tree Institute, University of Sfax, Hédi Karray Street, Tunis 1004, Tunisia; ali.rhouma@prima-med.org
* Correspondence: angel.carbonell@umh.es

Abstract: Pistachio (*Pistacia vera* L.) cultivation under rainfed conditions of the South Mediterranean has not been studied in depth. Under changing climate conditions, knowledge of cultivar and rootstock performances allows for the optimal use of genetic resources to improve yield and nut quality. This study aimed to evaluate growth, yield, and fruit characteristics of 'Mateur' and 'Achouri' pistachio cultivars grafted on *P. vera* and *P. atlantica* rootstocks grown in northeastern Tunisia. The research was based on three years worth of monitoring of growth, yield and fruit quality. Results showed that annual bio-climatic features had significant effects on all studied parameters. The 'Mateur' cultivar showed higher vigor and fruit yield than 'Achouri'. Fruit production was higher in *P. atlantica* rootstock for 'Mateur' but similar in both studied rootstocks for 'Achouri'. Kernel fat contents (40.7–46.8%) and fatty acid profiles were low or not affected by cultivar, rootstock and their combination. Oleic acid (C18:1) was the major fatty acid (68.94–69.22%) in kernel oil. The unsaturated/saturated fatty acid and oleic/linoleic acid ratios indicated that nuts obtained had high quality. The lower performances of 'Achouri' cultivar in the studied conditions may be related to low acclimation ability of this cultivar. These conditions, however, seem to be suitable for both studied rootstocks.

Keywords: *Pistacia vera*; yields; fruit characteristics; pistachio oil; rainfed conditions

Citation: Ouni, S.; Noguera-Artiaga, L.; Carbonell-Barrachina, A.; Ouerghui, I.; Jendoubi, F.; Rhouma, A.; Chelli-Chaabouni, A. Cultivar and Rootstock Effects on Growth, Yield and Nut Quality of Pistachio under Semi-Arid Conditions of South Mediterranean. *Horticulturae* **2022**, *8*, 606. https://doi.org/10.3390/horticulturae8070606

Academic Editor: Cristina Moniz Oliveira

Received: 6 June 2022
Accepted: 1 July 2022
Published: 5 July 2022

1. Introduction

Pistachio nut species, belonging to the *Anacardiaceae* family, are native to arid zones of Central and West Asia and are spread throughout the Mediterranean Basin. Pistachio kernels are a nutritive nutrient which can be included in a healthy diet to prevent heart diseases [1]. Pistachios have high energetic content with sensorial characteristics highly appreciated by consumers and are a good source of minerals, vitamins, fatty acids (linoleic, linolenic and oleic acids) and phenolic compounds [2–8].

With climate evolution, the Mediterranean Basin is expected to become 20% warmer than the global average with less and more variable precipitations [9]. This leads to the loss of genetic diversity, a reduction in yield and high vulnerability of plants to pests and diseases. Agricultural productivity is more affected when crops are grown under rainfed conditions or with low water supply, often, of low quality. In this situation, a water saving strategy is adopted for more efficient use of the available water for agriculture. Due to its tolerance to drought and its low water requirements, the cultivation of pistachio (*Pistacia vera* L.) has aroused great interest among farmers [10]. In addition, farmers are very interested in its cultivation under rainfed conditions to enhance the value of arid and semi-arid areas and maximize profits [11,12].

The main pistachio-producer countries are Iran, USA, Turkey, Syria, Italy, Greece and Tunisia. Tunisia is the main producer country in the southern part of the Mediterranean Basin, with a total annual production in 2020 of 4662 t, covering an area of 30,196 ha [13]. Most of the plantations are spread in the southern and the central parts of the country with arid and semi-arid climate and no more than 300 mm of average annual rainfall. These regions are among the areas most affected by water stress and temperature rise in the last decades [14–16]. Moreover, the durability of this crop is threatened due to low genetic diversity, with the 'Mateur' cultivar and *P. vera* rootstock being the most used. Selecting adapted and efficient plant material and determining the areas most suitable for this crop are among proposed strategies for the sustainability of this crop in Tunisia [15].

The rootstock influences growth, fruit production and quality of many fruit species including pistachio [7,17–20]. The scion–rootstock interaction may lead to variable responses linked to genetic and physiologic factors and, also, to soil–rootstock and root–soil microbiom interactions [21]. Rootstock has particular importance in pistachio orchards as grafting is the main technique used for its propagation. In traditional orchards, native *Pistacia* species are used as rootstocks (*P. atlantica*, *P. terebinthus*, *P. vera*, *P. integerrima*, *P. khinjuk*, *P. palaestina*, etc.). Recently, selected vigorous and stress-tolerant rootstocks were used in modern orchards to improve crop adaptation and agronomic performance [22–25]. As the diversification of cultivars and rootstocks can be a good tool for enhancing crop adaptation and productivity, recent researches have paid attention to the physiological and agronomic evaluation of pistachio rootstocks and scion–rootstock interactions under different growing conditions and their effects on agro-food parameters [1,11,22,26–31]. It was found that the responses were not homogeneous, and were sometimes contradictory, depending on the different experimental conditions and the stem–rootstock combination. Considering these variable responses, supplemental knowledge of scion–rootstock response under specific orchard managements and environmental conditions may help for an optimal use of available genetic resources with low water supply. Despite the large extension of non-irrigated pistachio orchards in the Mediterranean, only a few studies have been performed in such conditions to assess crop yield and nut quality.

The aim of this work was to study the agronomic performance of four pistachio scion–rootstock combinations grown under rainfed conditions in northeastern Tunisia. The evaluation focused on the effects of the cultivar, the rootstock and their interaction on plant growth, yield and physical and biochemical nut characteristics.

2. Material and Methods

2.1. Experimental Orchard and Growing Conditions

This study was carried out under rainfed conditions in the northeast of Tunisia in the Mornag experimental research unit (36°38′ N–10°16′ E) of the National Institute of Agronomic Research of Tunisia (INRAT). This region belongs to the upper semi-arid bioclimatic stage characterized by mild winters and dry and hot summers. Monitoring was undertaken during three consecutive years, from 2014 to 2016. The total annual rainfall was 352.5 in 2014, 468.9 in 2015 and 482.1 mm in 2016. Almost 80% of the total precipitations occurred from October to March. The monthly average minima and maxima temperatures varied between 7 °C and 35 °C. The coldest and hottest months were, respectively, December to February and July and August. The soil of the orchard was highly calcareous (26.3% total $CaCO_3$) with a silty loam texture. The soil pH and the electrical conductivity (EC) were 7.4 and 1 mS/cm, respectively. The orchard management followed usual standards of pruning, fertilization and tillage. The cultural management practices were those conventionally used for this crop under these growing conditions. Those practices consisted of tillage (mechanical plowing and weed control), N-P-K fertilization spread over the soil, winter pruning and pest and disease control.

2.2. Plant Material

Experiments were performed on twenty-two-year-old trees of 'Mateur' and 'Achouri' pistachio cultivars grafted on *P. vera* and *P. atlantica* rootstocks. Grafting was realized at the second year after rootstock planting. The tree spacing in the orchard was 7 × 8 m. Male trees were placed in entire rows between each three female rows. For each treatment (scion–rootstock combination), at least five trees of comparable size and vigor were chosen for the monitoring.

2.3. Tree Vigor, Yield and Yield Efficiency

For the determination of scion-trunk-cross-sectional area (TCSA), the tree-trunk perimeter was measured at about 30 cm above the graft point when trees have reached the full dormancy (December). TCSA below the graft point was not determined because of the low position of grafting points. Fruit clusters were harvested manually at full maturity occurring in late August. The fruit is considered ripened when the hull becomes easily separated from the shell. For each tree, fruits were separated from clusters and weighed. Tree yield efficiency (g/cm^2) was calculated by dividing the fruit yield by the TCSA.

2.4. Physical Traits of Fruits

Fresh weights of fruit and nut as well as dry weights of nut and kernel were measured for 3 lots of 100 fruits per tree. Nut dehiscence rate was determined. The nut and kernel length (L), width (W) and thickness (T) were measured with a digital caliper for 25 fruits per tree following IPGRI descriptors for *Pistacia vera* L. [32]. Sphericity (ϕ) of fruits, nuts and kernels was calculated according to Equations (1) and (2), with Dg being the geometric diameter:

$$\phi = Dg/L * 100 \tag{1}$$

$$Dg = (L * W * T)^{1/3} \tag{2}$$

2.5. Kernel-Oil Content and Fatty-Acid Composition

The fat content of pistachio kernels, harvested in 2014, was extracted using hexane as solvent. A one-gram sample of grounded kernels was mixed with 3 mL of n-hexane. The mixture was sonicated in an ultrasonic bath (Model 3000512, JP Selecta S.A., Barcelona, Spain) with a constant frequency of 40 kHz at room temperature for 3 h before being centrifuged at 15,000 rpm for 10 min. The fat portion was recuperated by n-hexane evaporation using a stream of nitrogen. Fatty-acid methyl esters (FAMEs) were prepared according to the method described by Carbonell-Barrachina et al. [6] using identical chromatography set-up and conditions. Identification of FAMEs was carried out on 50 mg of extracting oil by comparison with authentic standards from Sigma-Aldrich. This analysis was run in triplicate, and results were expressed as percent of the total area.

Kernel oil extraction in 2015 and 2016 was performed using hexane as solvent and a VELP SCIENTIFICA soxhlet apparatus. Dried kernels were ground, weighed and introduced in soxhlet cartridges before immersion in hexane for 72 min at 130 °C, solvent washing for 20 min and finally solvent recover for 30 min. After complete evaporation of the solvent, the buckets containing the oil were weighed. The percent of fat content (FC) relative to dry matter was calculated.

2.6. Statistical Analyses

Data were subjected to one-way analysis of variance (ANOVA) using SPSS 20.0 software (Manugistics, Inc., Rockville, MD, USA). Duncan multiple-range test was used to compare the means. Differences were considered statistically significant at $p \leq 0.05$.

3. Results

3.1. Tree Vigor, Yield and Yield Efficiency

The TCSA showed no significant variation according to year and rootstock (Table 1). The cultivar and the scion–rootstock interaction, however, significantly affected tree vigor with higher trunk growth on *P. atlantica* and *P. vera* rootstocks for, respectively, 'Mateur' and 'Achouri' cultivars.

Table 1. Trunk-cross-sectional area (TCSA), average tree yield and yield efficiency (YE). Values (n = 5 to 8) followed by the same letter, within the same column, were not significantly different; NS: not significant; * Significant ($p \leq 0.05$).

	TCSA (cm^2)	Yield (kg tree^{-1})	YE (g cm^{-2})
ANOVA			
Year	NS	*	*
Cultivar	*	*	*
Rootstock	NS	NS	NS
Cultivar/Rootstock			
'Mateur'/*P. vera*	304.8 b ± 59.6	7.3 b ± 5.6	27.4 a ± 19.8
'Mateur'/*P. atlantica*	437.3 a ± 73.9	10.1 a ± 7.1	23.8 a ± 18.0
'Achouri'/*P. vera*	327.9 b ± 100.5	3 c ± 3.6	11.3 b ± 13.2
'Achouri'/*P. atlantica*	248.8 c ± 69.2	3.5 c ± 3.6	13.8 b ± 13.7

Average tree yield showed significant annual variation (Table 1). The highest yields were recorded in 2014 and 2016, while 2015 was marked by a significantly lower tree production, indicating 'off' year features. 'Mateur' was 2.6-fold more productive than 'Achouri' with a significantly higher yield for *P. atlantica* rootstock. No effect of rootstock was recorded for 'Achouri' with this parameter. Tree production (Figure 1) showed no significant effect of rootstock for both cultivars except a significantly higher yield of 'Mateur' on *P. atlantica* rootstock in 2016. Inter-annual variation of fruit yield differed with cultivar. 'Mateur' exhibited 'on' year behavior in 2014 and 2016. 'Achouri' only exhibited an 'on' year in 2014, while 2015 and 2016 were consecutively 'off' years. In the 'off' year (2015), the 'Mateur' production was, respectively, 55.8% and 71.2% lower than those of the 'on' years (2014 and 2016). 'Achouri' exhibited a yield reduction of 85.9% and 70.5%, respectively, in the 'off' years (2015 and 2016) in comparison with the 'on' year (2014). Yield efficiency of trees (Table 1) varied according to the year and the cultivar but was not influenced by the rootstock. 'Mateur' YE was about two-fold that of 'Achouri'.

3.2. Fruit Physical Traits

Data of Table 2 show significant effects of year and cultivar on fruit weight and nut dehiscence rate, but no effect of rootstock. Nut fresh and dry weights in 2014 and 2015 were significantly higher than in 2016, while nut dehiscence rate was significantly lower in 2014 than in 2015 and 2016 (data not shown). Fruit weight and almost all fruit size and shape parameters showed significant variation from year to year, with higher values in 2015 (Tables 2 and 3). Fruits and nuts of 'Mateur' were significantly heavier, longer, wider and thicker than those of 'Achouri'. The same was the case concerning the length and the width of 'Mateur' kernel. The average thickness of the 'Achouri' kernel, however, was significantly higher than that of 'Mateur'. A minor effect due to rootstock was recorded for these parameters. The thickness of fruits, the length and the width of nuts, and the thickness of kernels had more variation according to rootstock with significantly higher values on *P. vera* rootstock. The scion–rootstock combination data revealed few significant differences. The 'Mateur' cultivar produced shorter nuts and thinner kernels on *P. atlantica* rootstock. Fruit sphericity (ϕ) varied significantly from year to year with higher values in 2016.

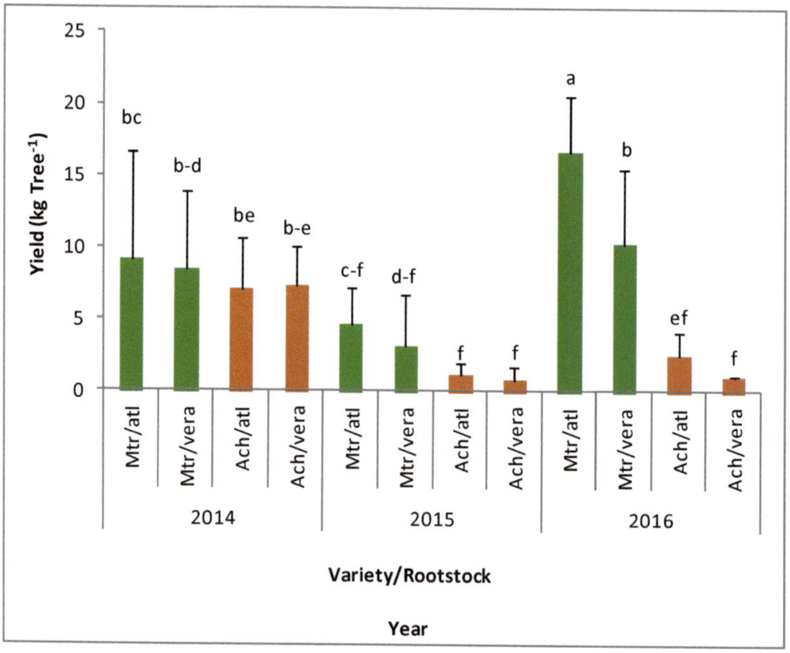

Figure 1. Effect of cultivar–rootstock interaction on annual fruit yield of pistachio. Mtr: Mateur; Ach: Achouri; atl: *P. atlantica*; vera: *P. vera*; Columns with the same letter are not significantly different (ANOVA, $p \leq 0.05$).

Table 2. Mean weights (g) of fresh fruit (FFW) and nut (NFW), dry weights of nut (NDW) and kernel (KDW), and nut dehiscence rate (DR) in percent (%).

Treatment	FFW	N FW	NDW	KDW	DR
ANOVA					
Year	*	*	*	NS	*
Cultivar	*	*	*	*	*
Rootstock	NS	NS	NS	NS	NS
Cultivar * Rootstock	*	*	*	*	*
Cultivar/Rootstock					
'Mateur'/*P. vera*	209.6 a ± 30.8	124.4 a ± 14.8	88.6 a ± 12.4	52.2 a ± 17.9	53.0 b ± 52.4
'Mateur'/*P. atlantica*	215.0 a ± 33.52	121.3 a ± 18.8	91.4 a ± 28.0	44.3 ab ± 10.1	63.5 ab ± 30.7
'Achouri'/*P. vera*	173.2 b ± 30.52	94.6 b ± 23.9	74.1 b ± 19.8	31.0 c ± 10.4	83.4 a ± 19.9
'Achouri'/*P. atlantica*	177.1 b ± 44.68	101.9 b ± 26.8	74.9 ab ± 17.4	39.0 bc ± 12.6	82.6 a ± 19.5

Values (mean of three lots of 100 fruits per tree) followed by the same letter, within the same column, are not significantly different ($p \leq 0.05$); Kernel weight values and corresponding statistical analysis were for 2015 and 2016 only; NS: not significant; * Significant.

3.3. Kernel Oil Content and Fatty-Acid Profile

The oil content of pistachio kernels ranged between 40.7% and 46.8% (Table 4). No significant variation was found according to year, cultivar, rootstock and scion–rootstock combination. However, a clear tendency of increasing kernel oil content on *P. atlantica* rootstock can be observed for both studied cultivars.

Table 3. Length, width, thickness (mm) and sphericity of fruit, nut and kernel. Values (mean of at least 25 fruits per tree) followed by the same letter, within the same column, are not significantly different ($p \leq 0.05$); NS: not significant; * Significant; ϕ: Sphericity.

	Fruit				Nut				Kernel			
Treatment	Length	Width	Thickness	ϕ	Length	Width	Thickness	ϕ	Length	Width	Thickness	ϕ
ANOVA												
Year	*	*	*	*	*	*	*	*	*	*	*	*
Cultivar	*	*	*	NS	*	*	*	NS	*	NS	*	NS
Rootstock	NS	NS	*	NS	*	*	NS	NS	NS	NS	*	NS
'Mateur'/*P. vera*	24.7 b ± 2.1	13.4 a ± 1.4	12.4 a ± 1.4	64.8 a	21.3 a ± 1.6	11.8 a ± 1.0	10.0 b ± 0.9	65.2 a	16.3 a ± 2.7	8.4 a ± 2.0	8.0 a ± 0.9	62.8 ab
'Mateur'/P. atlantica	24.8 a ± 2.1	13.6 a ± 1.5	12.3 a ± 1.2	64.6 a	20.3 b ± 1.4	11.8 a ± 0.9	10.4 a ± 0.9	66.6 a	16.3 a ± 1.4	8.5 a ± 3.6	7.8 b ± 0.6	62.8 ab
'Achouri'/*P. vera*	23.4 a ± 2.12	12.5 b ± 1.4	11.6 b ± 1.3	64.4 a	20.1 c ± 1.3	11.0 b ± 0.8	10.0 b ± 1.1	65.0 a	15.5 b ± 1.3	8.5 a ± 6.7	7.7 b ± 0.8	63.5 a
'Achouri'/*P. atlantica*	23.2 b ± 2.2	12.6 b ± 5.1	11.6 b ± 1.2	64.7 a	20.1 c ± 1.5	10.9 c ± 0.9	10.5 b ± 2.1	65.2 a	15.8 b ± 1.7	8.0 a ± 1.0	7.7 b ± 0.7	62.5 b

Table 4. Fat content and fatty-acid profile of pistachio kernels.

Treatment	Fat Content (%)	C16:0	C16:1	C18:0	C18:1	C18:2	C18:3	C20:0	C20:1	O/L	UFAs/SFAs
ANOVA											
Year	NS	*	*	*	*	NS	NS	NS	NS	NS	NS
Cultivar	NS	NS	NS	NS	NS	NS	NS	NS	NS	NS	NS
Rootstock	NS	NS	NS	NS	NS	NS	NS	NS	NS	NS	NS
'Mateur'/*P. vera*	43.2 a ± 4.7	9.88 a	0.78 a	1.83 b	68.94 a	18.03 a	0.27 a	0.16 ab	0.38 a	4.4 a	7.8 a
'Mateur'/*P. atlantica*	45.7 a ± 7.5	9.98 a	0.79 a	2.12 a	70.22 a	16.70 a	0.29 a	0.17 a	0.38 a	4.5 a	7.5 a
'Achouri'/*P. vera*	40.7 a ± 4.5	9.99 a	0.79 a	2.01 ab	69.41 a	17.28 a	0.26 a	0.14 b	0.35 a	4.4 a	7.6 a
'Achouri'/*P. atlantica*	46.8 a ± 4.8	9.81 a	0.74 a	2.04 ab	69.26 a	17.64 a	0.28 a	0.15 ab	0.37 a	3.8 a	7.8 a

Fat content values, Oleic/Linoleic acid ratio (O/L) and Unsaturated/Saturated fatty-acid ratio (UFAs/SFAs) for 2015 and 2016 only; Fatty acids: C16:0: Palmitic acid; C16:1: Palmitoleic acid; C18:0: Stearic acid; C18:1: Oleic acid; C18:2: Linoleic acid; C18:3: Linolenic acid; C20:0: Arachidic acid; and C20:1: Gadoleic acid; Values followed by the same letter, within the same column, are not significantly different ($p < 0.05$); NS: not significant ($p < 0.05$); * Significant ($p \leq 0.05$).

Eight fatty acids were found in pistachio kernels (Table 4): palmitic acid (C16:0), palmitoleic acid (C16:1), stearic acid (C18:0), oleic acid (C18:1), linoleic acid (C18:2), linoleic acid (C18:3), arachidic acid (C20:0) and gadoleic acid (C20:1). Variations on the composition of fatty acids were observed according to the year, especially palmitic, palmitoleic, stearic and oleic acids. Linolenic, arachidic and gadoleic acids were more stable and in minor proportions (0.14–0.38%). In all samples, unsaturated (UFAs) and saturated fatty acids (SFAs) showed little variation and ranged between 88.1% and 88.7% and 11.5% and 12.3%, respectively. The UFA–SFA ratios were between 7.5 and 7.8. The mono-saturated oleic acid was the major fatty acid with little variation (67.62% to 70.41%). Linoleic acid content varied from 16.7% to 18.7%. The oleic/linoleic acid ratios were between 3.8 and 4.5. Obtained results showed no effect of cultivar or rootstock on kernel fatty acid profile for all the experimental periods. The scion–rootstock combination, however, influenced the stearic acid profile for the 'Mateur' cultivar with significantly higher amounts on *P. atlantica* rootstock.

4. Discussion

4.1. Tree Vigor, Yield and Yield Efficiency

The scion–rootstock interaction has been demonstrated to have a significant impact on growth and yield of pistachio [26,31,33,34]. The results of this study clearly established this effect. 'Mateur' with higher vigor (TCSA) exhibited higher yield performance than 'Achouri'. The rootstock had no significant effect on TCSA, yield and yield efficiency.

Data of average tree yield of 'Mateur' cultivar in this study were consistent with those claimed by Oukabli [35] in rainy conditions (9.7 kg tree^{-1}). In the drought conditions of Sfax (southeast Tunisia), the average yield of this cultivar was 2.5 kg/tree [36]. The lower performance of 'Achouri' suggested low acclimation capacity to the tested bio-climatic conditions. Lacasta et al. [22,26] reported higher yield production of 'Achouri' compared to 'Mateur' under the rainfed conditions of central Spain. Rootstock had a significant effect on 'Mateur' fruit yield with higher performance for *P. atlantica* rootstock. These results were consistent with those of Carbonell-Barrachina et al. [6] on the Kerman variety that produced significantly higher yield on *P. atlantica* than on *P. integerrima* and *P. terebinthus* rootstocks. Similarly, Lacasta et al. [22] reported higher productivity of eight cultivars including 'Achouri' (8.4 kg/tree) on *P. atlantica* rootstock compared to *P. terebinthus*, *P. vera* and *P. integerrima*. This was not the case in our study for the 'Achouri' cultivar with any rootstock effect noted. The lower tree growth and productivity of the 'Achouri' cultivar compared to 'Mateur' may be explained by the lesser adaptive capacity of this cultivar to Mornag rainfed conditions.

An alternate bearing is a common trait of pistachio [37,38]. This phenomenon is accentuated by warm climate and rainfed conditions. Its intensity may vary according to the cultivar. Yield variation between the 'off' and the 'on' year can reach three- to five-fold for the Kerman pistachio variety [6]. In the case of this study, this variation did not reach two-fold. Before building up results from a longer monitoring period, these results cannot give exhaustive information on the alternate bearing intensity of studied cultivars.

4.2. Physical Traits of Fruits

The physical characteristics of fruits can be influenced by cultivar, rootstock and bioclimatic features [39–42]. Data of 'Mateur' dry weight in shell nuts in this study are comparable to those obtained by Vargas et al. [43] in the climatic conditions of the northern Mediterranean (IRTA-Mas Bové—Spain). Those of 'Achouri' were, however, lower (74.5 g) than those obtained by these authors (100 g).

Shell splitting is an attractive trait of pistachio nuts that determines the choice of variety in the market. Its occurrence depends on the growth of kernel and nutshell [44]. Factors that enhance kernel size concerning shell size, such as irrigation management and crop load, are thought to promote shell splitting [22]. Climatic conditions in rainfed orchards could have a relevant impact on the seasonal growth of trees and consequently fruit quality. This may, partly, explain the annual variation of DR in this study, which was lower in 2014 than in 2015 and 2016. Shell splitting is closely linked to the cultivar [39]. In this study, significant differences of DR were shown between 'Mateur' and 'Achouri' cultivars. The nut split rate was significantly higher for 'Achouri' than for 'Mateur'. This was in accordance with Vargas et al. [35] in terms of variation, but the DR values in our experiments were higher (82.9% vs. 66% for 'Achouri' and 58.5% vs. 43% for 'Mateur'). It was reported that the DR of 'Achouri' nuts ranged between 55 and 90% [43,45]. For both studied cultivars, no effect of rootstock was observed on fruit weight and DR. Similarly, the 'Bianca' pistachio cultivar showed no variation of DR with rootstock [33]. Other pistachio cultivars, however, have been reported to exhibit significant differences in nut splitting rate according to the rootstock [41,46].

Fruit size and shape are important parameters that are considered in marketing and in post-harvest processing, such as design and setting of equipment [47]. Fruit sizes of 'Mateur' and 'Achouri' in this study were comparable to those of Vargas et al. [43]. Sphericity of fruits, nuts and kernels of both studied cultivars varied from 60.2% to 66.7%, indicating an elongated shape. The most ovoid-shaped Iranian pistachios (Akbari, Badami, Kalle-Ghuchi, Momtaz and Ohadi) were reported to range between 68.2% and 79.8% [47]. Overall, the year of harvest and, to a lesser degree, the cultivar had the most important effects on fruit size and shape of pistachio in this study. The influence of rootstock on these parameters was lower despite a tendency of 'Mateur' and 'Achouri' cultivars to produce smaller fruits on *P. atlantica* rootstock. The 'Achouri' cultivar produced significantly lower spherical kernels on *P. atlantica* rootstock.

4.3. Fat Content and Fatty-Acid Profile

Kernel fat content of pistachio samples varied from 45.25% in 2015 to 42.92% in 2016 (Table 4). No significant variation was recorded between the two studied cultivars and rootstocks as well as their interactions. Fat content values recorded here are similar to those reported by Carbonell-Barrachina et al. [6] for the Kerman pistachio cultivar but lower than those obtained for other pistachio cultivars [4,5]. The fatty-acid composition of kernel samples of studied pistachio cultivars varied significantly over years, but not according to cultivar, rootstock or cultivar–rootstock interaction in our experimental conditions. Significant variation in fatty-acid profile with rootstock was found for immature pistachio kernels [48]. Noguera-Artiaga et al. [30] found no significant variation with rootstock of these components except for the polyunsaturated α-Linolenic acid (C18:3). The content of oleic acid (67.62–70.41%), the major mono-unsaturated fatty acid in pistachio kernel samples, had similar percentages as previously found in Italian (70.1–71.5%), Turkish (53.16–72.63%) and Iranian (51.8–71.23%) kernels [4,5,49]. Similarly, data of linoleic acid contents (16.7–18.7%) were in agreement with those reported for Iranian (17.36–35.16%) and Turkish (16.58 and 35.40%) varieties [48,49]. The comparison of fatty-acid composition of 'Mateur' cultivar with that reported by Chahed et al. [3] in the southern area of Tunisia (Sfax) revealed higher palmitic, palmitoleic, stearic and oleic acids in Sfax than in the Mornag (northern) area. Conversely, higher and equal rates of linoleic and linolenic acids were found in Mornag compared to Sfax (7.6% and 0.2% for linoleic and linolenic acids, respectively). Similar geographical variations were reported by Chahed et al. [3] in oil composition of pistachio kernels originating in 'Mateur' and Sfax areas. This may be attributed to the effect of plant water status on pistachio nut quality [6,12]. The quality index based on the oleic/linoleic (O/L) acid ratio is commonly used for fat-quality assessment to predict chemical stability and shelf life of the oil [1]. The higher this ratio, the greater the stability of the oil. None of the studied parameters here had an effect on this index. Recorded values (3.8–4.5) were comparable to Greek and Italian cultivars (3.9–5.9) but higher than those reported by Tsantili et al. [4] and Esteki et al. [1] for Iranian pistachio cultivars (1.9–3.7). The total rates of unsaturated (UFAs) and saturated fatty acids (SFAs) as well as the UFA/SFA ratio results were in agreement with other pistachio research findings [4,5,48]. The nut pistachio quality of these cultivar–rootstock combinations may contribute to compensating for the reduced financial impact of relatively low-productivity orchards under these rainfed conditions.

5. Conclusions

This work showed high annual variation for the studied parameter for 'Mateur' and 'Achouri' pistachio cultivars grown under rainfed conditions of the Mornag semi-arid climate. Seasonal climatic features were shown to play a major role in plant growth, yield and quality in these growing conditions. The low productivity of 'Achouri' compared to 'Mateur' seems to be linked to less adaptation capacity. All studied parameters appeared to be dependent on cultivar–rootstock combination except for the fatty-acid profile of kernel oil. Technological traits of kernel oils revealed a high quality product that should be valorized to promote higher farmer incomes in this region. Obtained results support the alternative use of *P. atlantica* rootstock in Tunisian pistachio orchards to increase biodiversity and crop adaptation. Pistachio is a great alternative for farmers in the Mediterranean Basin given its low environmental impact and the appreciation of its fruits by consumers.

Author Contributions: Conceptualization, A.C.-C. and A.R.; methodology and validation, A.C.-C. and A.C.-B.; investigation, S.O., L.N.-A., I.O. and F.J.; resources: A.C.-C. and A.C.-B.; formal analysis, S.O.; writing—original draft preparation, S.O.; writing—review and editing, A.C.-C. and L.N.-A.; visualization and supervision, A.C.-C., A.R. and A.C.-B.; funding acquisition, A.R., A.C.-C. and A.C.-B. All authors have read and agreed to the published version of the manuscript.

Funding: This work was funded by the Ministry of Higher Education and Scientific Research of Tunisia, Laboratory of Horticulture (LR16INRAT03), the Institution of Agricultural Research and Higher Education (IRESA) of Tunisia and the University Miguel Hernandez de Elche (UMH), Alicante, Spain.

Institutional Review Board Statement: Not applicable.

Informed Consent Statement: Not applicable.

Acknowledgments: The authors thank Mahdi Fendri, a researcher in the Olive Tree Institue of Tunisia for allowing the access to his laboratory and providing the equipments needed for kernel fat analysis. They are very grateful to late Nabil Soua and Hmida Ben Hamda for their technical assistances.

Conflicts of Interest: The authors declare no conflict of interest.

References

1. Esteki, M.; Ahmadi, P.; Vander Heyden, Y.; Simal-Gandara, J. Fatty Acids-Based Quality Index to Differentiate Worldwide Commercial Pistachio Cultivars. *Molecules* **2018**, *24*, 58. [CrossRef] [PubMed]
2. Küçüköner, E.; Yurt, B. Some Chemical Characteristics of *Pistacia vera* Varieties Produced in Turkey. *Eur. Food Res. Technol.* **2003**, *217*, 308–310. [CrossRef]
3. Chahed, T.; Bellila, A.; Dhifi, W.; Hamrouni, I.; M'hamdi, B.; Kchouk, M.E.; Marzouk, B. Pistachio (*Pistacia vera*) Seed Oil Composition: Geographic Situation and Variety Effects. *Grasas Aceites* **2008**, *59*, 51–56. [CrossRef]
4. Tsantili, E.; Takidelli, C.; Christopoulos, M.V.; Lambrinea, E.; Rouskas, D.; Roussos, P.A. Physical, Compositional and Sensory Differences in Nuts among Pistachio (*Pistachia vera* L.) Varieties. *Sci. Hortic.* **2010**, *125*, 562–568. [CrossRef]
5. Abdoshahi, A.; Mortazavi, S.A.; Shabani, A.A.; Elhamirad, A.H.; Taheri, M. Evaluation of protein, fat and fatty acids content of the pistachio (*Pistacia vera* L.) cultivars of Damghan, Iran. *Int. J. Nuts Relat. Sci.* **2011**, *2*, 15–24.
6. Carbonell-Barrachina, Á.A.; Memmi, H.; Noguera-Artiaga, L.; Gijón-López, M.d.C.; Ciapa, R.; Pérez-López, D. Quality Attributes of Pistachio Nuts as Affected by Rootstock and Deficit Irrigation. *J. Sci. Food Agric.* **2015**, *95*, 2866–2873. [CrossRef]
7. Ojeda-Amador, R.M.; Fregapane, G.; Salvador, M.D. Composition and Properties of Virgin Pistachio Oils and Their By-Products from Different Cultivars. *Food Chem.* **2018**, *240*, 123–130. [CrossRef]
8. Mahdavi, M.; Sharifnia, F.; Salimpour, F.; Esmaeili, A.; Larypoor, M. Evaluation of Nuts Morphology and Composition of Fatty Acids in Certain Iranian *Pistacia vera* L. (Anacardiaceae) Cultivars. *Acta Agric. Slov.* **2021**, *117*, 1–10. [CrossRef]
9. Lionello, P.; Scarascia, L. The Relation between Climate Change in the Mediterranean Region and Global Warming. *Reg. Environ. Chang.* **2018**, *18*, 1481–1493. [CrossRef]
10. Yarahmadi, J.; Amini, A. Determining Land Suitability for Pistachio Cultivation Development Based on Climate Variables to Adapt to Drought. *Theor. Appl. Climatol.* **2021**, *143*, 1631–1642. [CrossRef]
11. Moriana, A.; Memmi, H.; Centeno, A.; Martín-Palomo, M.J.; Corell, M.; Torrecillas, A.; Pérez-López, D. Influence of Rootstock on Pistachio (*Pistacia vera* L. Cv Kerman) Water Relations. *Agric. Water Manag.* **2018**, *202*, 263–270. [CrossRef]

12. Roxas, A.A.; Marino, G.; Avellone, G.; Caruso, T.; Marra, F.P. The Effect of Plant Water Status on the Chemical Composition of Pistachio Nuts (*Pistacia vera* L. Cultivar Bianca). *Agriculture* **2020**, *10*, 167. [CrossRef]
13. Ministry of Agriculture, Hydraulic Resources and Fishery of Tunisia; Tunis, Tunisia. Personal Communication, 2021.
14. Mouelhi, S.; Laatiri, L. Drought Conditions and Management Strategies in Tunisia. UNW-DPC_NDMP Country Report of Tunisia. Available online: https://www.droughtmanagement.info/literature/UNW-DPC_NDMP_Country_Report_Tunisia_2014.pdf (accessed on 17 March 2022).
15. Benmoussa, H.; Luedeling, E.; Ghrab, M.; Ben Mimoun, M. Severe Winter Chill Decline Impacts Tunisian Fruit and Nut Orchards. *Clim. Chang.* **2020**, *162*, 1249–1267. [CrossRef]
16. Soula, R.; Chebil, A.; McCan, L.; Majdoub, R. Water scarcity in the Mahdia region of Tunisia: Are improved water policies needed? *Groundw. Sustain. Dev.* **2021**, *12*, 100510. [CrossRef]
17. Castle, W.S. Rootstock as a Fruit Quality Factor in Citrus and Deciduous Tree Crops. *N. Z. J. Crop Hortic. Sci.* **1995**, *23*, 383–394. [CrossRef]
18. Boukid, F.; Abbattangelo, S.; Carini, E.; Marseglia, A.; Caligiani, A.; Vittadini, E. Geographical Origin Discrimination of Pistachio (*Pistacia vera* L.) through Combined Analysis of Physical and Chemical Features. *Eur. Food Res. Technol.* **2019**, *245*, 143–150. [CrossRef]
19. Balducci, F.; Capriotti, L.; Mazzoni, L.; Medori, I.; Albanesi, A.; Giovanni, B.; Giampieri, F.; Mezzetti, B.; Capocasa, F. The Rootstock Effects on Vigor, Production and Fruit Quality in Sweet Cherry (*Prunus avium* L.). *J. Berry Res.* **2019**, *9*, 249–265. [CrossRef]
20. Font i Forcada, C.; Reig, G.; Mestre, L.; Mignard, P.; Betrán, J.Á.; Moreno, M.Á. Scion × Rootstock Response on Production, Mineral Composition and Fruit Quality under Heavy-Calcareous Soil and Hot Climate. *Agronomy* **2020**, *10*, 1159. [CrossRef]
21. Warschefsky, E.J.; Klein, L.L.; Frank, M.H.; Chitwood, D.H.; Londo, J.P.; von Wettberg, E.J.B.; Miller, A.J. Rootstocks: Diversity, Domestication, and Impacts on Shoot Phenotypes. *Trends Plant Sci.* **2016**, *21*, 418–437. [CrossRef]
22. Ferguson, L.; Polito, V.; Kallsen, C. The pistachio tree; botany and physiology and factors that affect yield. In *Pistachio Production Manual*, 4th ed.; Ferguson, L., Beede, R.H., Freeman, M.W., Haviland, D.R., Holtz, B.A., Kallsen, C.E., Eds.; University of California Fruit & Nut Research & Information Center: Davis, CA, USA, 2005; pp. 31–39.
23. Aznarte-Mellado, C.; Sola-Campoy, P.J.; Robles, F.; Rejón, C.R.; de la Herrán, R.; Navajas-Pérez, R. Molecular Characterization of the Interspecific Hybrid Pistacia Vigros (*P. vera* L. × *P. atlantica* Desf.). *Sci. Hortic.* **2014**, *179*, 180–183. [CrossRef]
24. Akbari, M.; Hokmabadi, H.; Heydari, M.; Ghorbani, A. 'Arota': A New Interspecific Hybrid Pistachio Rootstock. *HortScience* **2020**, *55*, 965–966. [CrossRef]
25. Vahdati, K.; Sarikhani, S.; Arab, M.M.; Leslie, C.A.; Dandekar, A.M.; Aletà, N.; Bielsa, B.; Gradziel, T.M.; Montesinos, Á.; Rubio-Cabetas, M.J.; et al. Advanced in Rootstock Breeding of Nut Trees: Objectives and Strategies. *Plants* **2021**, *10*, 2234. [CrossRef] [PubMed]
26. Lacasta, C.; Vadillo, J.R.; Gómez, G.; Couceiro, J.F. El Pistachero I: Estudio de Variedades en Secano y en Manejo Ecológico. Comunicaciones: VI Congreso SEAE—Almería 2004 Agroecología: Referente para la Transición de los Sistemas Agrarios. 2004, pp. 1497–1512. Available online: https://digital.csic.es/bitstream/10261/16698/1/2004%20El%20pistachero%20I%20variedades.pdf (accessed on 11 February 2022).
27. Kallsen, C.E.; Parfitt, D.E. Comparisons of Scion/Rootstock Growth Rates among U.S. Pistachio Cultivars. *HortScience* **2011**, *46*, 197–200. [CrossRef]
28. Memmi, H.; Gijón, M.C.; Couceiro, J.F.; Pérez-López, D. Water Stress Thresholds for Regulated Deficit Irrigation in Pistachio Trees: Rootstock Influence and Effects on Yield Quality. *Agric. Water Manag.* **2016**, *164*, 58–72. [CrossRef]
29. Noguera-Artiaga, L.; Pérez-López, D.; Burgos-Hernández, A.; Wojdilo, A.; Carbonell-Barrachina, Á.A. Phenolic and triterpenoid composition and inhibition of α-amylase of pistachio kernels (*Pistacia vera* L.) as affected by rootstock and irrigation treatment. *Food Chem.* **2018**, *261*, 240–245. [CrossRef]
30. Noguera-Artiaga, L.; Sánchez-Bravo, P.; Pérez-López, D.; Szumny, A.; Calin-Sánchez, Á.; Burgos-Hernández, A.; Carbonell-Barrachina, Á.A. Volatile, Sensory and Functional Properties of HydroSOS Pistachios. *Foods* **2020**, *9*, 158. [CrossRef]
31. Surucu, A.; Acar, I.; Demirkiran, A.R.; Farooq, S.; Gokmen, V. Variations in Nutrient Uptake, Yield and Nut Quality of Different Pistachio Cultivars Grafted on *Pistacia Khinjuk* Rootstock. *Scient. Hortic.* **2020**, *260*, 108913. [CrossRef]
32. International Plant Genetic Resources Institute. *Descriptors for Pistachio (Pistacia vera L.)*; International Plant Genetic Resources Institute: Rome, Italy, 1997; pp. 27–28. Available online: https://www.cgiar.org/ipgri/ (accessed on 11 February 2022).
33. Caruso, T.; Barone, E.; Marra, F.P.; Sottile, F.; La Mantia, M.; De Pasquale, C. Effect of Rootstock on Growth, Yield and Fruit Characteristics in Cv. Bianca Pistachio (*Pistacia vera* L.) trees. In Proceedings of the XIII GREMPA Meeting on Pistachios and Almond, Mirandela, Portugal, 1–5 June 2003; Oliveira, M.M., Cordeiro, V., Eds.; Options Méditerranéennes, Série A, Séminaires méditerranéens, N° 63. CIHEAM-IAMZ: Saragoza, Spain, 2005; pp. 117–122.
34. Gijón, M.d.C.; Gimenez, C.; Perez-López, D.; Guerrero, J.; Couceiro, J.F.; Moriana, A. Rootstock Influences the Response of Pistachio (*Pistacia vera* L. Cv. Kerman) to Water Stress and Rehydration. *Sci. Hortic.* **2010**, *125*, 666–671. [CrossRef]
35. Oukabli, A. Phénologie et Caractérisation Pomologique de Quelques Variétés De Pistachier. *Rev. Res. Amélior. Prod. Milieu Aride* **1995**, *7*, 11–18.

36. Ghrab, M.; Gouta, H. "Mateur" and "Ohadi" Cultivars Characteristics over Fifteen Years of Production in Sfax Semi-Arid Region. In Proceedings of the XIII GREMPA Meeting on Pistachios and Almond, Mirandela, Portugal, 1–5 June 2003; Oliveira, M.M., Cordeiro, V., Eds.; Options Méditerranéennes, Série A, Séminaires méditerranéens, N° 63. CIHEAM-IAMZ: Saragoza, Spain, 2005; pp. 39–42.
37. Rosecrance, R.C.; Weinbaum, S.A.; Brown, P.H. Alternate bearing affects nitrogen, phosphorus, potassium and starch storage pools in mature pistachio trees. *Ann. Bot.* **1998**, *82*, 463–470. [CrossRef]
38. Spann, T.M.; Beede, R.H.; Dejong, T.M. Seasonal carbohydrate storage and mobilization in bearing and non-bearing pistachio (*Pistacia vera*) trees. *Tree Physiol.* **2008**, *28*, 207–213. [CrossRef] [PubMed]
39. Parfitt, D.; Kallsen, C.; Maranto, J. The orchard. Pistachio cultivars. In *Pistachio Production Manual*, 4th ed.; Ferguson, L., Beede, R.H., Freeman, M.W., Haviland, D.R., Holtz, B.A., Kallsen, C.E., Eds.; University of California Fruit & Nut Research & Information Center: Davis, CA, USA, 2005; pp. 62–66.
40. Seferoglu, S.; Seferoglu, H.G.; Tekintas, F.E.; Balta, F. Biochemical Composition Influenced by Different Locations in Uzun Pistachio Cv. (*Pistacia vera* L.) Grown in Turkey. *J. Food Compos. Anal.* **2006**, *19*, 461–465. [CrossRef]
41. Rahemi, M.; Tavallali, V. Effects of rootstock on Iranian pistachio scion cultivars. *Fruits* **2007**, *62*, 317–323. [CrossRef]
42. Zhang, L.; Laca, E.; Allan, C.J.; Mahvelati, N.M.; Ferguson, L. Nonlinear Model Selection for Fruit and Kernel Development as a Function of Heat in Pistachio. *HortScience* **2021**, *56*, 769–779. [CrossRef]
43. Vargas, F.J.; Romero, M.A.; Monastra, F.; Mendes Gaspar, A.; Rouskas, D. Sélection de Variétés de Pistachier Adaptées à l'aire Nord Méditerranéenne. In *Amélioration D'espèces à Fruits à Coque: Noyer, Amandier, Pistachier*; Germain, E., Ed.; Options Méditerranéennes: Série B. Etudes et Recherches; CIHEAM: Zaragoza, Spain, 1997; Volume 16, pp. 93–119.
44. Polito, V.S.; Pinney, K. Endocarp Dehiscence in Pistachio (*Pistacia vera* L.). *Int. J. Plant Sci.* **1999**, *160*, 827–835. [CrossRef]
45. Hadj-Hassan, A.I. Pistacia in West and Central Asia Cultivated Syrian pistachio varieties. In *Pistacia: Towards a Comprehensive Documentation of Distribution and Use of Its Genetic Diversity in Central and West Asia. North Africa and Mediterranean Europe*; Padulosi, S., Hadj-Hassan, A., Eds.; Project on Underutilized Mediterranean Species. Report of the IPGRI Workshop; IPGRI: Irbid, Jordan, 1998; pp. 1–12. Available online: https://www.yumpu.com/en/document/view/13816907/ (accessed on 10 May 2022).
46. Turker, S.; Ak, B.E. Effects of Different Rootstocks on Physical Traits of Siirt and Ohadi Pistachio Cultivars. *Options Méditerranéennes* **2010**, *94*, 245–250.
47. Razavi, S.M.A.; Emadzadeh, B.; Rafe, A.; Mohammad Amini, A. The Physical Properties of Pistachio Nut and Its Kernel as a Function of Moisture Content and Variety: Part I. Geometrical Properties. *J. Food Eng.* **2007**, *81*, 209–217. [CrossRef]
48. Acar, I.; Kafkas, S.; Kapchina-Toteva, V.; Ercisli, S. Effect of rootstock on fat content and fatty acid composition of immature pistachio kernels. *Comptes Rendus l'Académie Bulg. Sci.* **2017**, *70*, 1049–1060.
49. Agar, I.T.; Kafkas, S.; Kaska, N. Lipid characteristics of Turkish and Iranian pistachio kernels. *Acta Hortic.* **1998**, *470*, 378–386. [CrossRef]

horticulturae

MDPI

Article

Effect of Combination of $KMnO_4$ Oxidation and UV-C Radiation on Postharvest Quality of Refrigerated Pears cv. 'Ercolini'

Ramiro Alonso-Salinas [1], José Ramón Acosta-Motos [2,3,*], Antonio J. Pérez-López [1], Luis Noguera-Artiaga [4], Estrella Núñez-Delicado [2], Francisco Burló [4] and Santiago López-Miranda [1]

1 Department of Food Technology and Nutrition, UCAM Universidad Católica de Murcia, Avenida de los Jerónimos 135, Guadalupe, 30107 Murcia, Spain
2 Chair of Entrepreneurship in the Agri-Food Sector UCAM-Santander, UCAM Universidad Católica de Murcia, Avenida de los Jerónimos 135, Guadalupe, 30107 Murcia, Spain
3 Group of Fruit Tree Biotechnology, Department of Plant Breeding, CEBAS-CSIC, Campus Universitario de Espinardo, 30100 Murcia, Spain
4 Research Group "Food Quality and Safety", Centro de Investigación e Innovación Agroalimentaria y Agroambiental, Miguel Hernández University of Elche, Carretera de Beniel km 3.2, 03312 Orihuela, Spain
* Correspondence: jracosta@ucam.edu; Tel.: +34-968-278756

Abstract: This present study proposes an improvement for the postharvest preservation of the 'Ercolini' pear, a fruit that is little tested in the field, using a combination of ethylene elimination methods. The techniques used were potassium permanganate filters in devices with ultraviolet radiation and constant air flow to favour the contact of ethylene with the oxidising agents. The analysis carried out included weight, diameter, firmness, soluble solids content, total acidity, maturity index, ascorbic acid concentration, total phenolic compounds, antioxidant capacity via the ORAC method and a descriptive sensory analysis using experts. In addition, the ethylene removal method was tested at two storage temperatures: 1 °C, near optimal temperature, and 8 °C, the standard temperature for transport and storage of fruit on a commercial scale. The results showed a marked improvement in the maintenance of postharvest physicochemical quality using the proposed combination of methods. The sensory analysis confirmed what was observed in the laboratory, with higher organoleptic quality values observed in pears treated with the complete system under study consisting of filter and machine, highlighting the greater presence of flavours and odours related to green fruit. Ultimately, this innovation could be highly relevant for the food industry.

Keywords: climacteric fruit; fruit storage; potassium permanganate; *Pyrus communis*; UV-C

Citation: Alonso-Salinas, R.; Acosta-Motos, J.R.; Pérez-López, A.J.; Noguera-Artiaga, L.; Núñez-Delicado, E.; Burló, F.; López-Miranda, S. Effect of Combination of $KMnO_4$ Oxidation and UV-C Radiation on Postharvest Quality of Refrigerated Pears cv. 'Ercolini'. *Horticulturae* **2022**, *8*, 1078. https://doi.org/10.3390/horticulturae8111078

Academic Editor: Hongliang Zhu

Received: 14 October 2022
Accepted: 12 November 2022
Published: 16 November 2022

1. Introduction

The presence of ethylene in preservation atmospheres has been shown to be detrimental to the quality and shelf life of fruit and vegetables. In the case of climacteric fruits, ethylene accelerates the ripening processes, a highly undesirable aspect for their optimal commercialisation, leading to waste in a world that is becoming increasingly populated and more demanding of high-quality food. Fruit ripening is a complex process that promotes both physical and physiological changes, leading to a progressive deterioration of the products. Postharvest ripening cannot be stopped but it can be slowed down [1–3].

According to the report on the "DOP Pera Ercolina de Jumilla" carried out by Jesús García Brunton in 2011, pears of the variety 'Ercolini' belong to the group of climacteric fruits. This variety is characterised for having a medium size, as compared to other varieties such as 'Bosc Kobak' [4,5], with white and juicy flesh. Its skin is green and turns yellow when it ripens, a process that takes place very quickly. For this reason, it is essential to store them correctly in atmospheres without ethylene. Its maximum storage time is 3 weeks. The annual production of the 'Ercolini' pear in the Region of Murcia is approximately

22,000 tonnes, 48% of the Spanish national production, and 24% of the European production of this variety, and is, therefore, a crop of national importance [5].

It is known that the exposure to ethylene produces undesirable effects on fruit. In stone fruits such as apricot or peach, other authors have shown that their conservation without ethylene elimination leads to loss of weight, firmness, total acidity (TA), organoleptic qualities, and an increase in soluble solid content (SSC) [6]. In tomato, Mansourbahmani and collaborators showed a similar effect when comparing the application of various ethylene removal methods; these authors concluded that ethylene removal treatments could be a useful tool for reducing spoilage and maintaining fruit quality [7]. In pears, Charoenchongsuk et al., showed a relationship between colour loss, chlorophyll degradation, and softening of Russet pear, due to high ethylene exposure [8]. This effect may be related to the production of reactive oxygen species (ROS) in the degradation processes associated with ripening. Therefore, it would be interesting to study the evolution of the antioxidant activity in this product.

According to Alonso-Salinas [9,10], Kim [11], and Wei [12], ethylene removal by potassium permanganate ($KMnO_4$) oxidation is the most interesting method in terms of cost-effectiveness. This oxidising agent is anchored in active adsorbent materials such as zeolites, activated carbon, carbon nanospheres, and silica gel, to keep $KMnO_4$ and ethylene in contact [13]. These metal-coated porous materials are often used as fillers or active ingredients to be added to packaging films or paper, or as carriers where other ethylene scavengers can be incorporated [14].

Photocatalysis is also a suitable technique for ethylene removal [11,15,16]. UV light has been extensively studied as part of an ethylene degradation system, which is mainly attributed to its photochemical reactivity. Ethylene photo-degradation starts with the radiation of UV-C light, which generates oxidizing agents [11,17]. Although it is true that the efficiency of this ethylene removal method is not the best [18–20], it is sufficiently versatile to be incorporated as a support to other methods, to improve their overall effectiveness.

However, the above options to remove ethylene and avoid its action on fruits are not the only methods used. For example, the most widely studied in recent years is the treatment with 1-methylcyclopropene (1-MCP), a chemically synthesised molecule formed by a small hydrocarbon very similar to ethylene that competes with this gas for its receptor binding points (Ad-ERS1a, Ad-ETR2 and Ad-ETR3) inhibiting the expression of several transcription factors associated with ethylene (Ad-ERF4, Ad-ERF6, Ad-ERF10 and Ad-ERF14). This largely prevents the ripening of climacteric fruits by avoiding the action of ethylene even if it is still in the preservation atmosphere [12,21]. Nevertheless, not a few authors have doubts about its effectiveness compared to other methods such as $KMnO_4$ or palladium [7,10,12,22]. Another method of ethylene removal briefly studied is the application of palladium as an ethylene oxidising agent, acting in a similar way to $KMnO_4$. Nevertheless, although Smith et al. [23] indicate that it is more efficient than $KMnO_4$, its industrial application is complicated, since the cost of this metal is very high in the current market.

The aim of this study was to determine the effect of a novel combined ethylene removal method ($KMnO_4$ and UV-C radiation) on the postharvest quality and sensory analysis of pear cv 'Ercolini' preserved at two refrigeration treatments (1 °C and 8 °C).

2. Materials and Methods

2.1. Plant Material

Forty kilograms of 'Ercolini' pears (*Pyrus communis* L.) were supplied by "Cooperativa Hortofrutícola Campos de Jumilla" (Jumilla, Murcia, Spain). This variety has D.O.P. certification in Jumilla. The pears were harvested in the traditional way and preserved at 1 ± 1 °C for a day until laboratory transport for subsequent analysis. On the day of harvesting (26 July 2022), the supplying company classified the produce by calliper, and on the same day, the harvest index analyses were carried out to check the homogeneity of the pears. All the harvest index analyses were performed on 15 pears randomly selected from those supplied by the company and in the same way as the subsequent studies. The harvest indexes are shown in Table 1. Then, the pears were cooled and transported the following day (27 July 2022) to the laboratory to start the study.

Table 1. Harvest indexes. The means ± standard error of the means (SEM) are shown. n = 15.

Parameters	Weight (g)	Calliper (mm)	Firmness (N)	Soluble Solid Content (SSC) (%)	Total Acidity TA (%)	Colour
Data	116 ± 10	54.3 ± 2.5	52.6 ± 4.5	11.8 ± 0.9	0.35 ± 0.06	a *: −10.9 ± 2.8 b *: 40.3 ± 4.8 L *: 70.1 ± 3.6
Method	Navigator Balance, Ohaus Europe Gmbh (Nänikon, Switzerland).	Mitutoyo 530-122, Mitutoyo Spain (Guipúzcoa, Spain).	CT3 texturometer, AMETEK Brookfield (Middleboro, MA, USA).	Pocket Brix-Acidity meter, Atago (Tokyo, Japan).	Pocket Brix-Acidity meter, Atago (Tokyo, Japan).	Colourpin II, Natural Color System (Stockholm, Sweden).

2.2. Experimental Design

A total of 340 pears (40 kg) were randomly distributed into six 150 L (volume) conservation chambers (CCs) (Eurofred Cool Head RCG200, Eurofred S.A., Barcelona, Catalonia, Spain) for ethylene removal and temperature treatments.

According to Alonso-Salinas [9], the filters used were composed of $KMnO_4$ anchored to the active centre of zeolite, which allowed for a better interaction of this oxidizing substance with ethylene. The composition of the filters in terms of granulometry and other adsorbent substances was patented in Spain by the company "Nuevas Tecnologías Agroalimentarias KEEPCOOL" (Molina de Segura, Spain), patent No. 2548787 (2016). The adsorbing material was covered by a semi-permeable paper, which enables the entry of ethylene-rich air and the output of air clean of this phytohormone. Conversely, this kind of paper prevented the intrusion of water or other particles that could interfere with the process. Ethylene filters were installed inside an M-CAM 50 device (KEEPCCOL, Molina de Segura, Spain), which is an air-flow-forcing machine, to ensure that all the air in the CC passes through the filter. The volume of air moved inside the system is 750 L/min, which means that all the ethylene inside the CC is removed in 12 s, since the chamber has a capacity of 150 L.

In addition, this system incorporates a photocatalytic ultraviolet light system UV-C (TUV 254 nm, Philips, Amsterdam, Netherland) to aid the $KMnO_4$ filters in the removal of ethylene. The ultraviolet light is focused on the air coming out of the filters, not on the fruit. Throughout the article, the machine, filter and UV-C light combination will be referred to as the filter-device (FD).

According to Yildirim [24], UV-C exposure may have a negative effect on food quality. To avoid this possible adverse effect, the light beam was focused on the ethylene and not on the fruit since the device is completely closed, except for two air inlet and outlet openings, the UV-C radiation does not leave the system. To clarify the operation of this process, a diagram of the system is shown in Figure 1.

The combination of KMnO4 and UV-C radiation was chosen due to $KMnO_4$ being more effective than 1-MCP according to the literature [7,10,12,22] and easier to implement in the food industry than palladium due to its low cost. In addition, UV light was added because of its easiness of application and its support to $KMnO_4$.

Two treatments at 1 °C and 8 °C were set-up. These temperatures were selected because they are the standard storage temperatures utilised by fruit distribution companies.

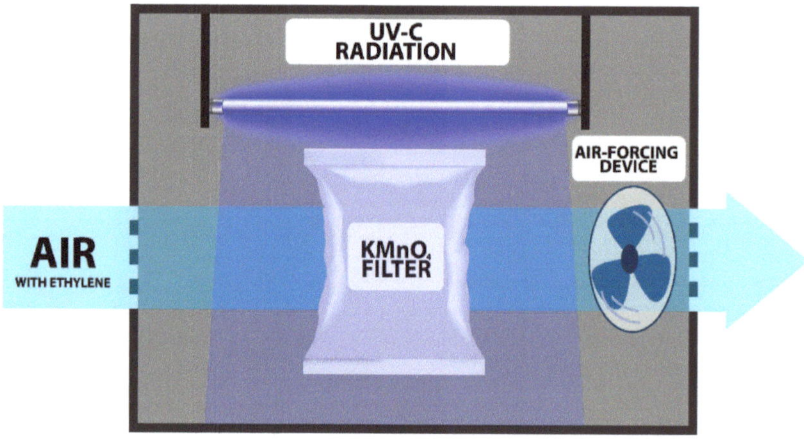

Figure 1. Ethylene scavenger diagram. Own source.

In terms of ethylene removal, preservation temperature and relative humidity the treatments were classified as follows (Table 2):

Table 2. Classification of the treatments applied according to storage temperature (°C), relative humidity (%) and the presence or absence of ethylene scavengers.

Treatments	**1 °C-C**	**1 °C-F**	**1 °C-FD**	**8 °C-C**	**8 °C-F**	**8 °C-FD**
Temperature	1 °C	1 °C	1 °C	8 °C	8 °C	8 °C
Relative humidity	90%	90%	90%	90%	90%	90%
Ethylene scavenger	None	Filter	Filter + Device	None	Filter	Filter + Device

2.3. Physicochemical Variables

All physicochemical analyses were carried out in triplicate on each pear, on five pears per treatment and per day (n = 5) throughout the entire storage period on the following days: 0, 7, 14, 21 and 28 (from 27 July 2022 to 24 August 2022). The shelf life (28 days) was established according to the optimum commercial life of the 'Ercolini' pear suggested by the supplying company. Similar storage times were also observed in other studies [3,25–27].

The ethylene (C_2H_4) concentration was measured using a Gas Analyzer (Felix Three F-950, Felix Instruments, Camas, WA, USA) and expressed as nmol kg^{-1} h^{-1}. The measuring flow rate of the Gas Analyser was 1 mL s^{-1} and 5 measurements were carried out for each day of analysis and treatment. A sealed access to the CCs was opened so that a sonde could be inserted to measure the ethylene concentration without disturbing the internal atmosphere of the chambers. The resolution of the Gas Analyzer was 0.1 ppm and the lower limit of detection is 0.15 ppm. Since at the time of arrival of the pears in the laboratory the ethylene concentration inside the CCs was 0, in order to observe possible differences, ethylene measurements on day 0 were made 6 h after the start of the study.

The weight was measured using a precision balance (Navigator Balance, Ohaus Europe Gmbh Nänikon, Switzerland), expressed in grams. The calliper of the pears was measured with a vernier calliper Mitutoyo 530-122, Mitutoyo Spain (Guipúzcoa, Spain) and expressed in millimetres. The calliper was considered as the equatorial diameter of the pears.

The firmness of the pears was measured with a CT3 texturometer (AMETEK Brookfield, Middleboro, MA, USA) equipped with a cylindrical probe measuring 35 mm height and 6 mm in diameter, which penetrated into the fruit 10 mm at a speed of 0.5 mm s^{-1}. Pear firmness was considered as the maximum force (N) measured during probe penetration.

The soluble solid content (SSC), pH, and total acidity (TA) were measured on fruit samples using the method adapted from Zhang [28]. Twenty grams of pear (without differentiating between skin and flesh) were taken and added to 20 mL of distilled water, then homogenised with a mixer (Ultra turrax T25, LabWare Wilmington, DE, USA) for 30 s. The homogenate was centrifuged at 3600× *g* for 10 min in a centrifuge at 4 °C (Eppendorf Centrifuge 5810, Hamburg, Germany), and the supernatant, mentioned in the rest of the manuscript as pear extract, was used to obtain SSC, pH, TA, ascorbic acid, TPC and antioxidant capacity according to ORAC.

The SSC of the pear extract was determined with a digital refractometer (Pocket Brix-Acidity meter, Atago Tokyo, Japan).) at 20 °C and expressed as a percentage (sugar equivalents in g 100 g^{-1}). The pH of the pear extract was determined with a pH-meter (Testo 206-pH2, Testo, Barcelona, Spain).

The determination of TA of the pear extract was made according to [28] with a Pocket Brix-Acidity meter, Atago (Tokyo, Japan). The results were expressed as g L^{-1}.

The maturity index (MI) was determined by dividing SSC (%) by TA (%). The expression of this parameter is dimensionless.

The ascorbic acid analysis was adapted from the Nielsen [29] method. First, two different solutions were prepared:

- Acid solution: 30 g of metaphosphoric acid (Acrós Organics, Geel, Belgium) and 80 mL of acetic acid (Panreac, Castellar del Vallés, Barcelona, Spain) were added in a 1 L flask and levelled.
- Dichlorophenol solution: 250 mg 2,6-dichlorophenol indophenol (Scharlab S.L., Barcelona, Spain), 210 mg sodium hydrogen carbonate (Panreac, Castellar del Vallés, Barcelona, Spain) were weighed, dissolved and levelled in a 1 L flask.

Ascorbic acid reduces 2,6-dichlorophenol indophenol from purple to a colourless solution. Thus, a 2 mL aliquot of the pear extract was taken and 5 mL of the acid solution was added. Subsequently, the resulting mix was titrated with the 2,6-dichlorophenol indophenol solution under constant stirring until a change in colour to pink was observed. The amount of ascorbic acid was determined by the following formula (F = 0.1):

$$Ascorbic\ acid\ \left(\mathrm{mg\ L^{-1}}\right) = \frac{F * \mathrm{mL}\ used\ of\ 2,6 - DCF * 1000\ \mathrm{mL}\ of\ juice}{Sample\ (\mathrm{mL})} \quad (1)$$

F = titter of dye (0.1 = mg ascorbic acid equivalent to 1.0 mL indophenol standard solution). Ascorbic acid was expressed as milligrams per 100 mL of pear juice (mg 100 mL^{-1}).

The total phenolic content (TPC) of the pear was determined colorimetrically at 765 nm using the Folin–Ciocalteau reagent according to a modification of the Kidron [30] method. The Folin–Ciocalteau reaction was performed by mixing 100 μL of the pear extract, 150 μL of Folin–Ciocalteau reagent, 450 μL of 20% Na_2CO_3, and 2300 μL of distilled water. After 2 h of reaction in dark, the absorbance of the sample was measured against a blank with a spectrophotometer (Shimadzu model UV-1603, Japan). Three measurements were made per pear and 5 pears were analysed for each treatment and day. The calibration curve (y = 0.5206x + 0.0899; R^2 = 0.998) was made using gallic acid as the standard at the range of 25–250 μg mL^{-1}. TPC was expressed in grams of gallic acid equivalents per kilograms of fresh pear (g kg^{-1}).

The antioxidant capacity of the pear extract was measured with ORAC (Oxygen Radical Absorbance Capacity) method following the one described by [31]. It was carried out with a SpectraMax ID3 multidetector microplate reader, from Bio-Tek Instruments, Inc. (USA), using 96-well polystyrene microplates with black sides and clear bottoms. To each well were added 100 mL of fluorescein (from a solution of 1.32 mg fluorescein in 1 L of distilled water), 50 mL of phosphate buffer (1%) and 20 uL of the pear extract diluted 1-10. After 30 min incubation in the dark at 37 °C, 30 uL of 2,2′-azobis(2-methylpropionamidine) dihydrochloride was added and the reaction started and finished after 2 h. Fluorescence was read through the clear bottom every minute of the reaction, with an excitation wavelength of

485 nm and an emission filter of 528 nm. The plate reader was controlled by SoftMax Pro 7.1 software. All reaction mixtures were prepared in triplicate and at least three independent assays were performed for each pear. A total of 5 pears were analysed for each treatment per day. The results were expressed in μmol of Trolox equivalents per kilograms of fresh pear (μmol kg^{-1}). The net area under the curve (AUC) for each well was calculated by subtracting the AUC for the blank from its AUC.

2.4. Descriptive Sensory Analysis

A trained panel consisting of 10 highly trained panellists (aged 25 to 55 years; 6 female and 4 male) from the Food Quality and Safety research group (Universidad Miguel Hernández de Elche, UMH, Orihuela, Spain) conducted the descriptive sensory analysis. Each panellist had more than 1000 h of experience with fruits. The methodology used for the descriptive sensory analysis was that previously described by Noguera-Artiaga [32] and the lexicon used was developed according to Gittins [33]. The scale used ranged from 10 (extremely high intensity) to 0 (no intensity) with 0.5 increments. The samples were served in odour-free disposable plates, at room temperature (~22 °C), and were coded using 3-digit numbers. Mineral water and unsalted crackers were provided to panellists to clean their palates between samples. The analyses were run in triplicate (n = 3). The descriptive sensory analysis was carried out at the beginning (establishing a control at day 0) and at the end of the study comparing the pears after 28 days (24 August 2022) of storage of the 6 treatments described.

2.5. Statistical Analysis

The descriptive statistics (mean and standard error of the mean [SEM] and the different tests described below were performed using the StatGraphics Centurion XV software (StatPoint Technologies, Warrenton, VA, USA). The Shaphiro-Wilk test was performed to check the normality of the data. In addition, to check the homogeneity of variance, Bartlett's test was applied. The six treatments were compared according to 10 variables analysed using a two-way analysis of variance (ANOVA) on days 7, 14, 21 and 28 of the experiment. Pearson's correlation coefficient (r) was calculated to measure the linear relationship between pairs of variables in the correlation matrix (at the end of the experiment, 28 days). A principal component analysis (PCA), followed by a partial least squares discriminant analysis, was conducted to assign the principal components displaying eigenvalues greater than or equal to 1.0, which led to the identification of two principal components that explained 78% of the variation within the data set (at the end of the experiment, 28 days). The sensory analysis was analysed using an analysis of variance (One-way ANOVA), comparing day 0 with day 28 of the experiment. Finally, Tukey's Multiple Range Test was utilised to separate the means and detect significant differences between the treatments (p-value < 0.05).

3. Results and Discussion

3.1. Ethylene

Pear is a climacteric fruit, which means that the ripening process continues once harvested, and this is highly affected by the presence of ethylene. Climacteric fruits increase the production of ethylene during post-harvest ripening, with this gas being responsible for the coordination of the ripening process. Ethylene has an autocatalytic feedback effect, i.e., a higher presence in the storage environment implies a higher ethylene production in the fruit. Therefore, it is crucial to maintain low ethylene levels to ensure an adequate shelf-life and quality of pears [9]. Furthermore, according to Hu [34], pear has a high ethylene sensitivity between 0.03 and 0.1 μL L^{-1}, which indicates that the presence of ethylene exceeding this threshold can cause significant damage to the product.

Figure 2 shows the ethylene concentration throughout preservation of 'Ercolini' pears. This parameter increased steadily in all treatments. However, it was affected by the different storage temperatures and by the use of the $KMnO_4$ filters or the complete ethylene removal system (filter + UV).

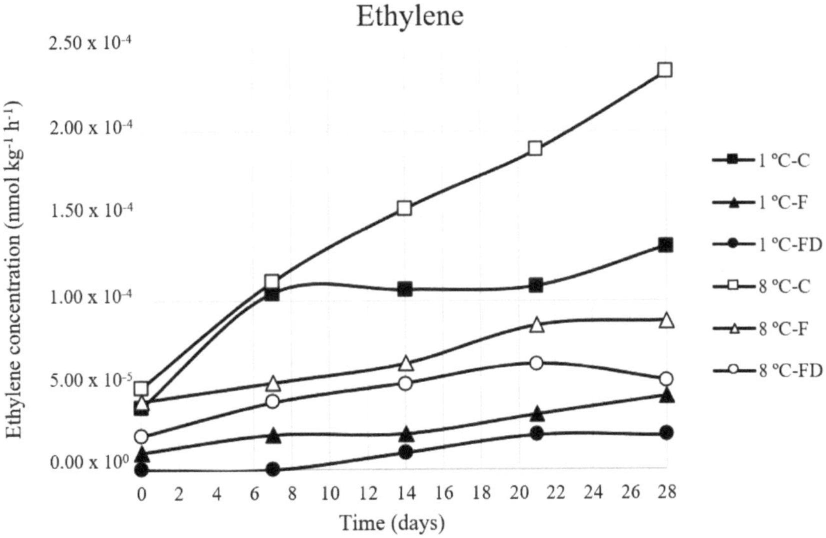

Figure 2. Ethylene concentration expressed as nmol kg^{-1} h^{-1} over the storage time in pears subjected to the different treatments: 1 °C-C (Control), 1 °C-F (Filter), 1 °C-FD (Filter + Device), 8 °C-C (Control), 8 °C-F (Filter) and 8 °C-FD (Filter + Device).

In the treatments kept at 1 °C, differences were observed from day 0 onwards. Treatment 1 °C-C (control) showed higher values than treatments 1 °C-F and 1 °C-FD. The 1 °C-C treatment reached its maximum value on day 28 with $1.32 \cdot 10^{-4}$ nmol kg^{-1} h^{-1}. The 1 °C-F treatment reached its maximum value of $4.37 \cdot 10^{-5}$ nmol kg^{-1} h^{-1} on day 28 as well, 3-fold lower than the control treatment. However, with the 1 °C-FD treatment, ethylene concentrations were obtained that barely exceeded $2 \cdot 10^{-5}$ nmol kg^{-1} h^{-1} throughout the entire storage time, between 5 and 7-fold lower than the control treatment.

In the treatments stored at a more stressful temperature (8 °C), higher levels of ethylene concentration were observed on average. The 8 °C-C treatment reached its maximum on the 28th day of analysis, $2.35 \cdot 10^{-4}$ nmol kg^{-1} h^{-1}. This value highly differed from the treatments in which different ethylene removal methods were used. On the one hand, for the treatment in which $KMnO_4$ filters were exclusively used (8 °C-F), the maximum ethylene concentration levels of $8.84 \cdot 10^{-5}$ nmol kg^{-1} h^{-1} were recorded on day 28, almost 3-fold lower than 8 °C-C. On the other hand, for the treatment in which the complete ethylene elimination system was used, the maximum concentration of this phytohormone was observed on day 28, with a value of $5.29 \cdot 10^{-5}$ nmol kg^{-1} h^{-1}, 2-fold lower than 8 °C-F and close to 10-fold lower than 8 °C-C.

From an overall point of view, the use of $KMnO_4$ filters (treatments 1 °C-F and 8 °C-F) achieved a reduction in ethylene concentration of 69.91% and 62.38% compared to the control treatments at 1 °C and 8 °C, respectively, on day 28. However, the use of the complete system (treatments 1 °C-FD and 8 °C-FD) reduced the concentration by 84.77% and 77.48% compared to the control treatments, at 1 °C and 8 °C, respectively, at the end of the trial.

These results are in agreement with the existing literature. According to Bower [1], it is certainly desirable to minimise ethylene concentrations around stored pears to reduce the incidence of scald and internal breakdown. Alonso-Salinas [9] reported a reduction in the ethylene concentration of 52% in peaches stored at 1 ± 1 °C, using the same ethylene elimination system as in the present study. Other researchers showed the effect of $KMnO_4$ and ultraviolet light as ethylene scavengers separately [19,35,36], nevertheless, their individual effectiveness is lower than that seen in this study as a combination.

Álvarez-Hernández [6] described the effect of applying $KMnO_4$-based ethylene scavengers on apricots stored at 2 °C, indicating a reduction of close to 100; these data are similar to those found in the present manuscript. However, Nguyen [4] reported that using 0.14% of 1-MCP, (a synthetic molecule that competes with ethylene for its receptors), achieved a reduction in the ethylene concentration in 'Bosc Kobak' pears between 5% and 60% compared to the control, depending on the starting day of treatment; similar data have been seen in other studies [3,37]. The use of this molecule, 1-MCP, is shown to be less effective in ethylene removal than those based on potassium permanganate.

3.2. Physico-Chemical Variables Analysed

Table 3 shows the evolution of the physico-chemical variables of pears from the beginning to the end of the conservation period for each treatment. In this table, the 1 °C-FD treatment had a better preservation performance and/or longer shelf life than the rest of the treatments at the end of the experiment (day 28), with a higher weight percentage (97.1%), higher calliper percentage (92.5%), except for 8 °C-FD treatment (95.4%), higher firmness (44.9 N), lower SSC (12.0%), higher TA (3.27 g L^{-1}), lower MI (39.4), higher ascorbic acid concentration (4.4 mg 100 mL^{-1}), higher TPC concentration (0.39 g kg^{-1}), and higher antioxidant capacity measured with ORAC (3.14 μmol kg^{-1}). All of these results are indicative of a less advanced ripening stage and a consequent prolongation of pear shelf life. The raw data are presented in the Supplementary Materials Tables S1–S3 which include the results of the parameters analysed on days 0, 7, 14, 21, and 28 separated into physical variables (Table S1), biochemical variables (Table S2), and bioactive variables (Table S3).

The order of treatments in terms of effectiveness is as follows: 1 °C-FD, 1 °C-F, 8 °C-FD and 1 °C-C. This suggests that the increase in temperature followed by a correct elimination of ethylene (8 °C-FD) slows down the loss of post-harvest quality of the pear, as compared to the control treatment at normal refrigeration temperature (1 °C-C).

In regard to the weight and calliper variables, Charoenchongsuk [8] reported a 20% loss in 'Gorham' pear firmness after 20 days of storage at 20 °C, and a delay in the softening of Russet pears using 1-MCP (1 μL L^{-1}) related to water and weight loss. This effect has also been observed by other authors in various fruit species such as melon [38], apricot [6,39], peach [9,40,41] or kiwifruit [42,43]. In terms of firmness, Nguyen [4] showed that the use of 1-MCP (0.14%) was able to maintain the firmness of the pears cv 'Bosc Kobak' for 14 days without a significant variation. However, Argenta [44] observed a decrease of about 40% of firmness in 1-MCP-treated treatments (0.42 mmol m^{-3}) after 30 days of storage. Escribano [45] reported that after 24 days of application of 0.6 μL L^{-1} of 1-MCP, the firmness of treated 'Bartlett' pears was reduced by about 80%. Until day 12 of the trial, the firmness was maintained. However, on day 14 and subsequent days, the firmness dropped sharply from 80 N on day 12 to 15 N on day 24 of storage.

In relation to the biochemical variables, both of those directly related to maturity and those related to bioactive compounds, Nguyen [4] observed significant differences through the application of 1-MCP (0.14%) to inactivate the action of ethylene on 'Bosc Kobak' pears; Chiriboga [46] showed that, by applying 1-MCP (300 nL L^{-1}), it is possible to preserve the ability to remove reactive oxygen species (ROS) in 'Conference' pears by blocking the autocatalytic feedback effect of ethylene. These researchers analysed electrolyte leakage (EL), total peroxidase (POX), superoxide dismutase (SOD) and catalase activity (CAT) to reach this conclusion. Using $KMnO_4$ as ethylene scavengers, Álvarez-Hernández [47] and Salamanca [48] observed similar effects, but on apricot and in 5 different varieties of tomato, respectively. Alvarez-Hernandez et al. [6] and Salamanca et al. [48] obtained alike results to those presented in this manuscript, supporting the findings reported here. However, Chiriboga et al. [46] obtained lower oxidation protection results than those observed in this study, corroborating that potassium permanganate-based ethylene scavengers are more effective in protecting the bioactive activity of pears than 1-MCP.

Table 3. Evolution from day 0 to day 28 of the physicochemical variables in pears subjected to the different treatments: 1 °C-C (Control), 1 °C-F (Filter), 1 °C-FD (Filter + Device), 8 °C-C (Control), 8 °C-F (Filter) and 8 °C-FD (Filter + Device). The variables measured were: weight expressed as average percentage compared to day 0 (being day 0 the 100% of average weight value of each treatment individually); calliper expressed as percentage compared to day 0 of each treatment; firmness expressed in Newtons; SSC expressed as percentage; pH; TA expressed as g L^{-1}, MI as the SSC (%)/TA (%) ratio; ascorbic acid content expressed as mg 100 mL^{-1}; total phenolic compounds expressed as $g_{galic\ acid}$ kg^{-1} and antioxidant capacity measured with the ORAC method expressed as $\mu mol_{Trollox.Eq.}$ kg^{-1}. The means ± standard error of the means (SEM) are shown. Different letters for each treatment represent statistically significant differences according to Tukey's test, n = 5 per treatment and day. The parameters have also been analysed by factors: ethylene (E), temperature (T) and the interaction of both ethylene × temperature (E × T).

Treatments	Weight (%)		Calliper (%)		Firmness (N)	
	Day 0	Day 28	Day 0	Day 28	Day 0	Day 28
1 °C-C	100	82.9 ± 3.5 b	100	82.5 ± 1.5 b	48.9 ± 1.7	34.4 ± 1.6 c
1 °C-F		91.7 ± 6.3 a		91.6 ± 1.5 a		43.4 ± 1.6 ab
1 °C-FD		97.1 ± 8.3 a		92.5 ± 0.7 a		44.9 ± 2.1 a
8 °C-C		89.9 ± 4.4 a		83.3 ± 2.0 b		26.1 ± 2.6 d
8 °C-F		91.1 ± 2.8 a		89.9 ± 1.7 a		33.0 ± 1.8 c
8 °C-FD		93.2 ± 1.6 a		95.4 ± 2.7 a		37.4 ± 2.1 bc
Ethylene (E)	-	**	-	***	-	***
Temperature (T)	-	n.s.	-	**	-	***
E × T	-	**	-	n.s.	-	n.s.

Treatments	SSC (%)		TA (g L^{-1})		MI	
	Day 0	Day 28	Day 0	Day 28	Day 0	Day 28
1 °C-C	12.5 ± 0.42	15.5 ± 0.65 ab	3.69 ± 0.47	1.76 ± 0.29 c	34.4 ± 2.44	95.9 ± 16.2 ab
1 °C-F		12.5 ± 0.29 c		2.93 ± 0.29 ab		43.7 ± 3.7 c
1 °C-FD		12.0 ± 0.41 c		3.27 ± 0.46 a		39.4 ± 6.3 c
8 °C-C		17.5 ± 0.65 a		1.51 ± 0.22 c		125.5 ± 22.0 a
8 °C-F		15.3 ± 1.10 b		1.93 ± 0.29 bc		84.1 ± 12.1 abc
8 °C-FD		15.0 ± 0.41 b		2.09 ± 0.29 bc		75.7 ± 10.1 bc
Ethylene (E)	-	***	-	***	-	***
Temperature (T)	-	***	-	***	-	***
E × T	-	n.s.	-	n.s.	-	n.s.

Treatments	Ascorbic Acid (mg 100mL^{-1})		TPC (g kg^{-1})		Antioxidant capacity (µmol kg^{-1})	
	Day 0	Day 28	Day 0	Day 28	Day 0	Day 28
1 °C-C	5.0 ± 0.47	2.9 ± 0.37 bcd	0.49 ± 0.03	0.26 ± 0.02 cd	4.17 ± 0.08	2.18 ± 0.14 c
1 °C-F		3.8 ± 0.28 ab		0.35 ± 0.04 ab		2.83 ± 0.06 ab
1 °C-FD		4.4 ± 0.21 a		0.39 ± 0.03 a		3.14 ± 0.22 a
8 °C-C		1.7 ± 0.22 d		0.21 ± 0.04 d		1.78 ± 0.14 d
8 °C-F		2.0 ± 0.39 cd		0.29 ± 0.05 bc		2.21 ± 0.05 c
8 °C-FD		3.1 ± 0.48 abc		0.36 ± 0.05 ab		2.56 ± 0.19 bc
Ethylene (E)	-	***	-	***	-	***
Temperature (T)	-	***	-	**	-	***
E × T	-	n.s.	-	n.s.	-	n.s.

Levels of statistical significance are: * $p < 0.05$, ** $p < 0.01$ and *** $p < 0.001$. n.s.: no significant differences.

However, Mansourbahmani [7] observed no significant differences between the application of 1-MCP (0.14%), $KMnO_4$ (20%) or 5% palladium as ethylene scavengers in AA, TPC, and antioxidant capacity (ORAC) during 35 days of tomato preservation. The data presented in this study contradict the findings of Mansourbahmani et al. [7] since we have seen differences in the application of $KMnO_4$ and UV-C radiation in these parameters.

3.3. Sensory Analysis

After the descriptive sensory analysis of pear samples, statistically significant differences were found in 18 of the 23 sensory descriptors studied (Table 4).

Table 4. Descriptive sensory analysis of pears. Number of panellists: 10. n = 3.

Sensory Descriptor	ANOVA	Day 0	1 °C (Day 28)			8 °C (Day 28)		
			C	F	FD	C	F	FD
COLOUR								
External	**	9.0 a	6.0 c	7.0 bc	8.5 b	2.0 e	3.0 de	4.0 d
Internal	***	8.0 a	6.0 b	6.0 b	7.5 a	2.0 c	2.0 c	5.0 b
Spotting	*	1.0 c	8.0 a	5.5 b	3.0 c	8.5 a	9.0 a	6.0 b
ODOUR								
Pear	***	7.0 a	4.0 c	5.5 b	7.0 a	1.5 d	2.0 d	2.0 d
Fruity (Green)	***	8.0 a	4.0 bc	5.0 b	8.0 a	1.0 c	1.0 c	3.0 c
Fruity (ripe)	***	4.0 d	8.0 bc	8.0 bc	5.0 d	9.0 a	9.0 a	7.0 b
Floral	**	3.0 a	1.0 b	1.0 b	2.5 a	1.0 b	1.0 b	2.5 a
Earthy	n.s.	1.5	3.0	3.0	1.0	3.5	4.0	3.0
FLAVOUR								
Pear	**	8.5 a	4.0 b	4.5 b	7.0 a	1.5 d	1.5 d	2.5 c
Fruity (Green)	***	8.5 a	5.0 b	6.0 b	7.0 a	0.5 d	1.0 d	3.0 c
Fruity (ripe)	***	2.0 d	4.0 c	4.0 c	2.5 d	9.5 a	9.5 a	7.0 b
Floral	*	3.0 a	1.5 b	2.0 b	3.0 a	1.0 b	1.0 b	2.0 b
Earthy	*	3.0 a	4.5 a	4.0 a	2.5 bc	1.0 b	2.0 b	2.0 b
Sweet	*	4.0 b	3.0 c	3.5 c	5.0 a	4.5 b	5.5 a	5.0 a
Sour	n.s.	1.0	1.0	1.0	1.5	2.0	2.5	2.0
Bitter	*	0.5 c	0.5 c	0.5 c	0.5 c	3.0 a	2.0 b	2.0 b
Astringent	n.s.	1.0	1.0	1.0	1.0	2.0	1.5	1.5
Aftertaste	**	4.0 a	2.0 b	2.5 b	4.5 a	1.0 b	1.5 b	2.5 b
TEXTURE								
Hardness	***	9.0 a	6.0 c	6.0 c	7.5 b	2.0 e	2.5 e	4.5 d
Crunchiness	***	9.0 a	6.0 c	5.5 c	7.5 b	1.0 d	1.5 d	3.0 d
Solubility	***	9.0 a	7.0 bc	6.5 bc	8.0 b	6.0 c	7.5 b	8.0 b
Residual particles	n.s.	3.0	5.0	4.5	2.5	3.5	4.0	3.5
Fibrousness	n.s.	1.0	3.0	3.0	1.0	3.0	3.0	3.0

Levels of statistical significance are: * $p < 0.05$, ** $p < 0.01$ and *** $p < 0.001$. n.s.: no significant differences. Values (mean of 3 replications) followed by the same letter, within the same sensory descriptor, were not significantly different ($p > 0.05$), according to Tukey's least significant difference test.

The colour of the fruits was affected by the ethylene treatments studied, both externally and internally. Samples from the 1 °C-FD treatment were the most preserved as compared with the initial colour values (control sample, day zero), while samples from the 8 °C-C treatment changed the most. Initially, the fruits had an intense green external colour, which gradually turned brown in the samples with less intensity. Similar changes were observed in the internal colour of the fruit, but in this case, the colour changed from a characteristic white to yellow-brown. Regarding staining, it was intense in the 8 °C-C, 8 °C-FD, 8 °C-F and 1 °C-C samples, while it remained practically null in the rest of the samples.

Regarding the odour, again, the 1 °C-FD sample obtained intensity values close to the control sample, demonstrating that this treatment is capable of very effectively preserving the characteristic odour of the product for a longer period. In the rest of the treatments, a change was observed in terms of the perception of the fruity odour, with the appearance of high intensities of fruitiness characteristic of ripe fruits. These perceptions were maintained when the same descriptors were analysed via the retro nasal route. Fruit aroma plays a well-established role in determining the final sensory quality, and is a strong determinant of consumer preference for a fruit. Thus, the longer the initial aroma of the fruit is preserved, the more market possibilities it has [49]. Volatile esters are a mayor group of volatile organic compounds contributing to the aroma of pears and, although ripening is inhibited during refrigeration, the release of volatile compounds still proceeds [50]. The results reported here indicate that the 1 °C-F treatment slowed the loss of these compounds.

If we analyse the results related to sweetness, sourness, astringency, and aftertaste, we again observed that the 1 °C-FD treatment maintained the intensities of the control fruit, while the 8 °C-C treatment had the most altered organoleptic properties.

Finally, when analysing the results obtained after the texture analysis, no differences were found in the content of residual particles and the fibrousness of the samples. However, differences appeared in hardness, crunchiness, and solubility in the mouth. With respect to hardness and crunchiness, the 1 °C-FD treatment presented intensities closest to the control, although slightly lower, followed by the 1 °C-F and 1 °C-C treatments. As for the solubility in the mouth, all the treatments were affected in their intensity with respect to the control sample.

Some authors have shown that the synthesis of esters may be related to the production of ethylene [50–52]. In the case of pears, this connection occurs through the expression of the PuAAT1 gene, which is responsible for the synthesis of these aromatic compounds [53]. The results obtained after the use of $KMnO_4$ and UV light, may indicate that as the fruit is producing ethylene naturally, the production of esters is also being activated. However, by eliminating the excess of ethylene in the chamber, the ripening of the fruit is stopped, but the synthesis of esters continues, which is why the conservation is prolonged, maintaining the typical flavour of the fruit.

These sensory results allow us to conclude that the 1 °C-FD treatment reduced the ripening of the pears for a longer period of time, preserving the odour, flavour, and texture of samples.

3.4. Correlation Matrix and Principal Component Analysis

In order to study the association between the variables studied, a correlation matrix was created (Table 5). The results can be separated into physical variables (weight, calliper, and firmness), biochemical variables (SSC, pH, TA, and MI) and bioactive variables (ascorbic acid, total phenolic compounds, and antioxidant capacity measured with ORAC).

Table 5. Pearson's correlation matrix (r) for analysed. Significant interactions are highlighted in bold. The parameter r represented in this table ranges from 1 to −1 depending on whether the correlations between parameters are positive or negative respectively. The variables analysed were: weight (W), calliper (C), firmness (F), soluble solid content (SSC), pH, total acidity (TA), maturity index (MI), ascorbic acid (AA), total phenolic compounds (TPC) and antioxidant capacity (ORAC). The data used are from the final day of the study (day 28).

	W	C	F	SSC	pH	TA	MI	AA	TPC
C	**0.5976** **	-	-	-	-	-	-	-	-
F	0.1449 n.s.	0.2914 n.s.	-	-	-	-	-	-	-
SSC	−0.1510 n.s.	**−0.3793** *	**−0.8344** ***	-	-	-	-	-	-
pH	−0.0620 n.s.	−0.2884 n.s.	**−0.7391** ***	**0.8209** ***	-	-	-	-	-
TA	0.2226 n.s.	0.2732 n.s.	**0.7369** ***	**−0.6806** **	**−0.6353** **	-	-	-	-
MI	−0.2499 n.s.	−0.3314 n.s.	**−0.8158** ***	**0.7661** ***	**0.7083** ***	**−0.8862** ***	-	-	-
AA	0.1693 n.s.	0.3262 n.s.	**0.7097** ***	**−0.6896** **	**−0.6577** **	**0.6204** **	**−0.6240** **	-	-
TPC	0.2170 n.s.	**0.5682** *	**0.7509** ***	**−0.6930** **	**−0.7238** ***	**0.6594** **	**−0.7129** ***	**0.6207** **	-
ORAC	0.1422 n.s.	**0.4062** *	**0.8681** ***	**−0.8010** ***	**−0.6596** **	**0.6980** **	**−0.7054** ***	**0.8058** ***	**0.7003** ***

Levels of statistical significance are: * $p < 0.05$, ** $p < 0.01$ and *** $p < 0.001$. n.s.: no significant differences.

Regarding the physical variables, a moderate correlation between weight and calliper was observed. In addition, a weak correlation was also observed between calliper and some of the biochemical and bioactive variables, such as SSC, ORAC and, TPC. Firmness stands out from the rest of the variables, as it maintained a strong relationship with all the internal variables, both biochemical and bioactive compounds.

As for the biochemical variables, a strong correlation was observed between them and with the compounds involved in antioxidant capacity. SSC had a strong correlation with pH, MI, and antioxidant capacity (ORAC), and a moderate correlation with TA, ascorbic acid (AA), and TPC. pH had a moderate correlation with TA, AA, and ORAC, and a strong correlation with MI and TPC. TA had a strong correlation with MI and a moderate correlation with all compounds related to antioxidant activity. All of these variables represent compounds involved in fruit biochemical processes that are more dependent on fruit maturity.

The bioactive compounds correlated amongst themselves. AA concentration showed a strong correlation with ORAC and a moderate correlation with TPC. ORAC was strongly correlated with both variables. This can be explained by the fact that the ORAC method is a generic way of determining antioxidant activity, which together with other compounds, depends on the concentration of AA and the action of TPC, and therefore its correlation with them was strong.

Tables 4 and 5 show a relationship between the data from both sets of analyses. Some examples are given below; the panellists of the descriptive sensory analysis concluded that those pears with higher flavour and fruity odour (green) were those that matched with the lowest maturity indexes (MI) in the treatments 1 °C-F and 1 °C-FD; the same happens with the treatments preserved at 8 °C. Also, panellists found that the highest hardness fits with those treatments that presented the highest firmness. This allows us to have two ways of confirming the efficacy of the applied treatments. Moreover, the sensory point of view (more related to commercial acceptance) is entirely linked to the physicochemical one in this manuscript. Therefore, this method of ethylene elimination can be very useful as a tool to be applied at industrial level. Álvarez-Hernández et al. [6], described a similar effect applying a $KMnO_4$-based ethylene scavenger on apricot.

In addition, to be able to indicate which set of variables explained the greatest variability in the experiment, and how the different treatments were separated, a principal component analysis (PCA) was carried out.

The purpose of the analysis is to first obtain a small number of linear combinations of the 8 variables studied (W, C, F, pH, MI, AA, TPC and ORAC) that explain the greatest variability in the data. In this case, 2 components were extracted, since these 2 components had eigenvalues greater than or equal to 1.0. These components are principal component 1 (PC1) which explains 61.09% of the variability of the experiment, and principal component 2 (PC2) which explains 17.41% of the variability of the experiment. Together they explained 78.51% of the variability in the original data (Table S4 and Figure S1). The second step is to indicate, for each extracted component, which variables had more weight or were the most important (variables with a higher absolute value). In PC1, the variables with the most weight, from high to low, were: F > ORAC > FC > MI > pH > AA. It can be concluded that PC1 contained all the internal quality variables that ultimately affect firmness according to the observed ethylene concentrations; this suggests that firmness, in this study, was the priority marker of the internal quality of the fruit, and was also notably important in the physical quality of the pears. Following the same criteria, in PC2 the variables with the most weight, from highest to lowest, were: W > C. It can be concluded that the PC2 included all variables related with water loss, and therefore those that were affected by the different temperatures applied (Table S5).

Although no references to other similar correlation studies have been found, the findings of this manuscript are indirectly corroborated by the existing literature. Many authors consider firmness [4,54,55] and antioxidant capacity [46,56–58] as the main markers of fruit quality, which supports what was observed in this study, as they are the two parameters with more weight in PC1 of the analysis by principal components.

As indicated above, the other objective is to be able to locate the treatments in a scatter diagram (Figure 3) or bigraphic (Figure S2). These figures are achieved through of the principal component table where for each treatment (5 per treatment for a total of 30 data), the scores obtained for each component are represented. In addition, the average score for each of the six treatments is added (Table S6a).

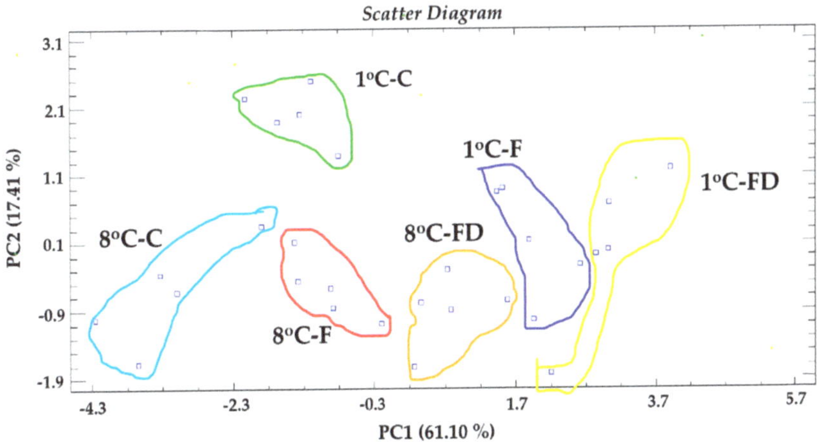

Figure 3. A principal component analysis applied to the different treatments (1 °C-C, 1 °C-F, 1 °C-FD, 8 °C-C, 8 °C-F and 8 °C-FD). Two principal components (PC1 and PC2) resulted in a model that explained 78.5% of the total variance.

The scatter plot shows that the treatments were well separated with the two PCs, but PC1 was the best, with a value of F = 75.72 ***, which allowed us to classify the treatments into four clusters (Table S6b): the first cluster is shaped by the 1 °C-FD and 1 °C-F treatments; the second cluster only includes the 8 °C-FD treatment; the third cluster is composed of the 8 °C-F and 1 °C-C treatments; and the fourth cluster only contained the 8 °C-C treatment (Table S6c). Regarding PC2, the treatments were not as well separated, with a value of F = 10.46 ***, which allowed us to classify the treatments into three clusters (Table S6d): the first cluster only formed by the 1 °C-C treatment; the second cluster included the 1 °C-F, 1 °C-FD, 8 °C-F and 8 °C-C treatments; and the third cluster only the 8 °C-FD treatment. These scores reconfirm the results described in the previous paragraph, which indicated that in PC1, the elimination of ethylene was essential in the preservation of the postharvest quality of the 'Ercolini' pear. Meanwhile, in PC2, a separation of the treatments was observed in relation to the temperature used for storage (Table S6e).

4. Conclusions

In this study, we carried out a complete evaluation of the effects of ethylene removal by $KMnO_4$, UV-C radiation and continuous air flow during the postharvest quality preservation, as well as the sensory analysis of 'Ercolini' pears at 1 °C and 8 °C. Based on the results described above, it can be concluded that the use of the combination of the aforementioned ethylene elimination methods maintained very low levels of this phytohormone, with values close to 0. As for the condition of the pears, those preserved using the complete ethylene elimination system (FD treatments) showed a higher physicochemical quality than the rest of the treatments, including increased weight, calliper, firmness, TA, ascorbic acid, TPC and antioxidant capacity retention and lower levels of SSC and MI, especially in the treatment with low temperature (1 °C-FD). In addition, that treatment had higher scores in the evaluation through sensory analysis. On the other hand, in the treatment with a more stressful preservation temperature, with the complete system of ethylene removal (8 °C-FD), the pears showed a state of maturity equal or lower than the control treatment at optimum temperature (1 °C-C). These results prove that the correct elimination of ethylene with the methods described here, delays the postharvest ripening of 'Ercolini' pears extending its shelf life.

Supplementary Materials: The following supporting information can be downloaded at: https://www.mdpi.com/article/10.3390/horticulturae8111078/s1, Figure S1: Sedimentation graph; Figure S2: Graphical representation of the principal components marking with lines for each variables and wit points for each score; Table S1: Evolution during storage time of the physical variables in pears subjected to the different treatments: 1 °C-C (Control), 1 °C-F (Filter), 1 °C-FD (Filter + Device), 8 °C-C (Control), 8 °C-F (Filter) and 8 °C-FD (Filter + Device); Table S2: Evolution during storage time of the biochemical variables in pears subjected to different treatments: 1 °C-C (Control), 1 °C-F (Filter), 1 °C-FD (Filter + Device), 8 °C-C (Control), 8 °C-F (Filter) and 8 °C-FD (Filter + Device); Table S3: Evolution during storage time of the bioactive compounds in pears subjected to different treatments: 1 °C-C (Control), 1 °C-F (Filter), 1 °C-FD (Filter + Device), 8 °C-C (Control), 8 °C-F (Filter) and 8 °C-FD (Filter + Device); Table S4: Principal Component Analysis; Table S5: Table of Component Weights; Table S6a: This table shows the scores of the principal components; Table S6b: ANOVA table for the component 1 scores according to the treatments; Table S6c: Multiple comparisons test for the component 1 scores by treatments using Tukey HSD method; Table S6d: ANOVA table for the component 2 scores according to the treatments; Table S6e: Multiple comparisons test for the component 2 scores by treatments using Tukey's HSD method.

Author Contributions: Conceptualization, R.A.-S., J.R.A.-M., S.L.-M. and A.J.P.-L.; methodology, R.A.-S., J.R.A.-M., S.L.-M. and A.J.P.-L.; software, R.A.-S., J.R.A.-M. and L.N.-A.; validation, E.N.-D. and F.B.; formal analysis, R.A.-S., J.R.A.-M., S.L.-M. and A.J.P.-L.; investigation, R.A.-S., J.R.A.-M., S.L.-M. and A.J.P.-L.; resources, J.R.A.-M., S.L.-M. and L.N.-A.; data curation, R.A.-S., J.R.A.-M., A.J.P.-L. and L.N.-A.; writing—original draft preparation, R.A.-S., J.R.A.-M., S.L.-M. and A.J.P.-L.; writing—review and editing, R.A.-S., J.R.A.-M., S.L.-M. and A.J.P.-L.; visualization, A.J.P.-L., J.R.A.-M. and S.L.-M.; supervision, J.R.A.-M., S.L.-M. and F.B.; project administration, S.L.-M. and E.N.-D.; funding acquisition, E.N.-D. and J.R.A.-M. All authors have read and agreed to the published version of the manuscript.

Funding: Nuevas Tecnologías Agroalimentarias (KEEPCOOL): CFE/KE/76-19; "Cátedra Emprendimiento en el Ámbito Agroalimentario" from Universidad Católica San Antonio de Murcia (UCAM) and Banco Santander: CFE/KE/76-19.

Acknowledgments: The authors gratefully acknowledge Antonio Cerdá (Cátedra UCAM-Banco Santander de Emprendimiento en el Ámbito Agroalimentario director), Cooperativa Hortofrutícola Campos de Jumilla from FECOAM ("Federación de Cooperativas Agrarias de Murcia"), especially Antonio Sanz and Lola Mondéjar for supplying the pears D.O.P "Ercolini" used in this study. Mario Fon for his help with English editing. and finally, Miriam Monje for her help with graphical abstract and figures.

Conflicts of Interest: The Doctoral Thesis (Ph.D.) of Ramiro Alonso Salinas, co-author of this work, is being co-financed by the company Nuevas Tecnologías Agroalimentarias (KEEPCOOL), within the UCAM Universidad Católica de Murcia Industrial Doctorate Program. The rest of the authors of the article declare that their contribution to this research was carried out in the absence of commercial or financial relationships that could be constructed as a possible conflict of interest.

References

1. Bower, J.; Biasi, W.; Mitcham, E. Effect of Ethylene in the Storage Environment on Quality of 'Bartlett Pears'. *Postharvest Biol. Technol.* **2003**, *28*, 371–379. [CrossRef]
2. Hiwasa, K.; Kinugasa, Y.; Amano, S.; Hashimoto, A.; Nakano, R.; Inaba, A.; Kubo, Y. Ethylene Is Required for Both the Initiation and Progression of Softening in Pear (*Pyrus communis* L.) Fruit. *J. Exp. Bot.* **2003**, *54*, 771–779. [CrossRef] [PubMed]
3. Horak, M.; Golias, J.; Balik, J.; Nemcova, A.; Snurkovic, P.; Koziskova, J. Changes in Asian and European Pear Cultivars Fruit after Harvest in Various Storage Conditions. *III Balk. Symp. Fruit Grow. Acta Hortic.* **2016**, *1139*, 609–615. [CrossRef]
4. Nguyen, L.L.; Pham, T.T.; Syium, Z.H.; Zsom-Muha, V.; Baranyai, L.; Zsom, T.; Hitka, G. Delay of 1-MCP Treatment on Post-Harvest Quality of 'Bosc Kobak' Pear. *Horticulturae* **2022**, *8*, 89. [CrossRef]
5. Jesús, G.B.; de Murcia, R. Informe sobre los parámetros de la calidad, contenido en Sólidos solubles (azucares) y Firmeza, 495 de los frutos de peral variedad 'Ercolini', cultivados en las condiciones medio ambientales de la comarca de Jumilla 'Altiplano' 496 (Murcia) en los años 2006, 2007 y 2008. *Región Murcia Cons. Agric. Y Agua* **2011**, 1–6.
6. Álvarez-Hernández, M.H.; Martínez-Hernández, G.B.; Avalos-Belmontes, F.; Miranda-Molina, F.D.; Artés-Hernández, F. Postharvest Quality Retention of Apricots by Using a Novel Sepiolite–Loaded Potassium Permanganate Ethylene Scavenger. *Postharvest Biol. Technol.* **2020**, *160*, 111061. [CrossRef]

7. Mansourbahmani, S.; Ghareyazie, B.; Zarinnia, V.; Kalatejari, S.; Mohammadi, R.S. Study on the Efficiency of Ethylene Scavengers on the Maintenance of Postharvest Quality of Tomato Fruit. *J. Food Meas. Charact.* **2018**, *12*, 691–701. [CrossRef]
8. Charoenchongsuk, N.; Matsumoto, D.; Itai, A.; Murayama, H. Ripening Characteristics and Pigment Changes in Russeted Pear Fruit in Response to Ethylene and 1-MCP. *Horticulturae* **2018**, *4*, 22. [CrossRef]
9. Alonso-Salinas, R.; Acosta-Motos, J.R.; Núñez-Delicado, E.; Gabaldón, J.A.; López-Miranda, S. Combined Effect of Potassium Permanganate and Ultraviolet Light as Ethylene Scavengers on Post-Harvest Quality of Peach at Optimal and Stressful Temperatures. *Agronomy* **2022**, *12*, 616. [CrossRef]
10. Alonso-Salinas, R.; López-Miranda, S.; Pérez-López, A.J.; Noguera-Artiaga, L.; Carbonell-Barrachina, Á.A.; Núñez-Delicado, E.; Acosta-Motos, J.R. Novel Combination of Ethylene Oxidisers to Delay Losses on Postharvest Quality, Volatile Compounds and Sensorial Analysis of Tomato Fruit. *LWT* **2022**, *170*, 114054. [CrossRef]
11. Kim, S.; Jeong, G.H.; Kim, S.-W. Ethylene Gas Decomposition Using ZSM-5/WO3-Pt-Nanorod Composites for Fruit Freshness. *ACS Sustain. Chem. Eng.* **2019**, *7*, 11250–11257. [CrossRef]
12. Wei, H.; Seidi, F.; Zhang, T.; Jin, Y.; Xiao, H. Ethylene Scavengers for the Preservation of Fruits and Vegetables: A Review. *Food Chem.* **2021**, *337*, 127750. [CrossRef]
13. Janjarasskul, T.; Suppakul, P. Active and Intelligent Packaging: The Indication of Quality and Safety. *Crit. Rev. Food Sci. Nutr.* **2018**, *58*, 808–831. [CrossRef]
14. Esturk, O.; Ayhan, Z.; Gokkurt, T. Production and Application of Active Packaging Film with Ethylene Adsorber to Increase the Shelf Life of Broccoli (*Brassica oleracea* L. Var. Italica). *Packag. Technol. Sci.* **2014**, *27*, 179–191. [CrossRef]
15. Ibhadon, A.; Fitzpatrick, P. Heterogeneous Photocatalysis: Recent Advances and Applications. *Catalysts* **2013**, *3*, 189–218. [CrossRef]
16. Keller, N.; Ducamp, M.-N.; Robert, D.; Keller, V. Ethylene Removal and Fresh Product Storage: A Challenge at the Frontiers of Chemistry. Toward an Approach by Photocatalytic Oxidation. *Chem. Rev.* **2013**, *113*, 5029–5070. [CrossRef]
17. Kaewklin, P.; Siripatrawan, U.; Suwanagul, A.; Lee, Y.S. Active Packaging from Chitosan-Titanium Dioxide Nanocomposite Film for Prolonging Storage Life of Tomato Fruit. *Int. J. Biol. Macromol.* **2018**, *112*, 523–529. [CrossRef]
18. Pathak, N.; Caleb, O.J.; Geyer, M.; Herppich, W.B.; Rauh, C.; Mahajan, P. V Photocatalytic and Photochemical Oxidation of Ethylene: Potential for Storage of Fresh Produce—A Review. *Food Bioprocess Technol.* **2017**, *10*, 982–1001. [CrossRef]
19. Pathak, N. *Photocatalysis and Vacuum Ultraviolet Light Photolysis as Ethylene Removal Techniques for Potential Application in Fruit Storage*; Technische Universität Berlin: Berlin, Germany, 2019.
20. Tytgat, T.; Hauchecorne, B.; Abakumov, A.; Smits, M.; Vinken, K.; Verbruggen, S.; Lenaerts, S. Photocatalytic Process Optimisation for Ethylene Oxidation. *Chem. Eng. J.* **2011**, *209*, 494–500. [CrossRef]
21. Liu, J.; Yang, J.; Zhang, H.; Cong, L.; Zhai, R.; Yang, C.; Wang, Z.; Ma, F.; Xu, L. Melatonin Inhibits Ethylene Synthesis via Nitric Oxide Regulation To Delay Postharvest Senescence in Pears. *J. Agric. Food Chem.* **2019**, *67*, 2279–2288. [CrossRef]
22. Álvarez-Hernández, M.H.; Artés-Hernández, F.; Ávalos-Belmontes, F.; Castillo-Campohermoso, M.A.; Contreras-Esquivel, J.C.; Ventura-Sobrevilla, J.M.; Martínez-Hernández, G.B. Current Scenario of Adsorbent Materials Used in Ethylene Scavenging Systems to Extend Fruit and Vegetable Postharvest Life. *Food Bioprocess Technol.* **2018**, *11*, 511–525. [CrossRef]
23. Smith, A.; Poulston, S.; Rowsell, L.; Terry, L.; Anderson, J. A New Palladium-Based Ethylene Scavenger to Control Ethylene-Induced Ripening of Climacteric Fruit. *Platin. Met. Rev.* **2009**, *53*, 112–122. [CrossRef]
24. Yildirim, S.; Röcker, B.; Pettersen, M.K.; Nilsen-Nygaard, J.; Ayhan, Z.; Rutkaite, R.; Radusin, T.; Suminska, P.; Marcos, B.; Coma, V. Active Packaging Applications for Food. *Compr. Rev. Food Sci. Food Saf.* **2018**, *17*, 165–199. [CrossRef] [PubMed]
25. Makkumrai, W.; Sivertsen, H.; Sugar, D.; Ebeler, S.E.; Negre-Zakharov, F.; Mitcham, E.J. Effect of Ethylene and Temperature Conditioning on Sensory Attributes and Chemical Composition of "Comice" Pears. *J. Agric. Food Chem.* **2014**, *62*, 4988–5004. [CrossRef] [PubMed]
26. Cruz-Hernández, A.; Paredes-López, O. Fruit Quality: New Insights for Biotechnology. *Crit. Rev. Food Sci. Nutr.* **2012**, *52*, 272–289. [CrossRef]
27. Gamrasni, D.; Ben-Arie, R.; Goldway, M. 1-Methylcyclopropene (1-MCP) Application to Spadona Pears at Different Stages of Ripening to Maximize Fruit Quality after Storage. *Postharvest Biol. Technol.* **2010**, *58*, 104–112. [CrossRef]
28. Zhang, B.; Peng, B.; Zhang, C.; Song, Z.; Ma, R. Determination of Fruit Maturity and Its Prediction Model Based on the Pericarp Index of Absorbance Difference (IAD) for Peaches. *PLoS ONE* **2017**, *12*, e0177511. [CrossRef]
29. Nielsen, S.S. *Vitamin C Determination by Indophenol Method BT—Food Analysis Laboratory Manual*; Nielsen, S.S., Ed.; Springer International Publishing: Cham, Switzerland, 2017; pp. 143–146. ISBN 978-3-319-44127-6.
30. Kidron, M.; Harel, E.; Mayer, A.M. Catechol Oxidase Activity in Grapes and Wine. *Am. J. Enol. Vitic.* **1978**, *29*, 30–35.
31. López-Miranda, S.; Serrano-Martínez, A.; Hernández-Sánchez, P.; Guardiola, L.; Pérez-Sánchez, H.; Fortea, I.; Gabaldón, J.A.; Núñez-Delicado, E. Use of Cyclodextrins to Recover Catechin and Epicatechin from Red Grape Pomace. *Food Chem.* **2016**, *203*, 379–385. [CrossRef]
32. Noguera-Artiaga, L.; Salvador, M.; Fregapane, G.; Collado, J.; Wojdyło, A.; López Lluch, D.; Carbonell-Barrachina, A. Functional and Sensory Properties of Pistachio Nuts as Affected by Cultivar. *J. Sci. Food Agric.* **2019**, *99*, 6696–6705. [CrossRef]
33. Gittins, C.; Calvo, P.; Miranda, M.J.; Barda, N. Developing a Sensory Descriptive Panel for Pear Quality Evaluation. *Acta Hortic.* **2011**, *909*, 617–624. [CrossRef]

34. Hu, B.; Sun, D.-W.; Pu, H.; Wei, Q. Recent Advances in Detecting and Regulating Ethylene Concentrations for Shelf-Life Extension and Maturity Control of Fruit: A Review. *Trends Food Sci. Technol.* **2019**, *91*, 66–82. [CrossRef]
35. Gaikwad, K.K.; Singh, S.; Negi, Y.S. Ethylene Scavengers for Active Packaging of Fresh Food Produce. *Environ. Chem. Lett.* **2020**, *18*, 269–284. [CrossRef]
36. Sammi, S.; Masud, T. Effect of Different Packaging Systems on the Quality of Tomato (*Lycopersicon esculentum* Var. Rio Grande) Fruits during Storage. *Int. J. Food Sci. Technol.* **2009**, *44*, 918–926. [CrossRef]
37. Shinozaki, Y.; Hao, S.; Kojima, M.; Sakakibara, H.; Ozeki-Iida, Y.; Zheng, Y.; Fei, Z.; Zhong, S.; Giovannoni, J.J.; Rose, J.K.C.; et al. Ethylene Suppresses Tomato (*Solanum lycopersicum* L.) Fruit Set through Modification of Gibberellin Metabolism. *Plant J.* **2015**, *83*, 237–251. [CrossRef]
38. Le Nguyen, L.P.; Zsom, T.; Dam, M.S.; Baranyai, L.; Hitka, G. Evaluation of the 1-MCP Microbubbles Treatment for Shelf-Life Extension for Melons. *Postharvest Biol. Technol.* **2019**, *150*, 89–94. [CrossRef]
39. Fan, X.; Shu, C.; Zhao, K.; Wang, X.; Cao, J.; Jiang, W. Regulation of Apricot Ripening and Softening Process during Shelf Life by Post-Storage Treatments of Exogenous Ethylene and 1-Methylcyclopropene. *Sci. Hortic.* **2018**, *232*, 63–70. [CrossRef]
40. Emadpour, M.; Ghareyazie, B.; Kalaj, Y.R.; Entesari, M.; Bouzari, N. Effect of the Potassium Permanganate Coated Zeolite Nanoparticles on the Quality Characteristic and Shelf Life of Peach and Nectarine. *Int. J. Agric. Technol.* **2015**, *11*, 1411–1421.
41. Crisosto, C.H.; Gugliuzza, G.; Garner, D.; Palou, L. Understanding the Role of Ethylene in Peach Cold Storage Life. In Proceedings of the Acta Horticulturae; International Society for Horticultural Science (ISHS), Leuven, Belgium, 30 June 2001; pp. 287–288.
42. Hu, Q.; Fang, Y.; Yang, Y.; Ma, N.; Zhao, L. Effect of Nanocomposite-Based Packaging on Postharvest Quality of Ethylene-Treated Kiwifruit (*Actinidia deliciosa*) during Cold Storage. *Food Res. Int.* **2011**, *44*, 1589–1596. [CrossRef]
43. Atkinson, R.G.; Gunaseelan, K.; Wang, M.Y.; Luo, L.; Wang, T.; Norling, C.L.; Johnston, S.L.; Maddumage, R.; Schröder, R.; Schaffer, R.J. Dissecting the Role of Climacteric Ethylene in Kiwifruit (*Actinidia chinensis*) Ripening Using a 1-Aminocyclopropane-1-Carboxylic Acid Oxidase Knockdown Line. *J. Exp. Bot.* **2011**, *62*, 3821–3835. [CrossRef]
44. Argenta, L.C.; Mattheis, J.P.; Fan, X.T.; Amarante, C.V.T. Managing "Bartlett" Pear Fruit Ripening with 1-Methylcyclopropene Reapplication during Cold Storage. *Postharvest Biol. Technol.* **2016**, *113*, 125–130. [CrossRef]
45. Escribano, S.; Lopez, A.; Sivertsen, H.; Biasi, W.V.; Macnish, A.J.; Mitcham, E. Impact of 1-Methylcyclopropene Treatment on the Sensory Quality of 'Bartlett' Pear Fruit. *Postharvest Biol. Technol.* **2016**, *111*, 305–313. [CrossRef]
46. Chiriboga, M.A.; Bordonaba, J.G.; Schotsmans, W.C.; Larrigaudiere, C.; Recasens, I. Antioxidant Potential of "Conference" Pears during Cold Storage and Shelf Life in Response to 1-Methylcyclopropene. *LWT* **2013**, *51*, 170–176. [CrossRef]
47. Álvarez-Hernández, M.H.; Martínez-Hernández, G.B.; Avalos-Belmontes, F.; Castillo-Campohermoso, M.A.; Contreras-Esquivel, J.C.; Artés-Hernández, F. Potassium Permanganate-Based Ethylene Scavengers for Fresh Horticultural Produce as an Active Packaging. *Food Eng. Rev.* **2019**, *11*, 159–183. [CrossRef]
48. Salamanca, F.A.; Balaguera-López, H.; Herrera, A. Effect of Potassium Permanganate on Some Postharvest Characteristics of Tomato "Chonto" Fruits (*Solanum lycopersicum* L.). *Acta Hortic.* **2014**, *1016*, 171–176. [CrossRef]
49. Willner, B.; Granvogl, M.; Schieberle, P. Characterization of the Key Aroma Compounds in Bartlett Pear Brandies by Means of the Sensomics Concept. *J. Agric. Food Chem.* **2013**, *61*, 9583–9593. [CrossRef]
50. Yao, M.; Zhou, X.; Zhou, Q.; Shi, F.; Wei, B.; Cheng, S.; Tan, Z.; Ji, S. Low Temperature Conditioning Alleviates Loss of Aroma-Related Esters of "Nanguo" Pears by Regulation of Ethylene Signal Transduction. *Food Chem.* **2018**, *264*, 263–269. [CrossRef]
51. Chai, J.; Liao, B.; Li, R.; Liu, Z. Changes in Taste and Volatile Compounds and Ethylene Production Determined the Eating Window of 'Xuxiang' and 'Cuixiang' Kiwifruit Cultivars. *Postharvest Biol. Technol.* **2022**, *194*, 112093. [CrossRef]
52. Yao, M.; Zhou, X.; Ji, Y.; Luo, M.; Sun, Y.; Zhou, Q.; Ji, S. Potential of Ethylene in Alleviating Cold-Induced Volatile Esters Loss of 'Nanguo' Pears by Regulating the Lipoxygenase Pathway. *Environ. Exp. Bot.* **2022**, *194*, 104723. [CrossRef]
53. Li, X.; Gao, S.; Yang, L.; Yin, M.; Li, J.; Zhang, H.; Ji, S. Ethylene Promotes Ester Biosynthesis through PuERF13/PuDof2.5 Synergically Activated PuAAT1 during Ripening of Cold-Stored 'Nanguo' Pear. *Postharvest Biol. Technol.* **2023**, *195*, 112108. [CrossRef]
54. Serra, S.; Anthony, B.; Masia, A.; Giovannini, D.; Musacchi, S. Determination of Biochemical Composition in Peach (*Prunus persica* L. Batsch) Accessions Characterized by Different Flesh Color and Textural Typologies. *Foods* **2020**, *9*, 1452. [CrossRef]
55. Álvarez-Hernández, M.H.; Martínez-Hernández, G.B.; Avalos-Belmontes, F.; Rodríguez-Hernández, A.M.; Castillo-Campohermoso, M.A.; Artés-Hernández, F. An Innovative Ethylene Scrubber Made of Potassium Permanganate Loaded on a Protonated Montmorillonite: A Case Study on Blueberries. *Food Bioprocess Technol.* **2019**, *12*, 524–538. [CrossRef]
56. Anton, D.; Bender, I.; Kaart, T.; Roasto, M.; Heinonen, M.; Luik, A.; Püssa, T. Changes in Polyphenols Contents and Antioxidant Capacities of Organically and Conventionally Cultivated Tomato (*Solanum lycopersicum* L.) Fruits during Ripening. *Int. J. Anal. Chem.* **2017**, *2017*, 2367453. [CrossRef]
57. Javanmardi, J.; Kubota, C. Variation of Lycopene, Antioxidant Activity, Total Soluble Solids and Weight Loss of Tomato during Postharvest Storage. *Postharvest Biol. Technol.* **2006**, *41*, 151–155. [CrossRef]
58. Rai, G.; Kumar, R.; Singh, A.; Rai, P.; Rai, M.; Chaturvedi, A.K.; Rai, A. Changes in Antioxidant and Phytochemical Properties of Tomato (*Lycopersicon Esculentum* Mill.) under Ambient Condition. *Pak. J. Bot.* **2012**, *44*, 667–670.

horticulturae

MDPI

Review

Strategies to Delay Ethylene-Mediated Ripening in Climacteric Fruits: Implications for Shelf Life Extension and Postharvest Quality

Ramiro Alonso-Salinas [1,†], Santiago López-Miranda [1,2,†], Antonio José Pérez-López [1,2] and José Ramón Acosta-Motos [1,2,*]

1 Plant Biotechnology for Food and Agriculture Group (BioVegA2), Universidad Católica San Antonio de Murcia, Avenida de los Jerónimos 135, Guadalupe, 30107 Murcia, Spain; ralonso4@alu.ucam.edu (R.A.-S.); slmiranda@ucam.edu (S.L.-M.); ajperez@ucam.edu (A.J.P.-L.)

2 Plant Biotechnology, Agriculture and Climate Resilience Group, UCAM-CEBAS-CSIC, Associated Unit to CSIC by CEBAS-CSIC, 30100 Murcia, Spain

* Correspondence: jracosta@ucam.edu

† These authors contributed equally to this work.

Abstract: Climacteric fruits undergo a characteristic ripening process regulated by ethylene, a key plant hormone. Extending the shelf life of these fruits while preserving their postharvest quality poses a significant challenge for the food industry. This review provides a comprehensive overview of physiological and molecular strategies to delay ethylene-mediated ripening in climacteric fruits and their impact on shelf life, postharvest quality, sensory attributes, and volatile compounds. Additionally, it examines the role of ethylene in fruit ripening, analysing various ethylene managing strategies including ethylene inhibitors, ethylene adsorbents, and ethylene scavengers by catalytic oxidation. This review concludes with future research directions including molecular and genetic approaches for reducing ethylene production or responsiveness in fruits, integrated strategies, environmental considerations, and commercial applications for improving postharvest handling and fruit quality.

Keywords: delayed ripening; fruit preservation; phytohormone

Citation: Alonso-Salinas, R.; López-Miranda, S.; Pérez-López, A.J.; Acosta-Motos J.R. Strategies to Delay Ethylene-Mediated Ripening in Climacteric Fruits: Implications for Shelf Life Extension and Postharvest Quality. *Horticulturae* **2024**, *10*, 840. https://doi.org/10.3390/horticulturae10080840

Academic Editor: Kazuo Ichimura

Received: 26 June 2024
Revised: 25 July 2024
Accepted: 5 August 2024
Published: 8 August 2024

1. Introduction

Ripening is a complex, genetically programmed, and irreversible phenomenon involving a series of physiological, biochemical, and organoleptic changes. Since this process is unstoppable, there is a finite time for commercialization and consumption where the product exhibits desirable quality attributes. Once this time has elapsed, ripening becomes an adversary as it leads to processes such as weight, size, and firmness reduction, organoleptic degradation, or loss in bioactive compounds, resulting in an inedible and unmarketable product, leading to economic and environmental losses.

According to the Ministry of Agriculture, Fisheries and Food of the Government of Spain, in its report entitled "More food, less waste" [1], global food losses and waste in 2020 amounted to one-third of the world's food production intended for human consumption. This translates into losses of approximately 1.3 billion tons per year, a quantity sufficient for feeding 2 billion people. In Europe, this figure stands at 89 million tons and that for Spain is 7.7 million tons. In economic terms, global food losses and waste would amount to throwing away more than €782.397 billion (excluding environmental and social costs).

According to De Laurentiis [2], households in the European Union waste about 17 billion kilograms of fresh fruit and vegetables each year, which is 35.3 kg per person per year. On average, 29% of fresh fruits and vegetables are wasted in households. Furthermore, according to the programme "Informe Semanal: Comer bien, tirar menos" by Radio Televisión Española (RTVE) [3], aired on 11 March 2023, more than two-fifths of all fruit, vegetable, and green crops are wasted because they are "ugly". This subjective criterion prevails over other objectives such as firmness, flavour, or internal composition of these foods.

Based on the aforementioned, proper management of ripening-related processes can provide more marketing and consumption time for the customer, thereby avoiding waste and subsequent economic losses and environmental problems for the entire production chain.

1.1. Ethylene: A Ripening Plant Hormone

Ethylene is a plant hormone produced by plants and fruits that remains in gaseous state under normal conditions (25 °C temperature and 1 atm pressure). It is biologically active in minimal amounts and its effects are commercially crucial in fruit and vegetable ripening [4]. The role of ethylene as a potent regulator of plant growth has been established over the last hundred years, but its effects have been known for several centuries. The use of ethylene to accelerate fruit ripening has been known since ancient times. Examples of this phenomenon include the ripening of apples in southern Italy using quinces for joint preservation or the ripening of mangoes in India using straw combustion. It is even known that in ancient Egyptian civilization, superficial cuts were made in the skin of figs to stimulate their ripening. Subsequently, it was discovered that these cuts or scarifications promote fruit stress, leading to increased respiration and ethylene production [5].

Phytohormones play a role in many aspects of plant development. Ethylene was one of the first plant hormones discovered. Its discovery arose from a remarkably curious fact. Although by the mid-19th century it was clear that the presence of gaseous materials in the air could modify plant growth, it was not until the late 19th century that the Russian researcher Dimitry Neljubow identified ethylene, an active component of street lamp gas, as the generator of a strange growth habit in pea seedlings suffering from etiolation [6]. The first evidence that plant material produces a gas that affects the growth of nearby plants was discovered by Cousins in 1910 [7], who hypothesized that gases emitted by oranges caused banana ripening in mixed commercial shipments. However, since healthy oranges produce very little ethylene because they are non-climacteric fruits, the origin of this ethylene was from oranges infected with fungi. In 1917, Sarah Doubt successfully correlated the presence of ethylene with the stimulation of leaf and fruit abscission [8].

In 1924, Frank E. Denny [9] observed that farmers cultivating Florida lemons stored their fruits in sheds with kerosene lamps, thinking that heat caused them to lose their green colour. Upon investigation, Denny discovered that it was the ethylene produced by those lamps that induced the colour change in lemons from green to yellow, a process later known as degreening. Gane [10] demonstrated in 1934 that fruits during ripening synthesize ethylene. He provided chemical evidence that ethylene was indeed mostly produced by ripest bananas, demonstrating that plants produce ethylene themselves, confirming Cousins' hypothesis. He later found that ethylene was also produced by other fruits and could promote seed germination [11]. Crocker reported in 1935 [12] that ethylene acts similarly to auxins, being involved in plant growth and the senescence of vegetative tissues in *Arabidopsis thaliana* L. Therefore, it was established that ethylene is a plant hormone.

Subsequently, between the 1940s and the early 1970s, methods for eliminating this phytohormone to extend the shelf life of plant products began to be proposed [13]. Southwick & Smock showed in 1943 [14] that by using activated charcoal with bromine as an adsorbent for ethylene, the shelf life of 'McIntosh' apples could be extended by a month. In 1971, Scott and colleagues [15] proposed the use of ultraviolet light to eliminate ethylene and thus extend the shelf life of fruits, although these methods were still far from industrial application.

In recent years, multiple procedures have been developed to eliminate or inactivate ethylene and its effects. Among them, oxidative ethylene removers (such as potassium permanganate or titanium dioxide) and ethylene inhibitors (such as salicylic acid or 1-MCP) stand out. Currently, this field faces the following challenges:

1. Observing the effect of different ethylene removal methods on as many foods as possible;
2. Discovering to what extent food quality is maintained through these methods;
3. Establishing which of them are truly applicable on an industrial scale and refining these methods to make them more effective.

Ethylene can be classified according to its origin as endogenous and exogenous. Endogenous ethylene is produced by a fruit as a result of its internal synthesis. In contrast, exogenous ethylene is produced by other agents, which can be biological, such as other adjacent fruits, or of another nature such as the combustion of vegetables like straw or stubble or combustion in vehicle engines [16].

Regarding the capacity of this gas to act, effects of ethylene have been recorded at very low concentrations, even below 0.001 μL per litre of air [17]. While it is true that in the initial stages of fruit development, a high presence of this gas can be beneficial as it promotes and accelerates their development, in later stages, especially during postharvest, it can be detrimental by accelerating senescence and reducing their commercial life [18,19]. This rapid ripening favoured by the presence of ethylene affects most of the qualitative parameters of fruits, from physical parameters such as weight or firmness to biochemical parameters such as antioxidant capacity, soluble solids, pH, or acidity.

1.2. Scope and Structure of This Review

This review aims to provide a comprehensive examination of the ripening dynamics of climacteric fruits, with a specific focus on ethylene control and removal. By synthesizing current research findings, theoretical frameworks, and practical applications, this review seeks to deepen our understanding of the interplay between ethylene and fruit ripening processes, elucidating the molecular mechanisms, physiological effects, and practical implications of ethylene biology in the horticultural industry.

To achieve these objectives, this review will be structured into several interrelated sections, each addressing specific aspects of climacteric fruit ripening and ethylene biology:

- Climacteric fruits and ethylene: This section will provide a comprehensive overview of climacteric fruits and their characteristic ripening process. It will define climacteric fruits, delineate the stages of their ripening process, and discuss the physiological changes associated with ripening. Additionally, it will highlight the pivotal role of ethylene as a master regulator of climacteric fruit ripening, emphasizing its multifaceted functions and importance in fruit development and maturation.
- Regulation of ethylene biosynthesis and signalling: This section will delve into the molecular mechanisms underlying ethylene biosynthesis, perception, and signal transduction in climacteric fruits. It will explore the regulatory pathways governing ethylene production, receptor-mediated signalling cascades, and downstream responses in fruit tissues. Additionally, it will discuss the environmental and hormonal factors that modulate ethylene biosynthesis and signalling pathways, providing insights into the complexity of ethylene regulation during fruit ripening.
- Physiological and molecular effects of ethylene on fruit ripening: This section will examine the specific molecular and physiological effects of ethylene on climacteric fruit ripening processes. It will elucidate how ethylene influences key ripening-related events, such as fruit softening, flavour development, colour changes, and aroma production. By integrating molecular biology, biochemistry, and physiology, this section will offer a comprehensive understanding of ethylene-mediated ripening processes at the cellular and tissue levels.
- Ethylene managing strategies: This section will be destined to go deep in different strategies to remove, inhibit, or reduce the effect that ethylene could have according to its interaction with climacteric and non-climacteric fruits.
- Practical implications and future directions: This final section will discuss the practical implications of ethylene biology for agricultural practices, postharvest management, and fruit quality enhancement. It will highlight potential strategies for manipulating

ethylene levels, optimizing postharvest handling techniques, and improving fruit quality and shelf life. Additionally, it will identify emerging research trends and future directions in ethylene research, pointing towards new opportunities for innovation and advancement in this field.

Overall, this review aims to provide a comprehensive synthesis of current knowledge on the interrelationship between climacteric fruit ripening and ethylene biology, offering valuable insights into the mechanisms, regulation, and practical applications of ethylene in fruit production and postharvest handling.

2. Climacteric Fruits and Ethylene

Fruit ripening has always been the subject of intense study due to its relevance to the nutritional characteristics that define its quality [20,21]. As mentioned earlier, depending on the respiration pattern displayed, fruits can be divided into climacteric fruits, which exhibit an increase in respiration rate mediated by a sudden rise in ethylene, and non-climacteric fruits, in which there is no increase in respiration rate or accumulation of ethylene [22].

Both climacteric and non-climacteric fruits show common pathways regarding the signal transduction pathway in response to ethylene [23] and accumulate abscisic acid (ABA), especially at the beginning of ripening [24,25]. The accumulation of ABA precedes and, therefore, modulates ethylene production in climacteric fruits and triggers ripening in non-climacteric ones while they are still in the plant [25,26]. Recently, the fundamental role of abscisic acid (ABA) in non-climacteric fruits, especially in *Fragaria × ananassa*, affecting cell wall modification was demonstrated. This is because ABA suppresses the production of pectin esterase and polygalacturonase enzymes, which degrade the cell wall and promote softening of the affected fruits [27,28].

The focus on the ripening process of climacteric fruits can be divided into several sequential stages, each characterized by specific physiological and biochemical changes (Figure 1):

- Pre-climacteric phase: At the beginning of fruit development, climacteric fruits are in the pre-climacteric phase. During this stage, ethylene production and respiration rates are relatively low. The fruits are typically firm, green, and physiologically immature. Although metabolic processes are occurring, they are not yet at levels indicative of ripening.
- Climacteric peak: As fruits reach maturity, they undergo a dramatic increase in ethylene biosynthesis and respiration, marking the climacteric peak. This peak is a pivotal event in the ripening process and triggers a cascade of biochemical and physiological changes. One of the most notable transformations is the conversion of starches into sugars, leading to increased sweetness. Additionally, the fruit softens as cell wall components break down, resulting in changes in texture and juiciness. Other changes include alterations in pigmentation, aroma development, and flavour enhancement.
- Climacteric phase: Following the climacteric peak, fruits enter the climacteric phase, characterized by sustained ethylene production and ongoing metabolic activity. Ripening processes initiated during the peak continue, albeit at a slower pace. This phase is crucial for the completion of ripening, as fruits continue to develop desirable sensory attributes and undergo structural modifications indicative of ripeness.
- Post-climacteric phase: Eventually, climacteric fruits enter the post-climacteric phase, marked by a decline in ethylene production and respiration rates. While fruits remain physiologically ripe during this phase, they may exhibit signs of senescence, such as loss in firmness, increased susceptibility to decay, and decline in sensory quality.

The ripening process of climacteric fruits is governed by a complex network of hormonal, genetic, and environmental factors. Ethylene, in particular, plays a central role in coordinating ripening-related processes by regulating gene expression, enzyme activities, and physiological responses. Additionally, interactions between ethylene and other hormones, such as auxins, abscisic acid, and gibberellins, further modulate ripening dynamics.

There are two ethylene production systems according to the developmental stages of plant tissues and fruits: system I and system II.

- System I of ethylene production is associated with vegetative tissues and fruits in the early stages of development and is characterized by low rates of ethylene production, auto-inhibitory character (a process by which ethylene induces and controls its own production), and absence of relevant peaks in the production of this phytohormone [29, 30].
- On the other hand, system II is present in more advanced development processes, especially in climacteric fruits, and is characterized by high rates of ethylene emission, feedback (higher ethylene concentration in the environment implies higher production of the same), and a high peak of ethylene production at the onset of physiological maturity [29].

The systems I and II of climacteric fruits are identical in the ethylene synthesis pathway. They exhibit different response patterns to exogenous ethylene, which may be related to the distinct properties of the ACS and ACO isoenzymes.

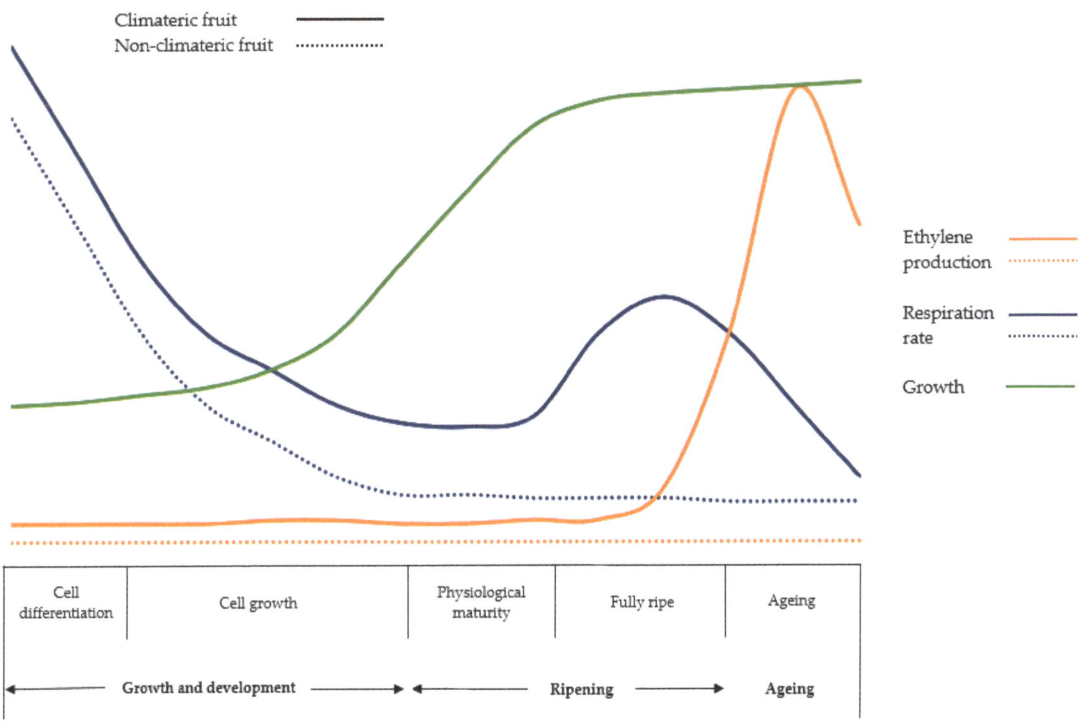

Figure 1. Respiration rate and ethylene production during postharvest (own source: http://hdl.handle.net/10952/6740 (accessed on 20 May 2023)).

As shown in Figure 1, climacteric fruits exhibit a peak in respiration and ethylene production during the ripening process and are capable of ripening even after harvest, whereas non-climacteric fruits do not show any peak in respiration and cannot ripen after harvest [31]. The differences between the two types of fruits are detailed below:

Firstly, climacteric fruits contain both the system I and system II of ethylene production, whereas non-climacteric fruits only emit ethylene through the so-called system I. This means that, at the onset of ripening, climacteric fruits experience a peak in respiration, followed by massive ethylene production [32]. As mentioned earlier, system II is characterized by feedback in ethylene production, where more ethylene in the environment leads to higher production. This maintains the ethylene peak until reaching an overripe state [30]. In contrast, for non-climacteric fruits, once the point of physiological maturity is reached, ethylene production remains, without significant changes, at a basal level [33,34]

Secondly, both climacteric and non-climacteric fruits show little difference during the developmental phase, as they only produce minimal amounts of ethylene. However, during the full ripening period, the amount of ethylene produced by climacteric fruits is much higher than that of non-climacteric fruits.

Thirdly, the application of exogenous ethylene is only effective in the early stages of development of climacteric fruits on the tree, which can lead to increased respiration and autocatalysis of endogenous ethylene. This reaction is irreversible because artificially applying ethylene accelerates the fruit ripening process, and as mentioned earlier, ripening cannot be stopped.

Finally, increasing the concentration of exogenous ethylene advances the appearance of the respiratory peak in climacteric fruits, but the intensity of the respiratory peak remains unchanged. However, for non-climacteric fruits, increasing the concentration of exogenous ethylene may increase respiratory intensity but not the duration of the respiratory peak [35].

The biochemical changes induced by ethylene, along with the microbiological damage caused by bacteria or fungi, are the main causes of deterioration in climacteric fruits, even under low-temperature storage conditions [36,37]. Consequently, an appropriate strategy must be adopted to prevent ethylene accumulation in order to prolong the postharvest life of such plant products and thereby reduce losses [38]. The effect of ethylene must be blocked or eliminated to improve its quality and extend its shelf life. Furthermore, its application should not only be limited to storage but should also be applied to handling and all stages after harvest [39]. To effectively eliminate or block the effect of ethylene, a thorough understanding of available methods, their industrial application, effectiveness, and cost is necessary.

The physiological maturity or ripening of a fruit is defined, according to the dictionary of the Royal Spanish Academy as "the state of development in which a fruit meets the requirements to be consumed or used by the consumer for a particular purpose". In these terms, two categories can be distinguished depending on whether, once harvested, they continue to ripen, known as climacteric fruits, or if their ripening is interrupted, known as non-climacteric fruits. For climacteric fruits, the ripening process is initiated by changes in their hormonal composition. It is not a gradual process; there is a peak of ripeness followed by ageing or overripening [5].

The onset of climacteric ripening is a well-defined process characterized by a rapid increase in the rate of respiration and ethylene production by the fruit. Climacteric fruits have the ability to continue ripening even when separated from the mother plant, provided they have reached a physiological state that ensures ethylene production. Examples of such fruits include apple, avocado, peach, or tomato [38,40–42].

Conversely, regarding non-climacteric fruits, the ripening process is gradual and continuous. They lack the ability to continue ripening once separated from the mother plant, so it must be ensured that they have reached an appropriate state of maturity for consumption at the time of harvest. Citrus fruits, grapes, pomegranates, or raspberries fall into this category of non-climacteric fruits [33,43].

About the effect of ethylene on non-climacteric fruits, although they do not show a clear increase in ethylene production rates during ripening, in certain cases, their exposure to exogenously applied ethylene can stimulate certain ripening-related processes, such as the inhibition of senescence, the inhibition of the development of physiological disorders, and the inhibition of colour changes [44]. Exogenous ethylene can react with ethylene receptors both in the early and late stages, leading to accelerated ripening. However, in this case, once exogenous ethylene application is stopped, the fruits will return to their pretreatment levels of respiration and ethylene production [35,45]. This process is widely applied in citrus fruits, for example, in lemons. These fruits are harvested without reaching their physiological ripeness (colour change from green to yellow). Subsequently, when they are ready for consumption, they undergo a process called "degreening" in which controlled ethylene is introduced into lemon storage chambers until they reach their optimal ripeness for consumption and then return to their pretreatment levels of respiration and ethylene production [46].

As an example, exposure to 1-MCP had an effect on delaying rachis browning in table grape varieties (Thompson Seedless'), thus suggesting the possible involvement of the ethylene signalling pathway in the regulation of the rachis browning process of non-climacteric fruit [47]. Another example can be found on pomegranates, where the main problems associated with their postharvest storage are related to weight loss and shrivelling, as well as peel discolouration. Previous studies indicated that exposure to ethylene blockers such as 1-MCP reduced skin shrivelling [48] and the incidence of peel browning [49].

Depending on the type of fruit and the ripening stage it is in, ethylene can have both positive and negative effects. The beneficial effects of ethylene include stimulating ripening in climacteric fruits until they reach the optimal consumption stage, developing colour through pigment synthesis (anthocyanin and lycopene), and chlorophyll degradation (a process known as degreening, commonly applied in the case of lemons). The negative effects of ethylene, especially during the postharvest of climacteric fruits, are numerous. Excessive softening, weight loss, loss in bioactive compounds, and emission of unpleasant volatile compounds in fruit [50]; leaf and flower abscission in higher plants; accelerated hardening of vegetables; increased susceptibility to bacterial or fungal pathogens; stimulation of sprouting; changes in shape; and appearance of reddish spots are some of them [43,51]. These undesirable changes often occur due to accelerated ripening caused by exposure to ethylene emitted by adjacent fruits and/or to ethylene generated as a contaminant in locations where the product is placed, whether in greenhouses, trucks, ships, or airplanes during storage and transport. Therefore, it is crucial to reduce surrounding ethylene as well as inhibit ethylene biosynthesis to minimize its impact on the product.

3. Regulation of Ethylene Biosynthesis and Signalling

3.1. Biosynthesis of Ethylene in Climacteric Fruits and the Enzymes Involved

As mentioned earlier, ethylene is involved in many aspects of plant development, including seed germination, root hair development in roots, seedling growth, leaf and petal abscission, fruit ripening, and plant organ senescence [4,52]. Ethylene production is regulated by internal signals during development and in response to biotic (pathogen attack) and abiotic environmental stimuli, such as wounds, hypoxia, ozone, cold, or freezing. Regulation can also occur at the level of perception or signal transduction [53,54].

Many plant tissues produce ethylene at concentrations that are mostly low. It was not until the discovery of gas chromatography that deeper knowledge of ethylene synthesis and metabolism began [55,56]. Bradford [57] clarified and explained the ethylene synthesis transduction pathway through a series of experiments in apples, summarizing it as a chain of reactions starting from the amino acid methionine and ending with ethylene synthesis, resulting in a series of intermediate products: methionine $\rightarrow$ S-adenosyl-L-methionine (SAM) $\rightarrow$ 1-aminocyclopropane-1-carboxylic acid (ACC) $\rightarrow$ ethylene.

Ethylene originates from the third and fourth carbon of methionine. During the cycle, methionine binds to adenosine to form SAM; subsequently, SAM, apart from giving rise to ACC, can produce 5′-methylthioadenosine (MTA), which, after the cycle repeat reaction and hydrolysis, regenerates methionine, leading to high rates of ethylene production without the need for high levels of intracellular methionine [55]. This ensures that the amino acid content in plants is ready for a new cycle of ethylene production [56–58]. This important discovery was made thanks to the work of Professor Shan Fa Yang [4]; therefore, this ethylene production cycle was named the "Yang cycle" (Figure 2).

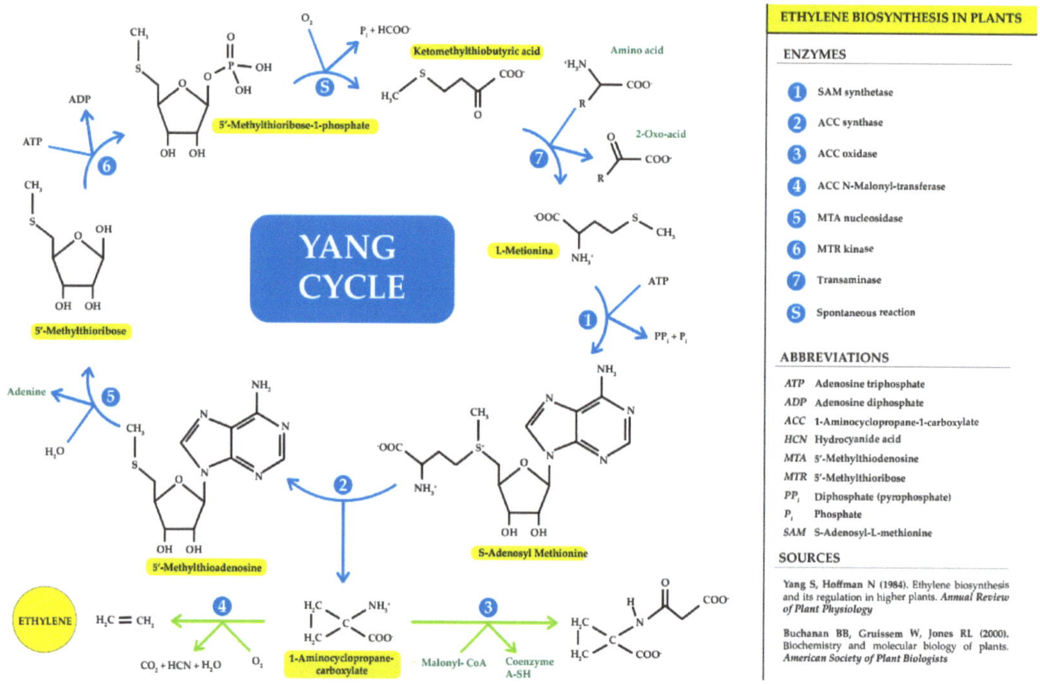

Figure 2. Yang cycle (own adaptation: http://hdl.handle.net/10952/6740 (accessed on 20 May 2023)) [4,59].

During ethylene synthesis, the activity of three enzymes is regulated: SAM synthetase (SAMS), ACC synthase (ACS), and ACC oxidase (ACO). SAM synthesized by the enzyme SAMS is also involved in other biochemical pathways such as that related to polyamine synthesis [60–62]. On the other hand, the ACS enzyme catalyses the breakdown of SAM into ACC and MTA, thus being the key point that limits the speed of the entire pathway. ACS is present in the cytoplasm forming monomers, dimers, or even trimers, with monomers exhibiting the highest catalytic activity. In biochemical assays carried out by Yamagami and colleagues in 2003 [63], the analysis of the amino acid sequence of the enzymes showed that all ACS isoforms depend on pyridoxal phosphate (PLP) coenzymes. Therefore, inhibitors of enzymes competing for PLP such as aminoethoxyvinylglycine (AVG) and aminooxyacetic acid (AOA) can be used to inhibit ethylene synthesis [64]. In addition, cobalt ion and low oxygen levels can also affect ethylene production by inhibiting the final step of the pathway, catalysed by ACC oxidase, to finally give ethylene. Finally, ACO forms ethylene, CO_2, and cyanide from ACC. This latter component (cyanide) is detoxified by beta-cyanoalanine synthase to beta-cyanoalanine to prevent toxicity of accumulated cyanide during high rates of ethylene biosynthesis [64]. In 2002, Chung and colleagues [65] discovered that the ACO enzyme is widely distributed in various plant tissues, and its function depends on the effects of ferrous ion (Fe^{2+}) and oxygen (O_2), and therefore various chelating agents can inhibit its activity. The activity of this enzyme also depends on ascorbic acid or vitamin C as a co-substrate, establishing a direct relationship between the concentration of this acid and ethylene production [66]. There is still no clear evidence about the cellular localization of the ACO enzyme, although several authors indicate that it could be found in the cytosol [67,68], but this is still unknown [69].

Recent advances have shown that ethylene synthesis and metabolism in plants are closely correlated with environmental factors such as light and temperature. There are also other processes in which the three key enzymes mentioned are involved that can regulate ethylene production. For example, Argueso and colleagues reported [70] that the expression of multiple ACS genes (ACS2, ACS4 and ACS5) is induced by wounds, increasing ethylene synthesis.

3.2. Ethylene Perception and Signalling

Researchers' understanding of the ethylene signal transduction pathways primarily stems from genetic and molecular biology studies of Arabidopsis thaliana L mutants [71]. The initial step in ethylene perception involves its binding to receptors for this molecule in plants and fruits. Cells possess a family of five receptors located in the endoplasmic reticulum (ER) membrane that are homologous to bacterial histidine kinases, namely, ETR1, ETR2, EIN4, ERS1, and ERS2, corresponding to ethylene receptors 1, 2, ethylene insensitive 4, and ethylene response sensor 1, 2, respectively, which bind to ethylene with the assistance of a copper cofactor in the transmembrane domain of the protein receptor [71–73].

While some receptor specialization has been recognized, it is believed that they primarily function by modulating the activity of the kinase called CTR1. Inactivation of this kinase in the presence of ethylene results in decreased levels of phosphorylation of the transmembrane protein located in the endoplasmic reticulum, EIN2 [74–76]. Two different responses have been characterized by studying EIN2 activity. On one hand, there is a rapid growth inhibition once the fruit is harvested, which occurs within minutes of hormone exposure [76,77]. On the other hand, many other changes, possibly slower, are induced by this hormone, including alterations at the transcriptional level in hundreds of genes [77,78]. Target genes include those involved in ripening, seed germination, wound response, senescence, and other phenomena whose functions are still unknown. This entire process is schematically illustrated in Figure 3.

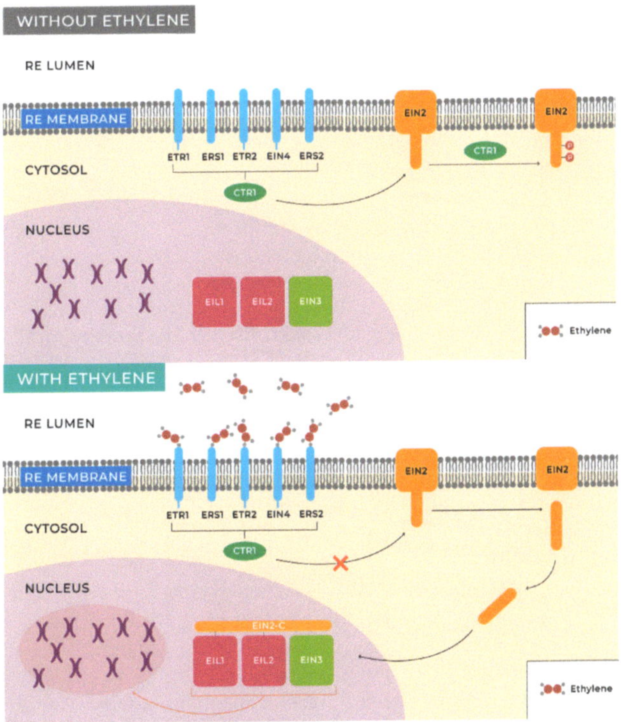

Figure 3. Ethylene perception and signalling (own source: http://hdl.handle.net/10952/6740 (accessed on 20 May 2023)).

Despite the growing evidence that CTR1 is directly regulated by ethylene receptors, the biochemical events controlling such regulation are not well understood. The involvement of a MAPK cascade and its role in ethylene signal transmission are still unclear, and more conclusive genetic evidence is needed. Another question to be addressed is the function of EIN2, a protein that plays an essential role in mediating all known ethylene responses. Regarding events occurring in the nucleus, determining how ethylene controls the stability/activity of primary transcription factors, such as EIN3, EIL1, or EIL2, represents another challenge, as well as determining the subsequent steps in gene regulation. Experiments using global gene expression profiling show that hundreds of genes are induced or repressed by ethylene. In the future, it will be important to discern how many of these genes are immediate targets of EIN3 and, in the long term, to categorize the transcriptional networks involved in ethylene responses.

Ethylene receptors, along with other two-component receptors—those having a part that receives the extracellular stimulus and another part that modifies a transcription factor regulating gene expression to rapidly adapt to environmental changes (such as phytochromes and cytokinin receptors)—are believed to have been acquired by plants from cyanobacteria, which allowed for the contribution of chloroplasts through endosymbiosis, as they are homologous to some bacterial receptors [79–84].

Data from recent phylogenetic analysis suggest a common origin for the ethylene-binding domain in cyanobacteria and plants [85]. Therefore, it is interesting to note that ethylene binding has been observed in at least one cyanobacterium, Synechocystis sp., which has a functional ethylene receptor regulating cell surface properties affecting biofilm formation and phototaxis, i.e., the ability of cells to orient in response to light intensity [86–88]. Furthermore, ethylene-binding affinities to some of these cyanobacteria and to the ethylene receptor of Synechocystis sp. are similar to those observed in plants [89], demonstrating conservation of this domain among these organisms. However, the primitive prokaryotic organism from which ethylene receptors first emerged remains unknown [71].

Plants contain multiple isoforms of ethylene receptors. Early studies identified ethylene-binding sites in plant endoplasmic reticulum (ER) membranes [90,91], a finding corroborated by subsequent research [92–97]. In Arabidopsis thaliana L., the five aforementioned isoforms (ETR1, ETR2, ERS1, ERS2, and EIN 4) have been identified [55,98–100]. Mutations in any of these receptors prevent ethylene binding to them and lead to a plant showing ethylene insensitivity [55,86,98,101]. Additionally, some mutations in these receptors do not affect ethylene binding but prevent signalling through the receptor, also resulting in ethylene insensitivity [86].

4. Physiological and Molecular Effects of Ethylene on Fruit Ripening

Ethylene-mediated fruit ripening involves a complex interplay of physiological and biochemical changes that occur in response to ethylene signalling. These changes are crucial for transforming immature, unripe fruit into fully mature, ripe fruit with desirable sensory attributes. Understanding the intricate mechanisms underlying ethylene-mediated ripening provides insights into the regulation of key ripening processes and offers opportunities for improving postharvest fruit quality and shelf life.

4.1. Effect of Ethylene Over Shelf Life, Quality and Disease Resistance

Phytohormone is also a key player in how long fruits and vegetables stay fresh. For example, ethylene speeds up the ripening process, which is great for achieving peak flavour and texture. However, beyond this point, ethylene can cause them to overripen and spoil much faster [51]. Ethylene also initiates the ageing process in plants, known as senescence, and induces abscission. Consequently, this results in a decline in the overall quality of the produce, affecting both its appearance and texture [53]. Another crucial role of the ethylene is stimulating the production of enzymes that degrade cell wall components, resulting in the softening of fruit tissue. This softening increases the susceptibility of the fruit to mechanical damage and microbial decay [102]. Lastly, ethylene increases the respiration rate in fruits, leading to a faster depletion of their energy reserves. This elevated metabolic activity accelerates the loss in firmness, flavour, and nutritional value, thereby contributing to a reduced shelf life [103].

Ethylene plays a crucial role in impacting the quality of fruits and vegetables, but its effects can be a double-edged sword if not carefully managed. Firstly, ethylene is known to induce the production of enzymes that break down cell walls, which leads to the softening of fruits. While this is essential for ripening, it can also result in a loss in texture and firmness, qualities highly valued by consumers [104]. Another effect of ethylene is the degradation of chlorophylls. This process affects the vibrant green colour of many fruits and vegetables, making them appear less fresh and less appealing. In addition, ethylene causes overripening, provoking a significant drop in overall quality and visual appeal. This deterioration can make fresh produce look unappetizing and less desirable [105]. Furthermore, the presence of ethylene during storage and transport is a significant issue. It can speed up the ripening process before the produce reaches consumers, resulting in a noticeable decline in quality by the time it reaches the shelves [106].

Ethylene also plays a multifaceted role in fruit disease resistance, acting as both a promoter and a suppressor depending on the context. Its influence is nuanced and varies with different factors. On one hand, ethylene can boost the fruit's defence mechanisms against pathogens. For example, research on 'Kyoho' grapes shows that treating them with ethephon, a compound that releases ethylene, before ripening can enhance the expression of genes related to both fruit colouration and disease resistance. This treatment led to improved resistance against Botrytis cinerea, a common fungal pathogen [107].

Firstly, the ethylene response involves a variety of genes that are differentially expressed when plants are challenged by pathogens. For instance, in a study on potato, a total of 1226 ethylene-specific differentially expressed genes (DEGs) were identified, including those encoding for transcription factors, kinases, defence enzymes, and disease resistance-related genes [108]. These genes are part of the plant's immune system, helping to activate defence responses. Secondly, transcription factors such as those from the APETALA2/ethylene response factor (AP2/ERF) family play a pivotal role in plant disease resistance. They act downstream of mitogen-activated protein kinase (MAPK) cascades and regulate the expression of genes associated with hormonal signalling pathways, biosynthesis of secondary metabolites, and formation of physical barriers [109,110]. For example, the ERF gene from the tomato is known to confer resistance to *Pseudomonas syringae* pv. tomato by activating genes encoding antifungal proteins and proteins involved in oxidative burst. Ethylene's role in disease resistance is closely linked to its biosynthesis and signalling pathways. In terms of disease resistance, the glutathione metabolism pathway, which includes key enzymes like glutathione S-transferase (GST), plays an important role in response to ethylene stimulus [111].

4.2. Effect of Ethylene on the Physical Characteristics of Fruits during Ripening

Ethylene is responsible for a series of physiological changes that transform the physical characteristics of fruits, directly affecting important physical parameters that impact their commercial value, such as weight, size, and texture. Controlling the presence of ethylene has a positive effect on maintaining the physical characteristics of fruits during postharvest storage.

Fruit weight is an important parameter for fruit producers, especially from an economic perspective, making the control of fruit weight loss crucial. This loss is associated with excessive water loss due to transpiration, which is related to low relative humidity (RH). Water loss after harvesting is an unavoidable phenomenon, resulting in weight loss, size reduction, wilting, abnormal textures, and quality degradation. The dehydration of peaches can be prevented by maintaining a high environment RH (90–95%), while also controlling air speed and using physical or chemical barriers [112,113].

Different authors have shown evidence that by controlling and eliminating ethylene, weight loss during postharvest storage can be reduced. Emadpour and colleagues [42] observed a weight loss reduction of only 3% during the storage of peaches using an ethylene eliminator like potassium permanganate. Mansourbahmani and colleagues [41] recorded a weight loss reduction of only 2% in 'Valouro' tomatoes when ethylene concentration in the storage chamber was controlled with potassium permanganate mixed with zeolite after 35 days of storage. Alonso-Salinas and colleagues [114] recorded weight reductions of between 3% and 7% in pears stored in an atmosphere with ethylene reduction systems, compared to weight losses of 10% and 17% in pears stored without ethylene control systems over a 28-day period.

Regarding fruit size during storage, it tends to show a similar trend as fruit weight [115]. These authors also observed that the decrease in the diameter of peaches, compared with weight loss, was less sensitive to the presence of ethylene in the storage atmosphere. The presence of ethylene promotes accelerated respiration, which causes a greater decrease in weight and size due to increased water loss [36]. In a study conducted with pears stored for 28 days, it was observed that the reduction in size was around 20% when ethylene was not removed from the storage chamber and between 5% and 10% when different ethylene removal techniques, such as potassium permanganate filters combined with ultraviolet light, were used [114].

Firmness is one of the main quality attributes that determine the acceptance of the product by the consumer. On one hand, weight loss influences firmness in a directly proportional relationship. On the other hand, ethylene significantly affects fruit firmness by triggering cell wall hydrolysis, which leads to fruit softening [116,117]. In addition, the expression of polygalacturonase-related genes is associated with ethylene production [118]. The action of this enzyme is considered key in the softening of fruits, increasing its activity in those treatments where the exposure to ethylene was higher [119]. Wu and colleagues [113] demonstrated that the effect of ethylene scavengers on apricots resulted in significantly higher firmness compared to control fruit.

Álvarez-Hernández and colleagues [120] showed that apricots with an ethylene removal system based on potassium permanganate and sepiolite sachets recorded 20% more firmness than the control treatment after 36 days of storage. Mansourbahmani and colleagues [41] recorded approximately 15% preservation of firmness on the final day when applying their ethylene removal method in tomatoes. In a study on peaches stored with and without an ethylene removal system, Alonso-Salinas and colleagues [115] observed that the firmness of the fruits from which ethylene was removed from the storage atmosphere maintained the same firmness during the first 7 days, tending then to equalize with the texture of the peaches stored without ethylene removers. These same authors observed, in a trial conducted with pears stored at 1 °C and 8 °C, that removing ethylene from the storage atmosphere reduced firmness by 10% and 28%, respectively, while not removing ethylene reduced firmness by 29% (1 °C) and 46% (8 °C) [114].

4.3. *Effect of Ethylene on the Biochemical Characteristics of Fruits during Ripening*

Ethylene is a crucial plant hormone that plays a significant role in the regulation of fruit ripening [121]. During ripening, ethylene significantly impacts the biochemical characteristics of fruits, enhancing attributes such as sweetness, maturity index, and antioxidant capacity, while modulating acidity, pH, and the concentration of bioactive compounds [122]. Understanding these changes during ripening is essential for optimizing postharvest handling and improving consumer satisfaction [123].

Ethylene significantly influences the accumulation of solid soluble content (SSC) in fruits. During postharvest ripening of climacteric fruits, such as peaches, sugars displace acids by certain metabolic processes, increasing SSC and giving the fruit a sweet taste. Therefore, SSC is a key indicator of the ripening stage of the fruit [124,125]. The increase in SSC primarily comprises an increase in sugars such as glucose, fructose, and sucrose [126]. This process is facilitated by the ethylene-induced activation of enzymes like invertase and amylase, which catalyse the breakdown of starches into simpler sugars [127]. Total acidity (TA) of fruits generally decreases during ethylene-induced ripening. This decline is attributed to the metabolic conversion of organic acids, such as citric and malic acids, into sugars and other metabolites [128]. The reduction in acidity not only affects the taste, making the fruit less sour, but also impacts the overall flavour profile and consumer preference [129]. The pH of fruits typically rises during ethylene-mediated ripening. This increase in pH correlates with the decrease in TA, reflecting the metabolic changes occurring within the fruit tissues. Monitoring pH changes is crucial for understanding the ripening dynamics and the biochemical environment of the fruit [130].

Maturity index (MI) defined as the ratio of SSC to TA, serves as an indicator of fruit ripeness. Ethylene accelerates the ripening process, thereby increasing the MI. A higher MI indicates a more advanced stage of ripeness, characterized by optimal sugar levels and reduced acidity, which are desirable traits for market quality and consumer acceptance, [131] but if the maturity index (MI) becomes too high due to the effect of ethylene, several negative outcomes can occur: (1) Overripening: Fruits may become too soft and mushy, leading to a loss in structural integrity and an undesirable texture for consumers [132]. (2) Flavour deterioration: Overripened fruits often experience changes in flavour, becoming overly sweet and losing the balance of acidity and sweetness that defines optimal ripeness [133]; (3) nutrient degradation: ascorbic acid (vitamin C), phenolic compounds, and antioxidant capacity (ORAC) initially can increase with ripening due to ethylene being able to stimulate the expression of genes involved in the ascorbate biosynthetic pathway and favouring the activity of phenylalanine ammonia-lyase (PAL), leading to increased synthesis of phenolic compounds (such as flavonoids and anthocyanins) and contributing to the accumulation of antioxidant capabilities. On the contrary, excessive ethylene exposure and overripening can lead to the degradation of these compounds due to oxidative degradation and metabolic consumption, reducing the fruit's nutritional value and antioxidant capacity [123,134,135]; (4) Increased susceptibility to diseases and pests: Overripened fruits are more prone to infections and infestations, leading to higher postharvest losses and a decrease in marketable yield [136]. (5) Economic losses: Fruits that are too ripe may have a shorter shelf life and become unmarketable, resulting in economic losses for producers and retailers [137,138].

Different studies performed by Alonso et al. in peaches, tomatoes, and pears [114,115,139] showed favourable responses to the preservation of the quality and shelf life of these fruits thanks to the different methodologies applied for the elimination of ethylene through oxidation reactions mediated by $KMnO_4$ and UV-C. Other studies have reported a less successful effect of ethylene scavengers on peach fruit conserved for 36 days at 0 °C [42,140] or in apricots stored at 15 °C [120].

4.4. *Effect of Ethylene on the Organoleptic Characteristics of Fruits during Ripening*

Fruit ripening brings out the best taste and scent, which influences the fruit's flavour. Fruits go through biochemical changes when they ripen, such as colour breakdown, starch hydrolysis, sugar and acid metabolism, volatile production, and cell wall disintegration, all of which affect flavour. These modifications add to the typical aroma of ripe fruits as well as their sweet and sour taste. Furthermore, when fruit ripens, its texture improves, becoming crisper, juicier, or melty, intensifying the taste release in the mouth. Ethylene gas regulates the ripening process and contributes to the softening and flavouring of fruit. Fruit ripening is an intricate process that involves a range of physiological and biochemical changes that have a big impact on the fruit's flavour [141–143].

The flavour changes associated with ethylene involve a network of genes, transcription factors, and metabolic pathways. Firstly, specific genes responsible for the synthesis of flavour compounds as genes encoding enzymes like lipoxygenases, hydroperoxide lyases, and alcohol dehydrogenases are involved in the formation of volatile compounds that contribute to fruit aroma [144]. Additionally, genes related to the biosynthesis of sugars, acids, and secondary metabolites like carotenoids and flavonoids are also ethylene-responsive and contribute to flavour [145]. Secondly, transcription factors such as the RIN (RIPENING

INHIBITOR), NOR (NONRIPENING), and CNR (COLOURLESS NONRIPENING) play crucial roles in the regulation of ripening-related genes. These factors can directly or indirectly influence the expression of genes involved in flavour compound biosynthesis [144]. For instance, the transcription factor HY5 has been shown to bind to the promoter of SWEET12c, a gene involved in sugar transport, thereby modulating sugar content and influencing sweetness in tomato fruit [145]. Finally, ethylene influences several metabolic pathways that are directly related to flavour development. The Lox pathway is responsible for the production of volatile compounds that contribute to aroma [146]. The glycolytic pathway and tricarboxylic acid (TCA) cycle are involved in sugar and acid metabolism, affecting sweetness and sourness [147]. Ethylene also affects the isoprenoid pathway, which includes the biosynthesis of carotenoids, contributing to both colour and flavour [145].

Ethylene is a highly volatile substance produced by many fruits during their ripening process, and it is thus considered a maturation hormone. These fruits ripen due to the action of ethylene, as increased concentrations accelerate the maturation process, altering the colour, firmness, flavour, and characteristic aromas of each fruit. During maturation, the starch in fruits is converted into sugars (fructose and glucose), tannins are reduced—these are compounds characteristic of unripe fruit that impart a bitter taste—and the pH increases, decreasing acidity. All these changes lead to modifications in the organoleptic characteristics of the fruit [148].

Volatile compounds, which undergo modifications during ripening, play a significant role in the flavour of foods. These compounds belong to various chemical families, with the most notable being hydrocarbons, aldehydes, ketones, alcohols, carboxylic acids, esters, and lactones [149,150].

Several authors have investigated the control of ethylene synthesis on the production of volatile compounds during fruit ripening. These changes in the aromatic profile have led to greater sensory acceptance by consumers of the final product [151,152].

Moreover, if ethylene control, through its inhibitors, is combined with other evaluative parameters such as different storage conditions—controlled or dynamic atmospheres, partial oxygen pressure, temperature, or fruit CO_2 production—it could offer commercial benefits. These conditions can slow down the respiration rate of the fruit, partially inhibit ethylene synthesis, and result in improvements in sensory parameters such as acidity and firmness, while also enhancing aroma formation and the synthesis of aromatic compounds [114,139,151,153–155].

Another crucial parameter in the sensory evaluation of fruits is colour, which influences flavour perception and is affected by ethylene synthesis and the presence of its inhibitors. This has led to positive evaluations from consumer panels analysing these products [29,105,153,156].

5. Ethylene Managing Strategies

Given that fruits and vegetables ripen faster regardless of whether ethylene exposure is exogenous or endogenous, it is essential to eliminate both types [51,157,158]. Therefore, strategies that allow for controlling ethylene levels in storage environments of climacteric fruits are crucial for reducing postharvest losses of products and maintaining the organoleptic quality of fruits [159].

Over the past two decades, various methods for ethylene control have been developed, becoming increasingly effective in achieving minimal ethylene concentrations in fruit storage and transportation environments. These methods can be classified into three well-defined categories:

- Ethylene inhibitors:
    - 1-MCP: 1-methylcyclopropene;
    - SA: salicylic acid;
    - AVG: aminoethoxyvinylglycine;
    - AOA: amino-oxyacetic acid.
- Ethylene adsorbents:

- Zeolite;
- Activated carbon;
- Metal–organic frameworks;
- Silica gel.

- Ethylene scavengers by catalytic oxidation:
  - $KMnO_4$: potassium permanganate;
  - UV radiation;
  - TiO_2: titanium dioxide;
  - O_3: ozone;
  - Palladium;
  - Cold plasma and other technologies.

5.1. Ethylene Inhibitors

Ethylene inhibitors act at some point in the biosynthesis, signalling, or perception pathways of ethylene as previously developed. By delaying or slowing down these metabolic processes, they hinder the natural progression of product ripening.

5.1.1. 1-Methylcyclopropene (1-MCP)

1-MCP is a chemically synthesized molecule belonging to the family of small-ring hydrocarbons (Figure 4). Under normal environmental conditions (25 °C and 1 atm), 1-MCP is gaseous; therefore, in its commercial form, it is encapsulated in γ-cyclodextrins that dissolve with the relative humidity of the environment, releasing this gas [160,161].

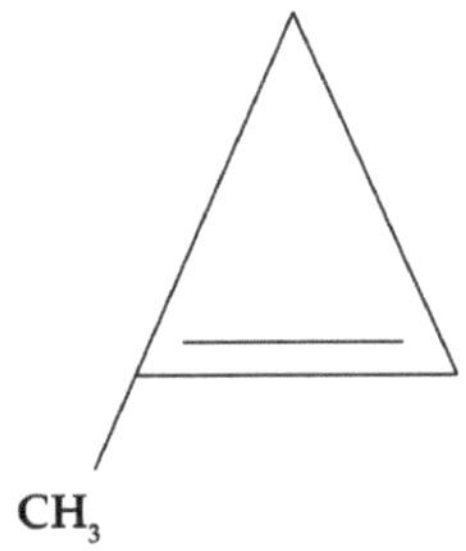

Figure 4. 1-MCP molecular structure.

This molecule is characterized by having a molecular structure similar to ethylene. This causes it to compete with this phytohormone for the ethylene membrane receptors (ETR1, ERS1, and EIN4) located on the plasma membranes of plant cells, inhibiting the ethylene response [29]. This mechanism can be observed in Figure 5.

Additionally, 1-MCP interacts with ethylene transmembrane receptors in a more intense and rapid manner than ethylene itself, facilitating the binding of 1-MCP to these receptors [162]. Other studies have shown that 1-MCP has other functions besides competing for ethylene reception; however, these mechanisms have not yet been discovered [163]. The effects of applying this molecule have been widely studied in the last 15 years (Table 1).

However, 1-MCP shows lower effectiveness than other ethylene scavengers or deactivators such as palladium compounds or the use of potassium permanganate as ethylene oxidants [29,41]. The reason for this is that 1-MCP is limited to competing with ethylene for cellular receptors, thereby partially preventing its reactivation but not eliminating it. Additionally, 1-MCP does not block all ethylene receptors in plant cells, thus its effectiveness is lower than expected [164].

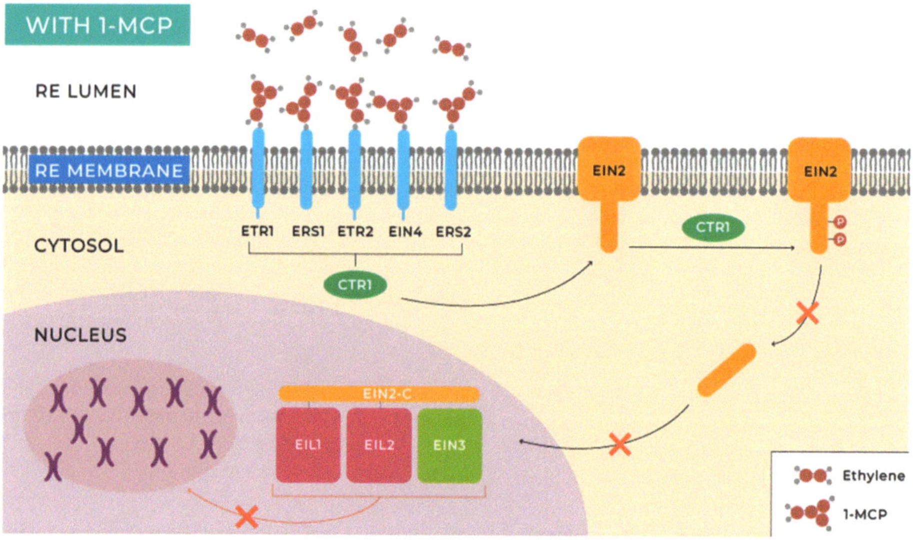

Figure 5. Ethylene perception and signalling with 1-MCP (own source: http://hdl.handle.net/10952/6740 (accessed on 20 May 2023)).

Table 1. Studies on the application of 1-MCP in different formats and concentrations on the preservation of various fruits.

Format	Concentration	Fruit	Conditions	Significative Results	Reference
Gas	0.5 μL L^{-1}	'Raf' Tomato	10 °C, 90% RH, 7 days 20 °C, 90% RH, 4 days	It reduces both ethylene production and respiration rate and, in turn, delays weight, soluble solids content, and total acidity changes.	[165]
Gas	1 μL L^{-1}	Pear cv: 'Gorham' 'Gran Champion' 'La France' 'Gold La France'	20 °C, 25 days	Reduction in ethylene production, significant delay in the expression of genes related to the change in skin colour of pears, and preservation of chlorophyll and fruit firmness.	[166]
Gas Gas	0.1 μL L^{-1} 0.035 μL L^{-1}	'Unicorn' Tomato	10 °C, 85% RH, 15 days	The treatment using a higher concentration of 1-MCP showed a higher conservation of lycopene and weight.	[167]
Gas	0.9 μL L^{-1}	'Hayward' Kiwi	20 °C, 95% RH, 20 days	Inhibition of ethylene production and respiration rate, delay of rot incidence, weight loss, increase in soluble solids content and total bacterial count. Improved preservation of firmness, chlorophyll, total acidity, ascorbic acid, and antioxidant capacity.	[168]
Micro-bubbles	100–400 ppb	'Khai' Banana	25 °C, 85% RH, 12 days	Reduced respiration rate and ethylene production. Higher preservation of total chlorophyll content, colour, firmness, total soluble solids, antioxidant capacity, and total phenolic compounds.	[169]
Gas	100 nL L^{-1}	'Gold' and 'Rainbow' Papaya	22 °C, 85% RH, 25 days 10 °C, 85% RH, 25 days	Fruits treated with high concentrations of 1-MCP showed increased firmness and delayed colour variation, meaning delayed ripening. The authors claim that commercial application could lead to a 30% reduction in papaya ripening.	[170]

5.1.2. Salicylic Acid (SA)

Salicylic acid (SA) (Figure 6) is a phenolic compound found in plants that is currently considered an endogenous plant hormone [171]. This molecule regulates many processes such as stress response, plant development, thermogenesis, photosynthesis, stomatal behaviour, transpiration, ion absorption and transport, seed germination, and glycolysis, among others [172,173].

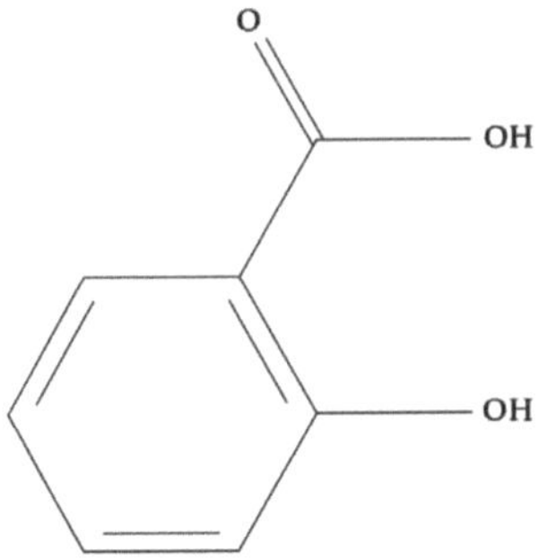

Figure 6. SA molecular structure.

It acts as an antagonist of ACC oxidase (ACO), inhibiting its action; this enzyme is crucial in the process of ethylene production. ACO oxidizes ACC to ethylene in its final stage [174]. It has been demonstrated that SA inhibits ethylene production in various plants. Leslie and Romani in 1988 [175] observed that applying a concentration of 80 μM of SA during washing of pears (*Pyrus communis* L.) reduced the ethylene production rate by 80% after 24 h. Srivastava & Dwivedi in 2000 [176] applied 100 μM of SA to bananas, achieving, approximately, a 50% reduction in respiration rate. These authors also observed a reduction in the activity of cellulase and polygalacturonase enzymes after SA application. Babalar and colleagues [173] applied SA at a concentration of 2 mM, achieving effective reduction in both fungal rot and respiration rate in 'Selva' variety strawberries. In 'Canino' apricots, Elmenofy et al. in 2021 [177] compared the application of two ethylene inhibitors such as SA (4 mM) and AVG (150 mg L^{-1}), recommending the application of SA (4 mM) which yielded optimal yield, quality, and preservation of this apricot variety when grown under Egyptian environmental conditions, where the research was conducted. Lastly, Mansourbahmani and colleagues [41] compared different ethylene removal methods, including SA. These authors concluded that applying a 1% SA solution to 'Valouro' variety tomatoes reduced ethylene production by 25%, which was lower than other investigated methods such as 1-MCP application or ethylene oxidation via the use of palladium, potassium permanganate, or UV-C radiation.

Recently, applications of SA as an elicitor have been discovered. It refers to a substance or signal that triggers a specific response in a cell, organism, or system. In plants, elicitors can be molecules like phytohormones, which activate specific receptors on the surface of cells, leading to a cascade of intracellular signalling events ultimately resulting in a cellular response. Gong and colleagues [178] reported in their review that SA application during fruit formation (lemon, pomegranate, mango, and strawberry among others) improved phenylpropanoid metabolism and carotenoid biosynthesis, increasing accumulation of flavonoids, ascorbic acid, and carotenoids, thereby enhancing antioxidant activity in the subsequently harvested fruit.

5.1.3. Aminoethoxyvinylglycine (AVG)

Similar to SA, aminoethoxyvinylglycine (AVG) (Figure 7) is a compound that inhibits ethylene production by intervening at some point in the metabolic pathways that leads to its production. Specifically, AVG is an inhibitor of ACC synthase (ACS) as it binds to its coenzyme, pyridoxal phosphate (PLP), preventing the conversion of SAM to ACC and thereby affecting the subsequent production of ethylene [179–181].

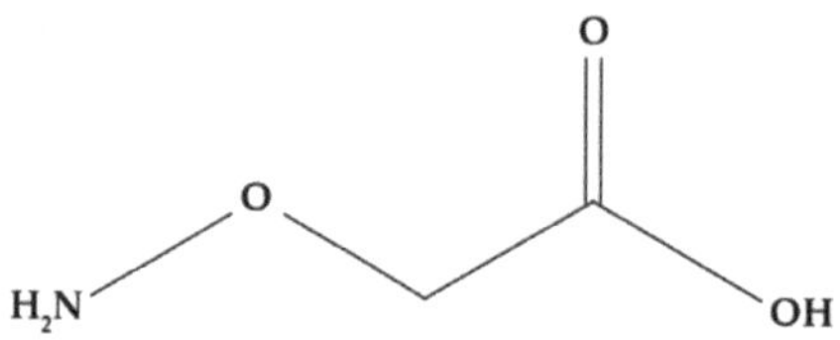

Figure 7. AVG molecular structure.

The postharvest application of AVG has been studied in various fruits: Salveit [182] applied a 10 mM solution of AVG in tomatoes (*Lycopersicon esculentum* Mill.) of the 'Castlemart' variety, resulting in a 30% reduction in respiration rate and a 70% reduction in ethylene production rate of the fruits. Ozturk and colleagues [183] applied 200 mg L^{-1} of AVG in plums two weeks before harvest in the 'Black Beauty', 'Black Amber', and 'Fibre' plum varieties. Results showed a 60% reduction in ethylene production and respiration rate. They also observed a delay in the colour change from green to red and prevented firmness loss (70% firmer fruits in treated fruits) and weight loss (15% more weight in treated fruits). The same author in 2019 applied 225 mg L^{-1} of AVG in kiwifruits (*Actinia deliciosa* L.) of the 'Hayward' variety, observing similar results, but no differences were observed with the control in the analysis of ascorbic acid and total flavonoids with the application of AVG [184]. Similar effects were observed by Yuan & Carbaugh [185] in apples of the 'Golden Supreme' and 'Golden Delicious' varieties, Muñoz-Robredo and colleagues [186] in apricots of the 'Patterson' variety, Xie and colleagues [187] in pears of the 'Starkrimson' variety, Kim and colleagues [188] in plums of the 'Formosa' variety, and Win and colleagues [189] in persimmons of the 'Sangjudungsi' variety.

However, possible negative effects on fruit aroma development have been reported, resulting in loss in intensity [190,191]. Additionally, its cost is too high for industrial application [180].

5.1.4. Aminooxyacetic Acid (AOA)

Unlike the compounds mentioned earlier (1-MCP, SA, and AVG), AOA (Figure 8) is a nonspecific inhibitor of all enzymes that use PLP as a coenzyme, including ACC synthase, so its application could decrease ethylene synthesis [180,192,193]. Due to its nonspecificity, it may interfere with other enzymatic reactions, negatively affecting other physiological processes [194].

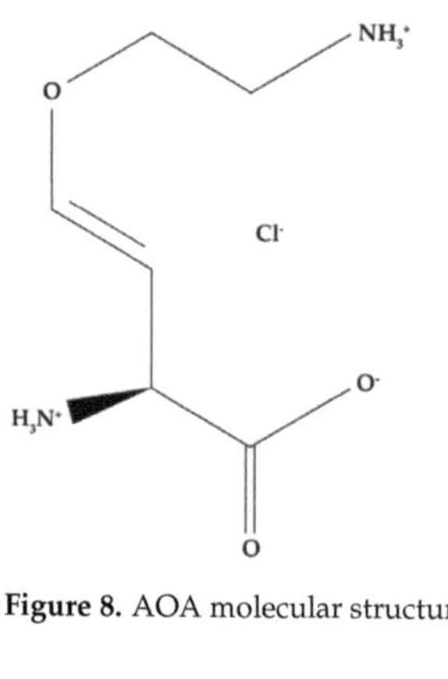

Figure 8. AOA molecular structure.

Bulantseva and colleagues [195] applied a solution of AOA at 10 mg L^{-1} to bananas harvested at three different ripening stages (green, green-yellow, and bright yellow). Those treated and harvested in the green and green-yellow stages had ethylene production rates about 30% lower than controls at the same ripening stages. Lima and colleagues [196] immersed the roots of sweet potatoes of the 'BRS Rubissol' variety in 1 mg L^{-1} of AOA, resulting in better preservation of weight, total sugars, and pH and a reduction in sprouting compared to untreated potatoes, thereby extending the shelf life of treated potatoes.

To date, the application of this compound has been mainly limited to the preservation of harvested flowers [197], while its application in fruits and vegetables has not been sufficiently tested.

5.2. Ethylene Adsorbents

Methods for trapping ethylene and minimizing its action are based on two physical phenomena: adsorption and absorption. Adsorption occurs when a particle adheres to the surface of a solid material. On the other hand, in absorption, the particle remains within the solid material. Many materials are capable of carrying out adsorption/absorption processes of ethylene, and they have been used to develop ethylene scavengers. Among them, zeolites, activated carbon, and metal–organic frameworks stand out.

5.2.1. Zeolite

Zeolites (Figure 9) [198] are three-dimensional microporous structures of crystalline aluminosilicates. Zeolites have negative charges in their structure that are balanced with alkali or alkaline earth ions [199]. The pore size of zeolites ranges from 3 to 12 Å, which gives them the capacity to adsorb many chemical substances with a certain degree of specificity, including ethylene [18].

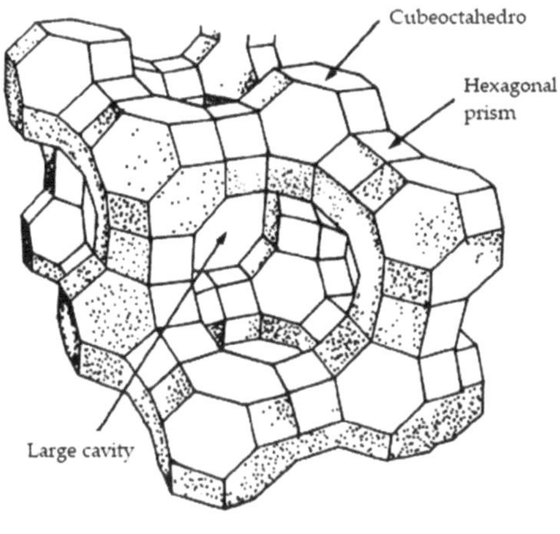

Figure 9. Zeolite typical structure (own adaptation from [199]).

Ethylene can be absorbed within the structure of zeolite and adsorbed on the surface of the zeolite structure [200]. In the case of natural zeolites, a large pore diameter (12 Å) favours ethylene (diameter 3.9 Å) to pass through the pore openings and be absorbed into the interior of the zeolite [201]. There are two theories for explaining this phenomenon. The first refers to a cation–π interaction (non-covalent molecular interaction between a cation and an electron-rich system) between the double-bond electrons of ethylene and metallic cations of the zeolite. The second theory involves a resulting CH-O interaction between the hydrogen atoms of ethylene and oxygen atoms on the surface of the zeolite [18,199,202].

De Bruijin and colleagues [203] applied zeolites (composed of 53% clinoptilolite, 40% mordenite, and 7% quartz) as adsorbents in the conservation of tomatoes of the 'Medano' variety and observed a 50% reduction in ethylene concentration measurements up to day

10 of storage at 10 °C. From that point on, ethylene levels between the control treatment and the treatment with zeolite adsorbents equalized. These authors concluded that it may be an effective method for ethylene capture but that it alone is not highly effective. Mariah and colleagues [204] reached a similar conclusion in their review work.

However, despite the fact that zeolites do not cause a significant environmental impact and are non-toxic, their ethylene retention capacity is low and nonspecific for this molecule [112].

5.2.2. Activated Carbon

The use of activated carbon dates back to 1940. Southwick and Smock as early as in 1943 [14] used brominated charcoal to remove ethylene from the storage atmosphere of 'McIntosh' apples, which allowed for a considerable extension of the fruit's life in that controlled atmosphere for a period of 2 to 3 months.

Activated carbon traps ethylene by adsorption (Figure 10) [205]. Its effectiveness depends on factors such as the contact surface, pore size, temperature, and relative humidity of the environment [193]. Granular powder activated carbon is the most common form used for ethylene capture with this methodology. Bailén and colleagues [206] confirmed that at an ethylene concentration ranging from 1 to 7.5 μL L^{-1} in the air, granular activated carbon of 20 to 60 meshes (a unit of measure for the pore diameter of a sieve in particle size analysis, in this case between 0.841 and 0.177 mm) could adsorb 70% of the available ethylene, while powdered activated carbon of 100 to 400 meshes (between 0.149 and 0.037 mm) could only adsorb 40%.

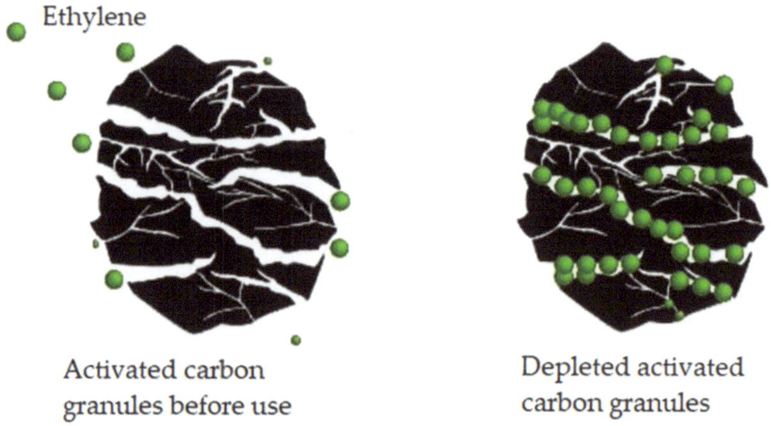

Figure 10. Active carbon acts as an ethylene adsorbent. When this phytohormone penetrates through the structural cracks in the active carbon, it remains adhered to the walls, thus preventing its effect (own adaptation from [205]).

Jaimun and Sabgsywan [207] applied activated carbon to chitosan-coated papers (aminopolysaccharide biopolymer) and vanillin to compare its effect with other ethylene scavengers such as zeolite in the transport boxes of 'Nam Dok Mai' mangoes preserved for 30 days at 13 °C. In their results, they observed that the application of zeolite in this context was more effective than activated carbon for ethylene adsorption, although the differences between these treatments were low (1.91 mL L^{-1} of ethylene for zeolite, 2.31 mL L^{-1} of ethylene for activated carbon, and 3.21 mL L^{-1} of ethylene for the control treatment at 28 days of trial). No differences were observed between applying activated carbon or zeolite in the rest of the analysed parameters (weight, firmness, titratable acidity, and soluble solids), but differences were observed with the control treatment, thus maintaining the quality of mangoes treated with both adsorbent substances during the storage time.

The main advantages of activated carbon as an ethylene remover are its environmental friendliness due to its carbon composition, low toxicity, availability, and low cost. However, its nonspecific adsorption nature is a significant limitation for its widespread use as an ethylene adsorbent alone, so it is usually used as an adjunct along with other ethylene scavengers and removers such as potassium permanganate ($KMnO_4$) or palladium [18,29].

5.2.3. Metal–Organic Frameworks

Metal–organic frameworks (MOFs) are a type of synthetic porous material formed by metal ions or groups of ions bonded to organic molecules. The combination of different organic and inorganic molecules provides flexibility in terms of pore size, shape, and structure [208]. MOFs have an exceptionally high adsorption surface area (from 1000 to 3000 $m^2\ g^{-1}$), greater than zeolites (320 $m^2\ g^{-1}$) and activated carbon (827 to 1120 $m^2\ g^{-1}$) [209]. MOFs can selectively adsorb volatile compounds such as ethylene [208,210].

Awalgaonkar [164] found that MOFs had better ethylene inactivation capacity than potassium permanganate under low and controlled relative humidity conditions. Chopra and colleagues [210] reported that MOFs do not adsorb ethylene as efficiently when there is water in the storage environment. This is indeed a problem for its implementation at an industrial level since the storage environments of plant products typically have relative humidity values between 90 and 95%. Although this method is effective for ethylene capture, it has not yet been approved by the U.S. Food and Drug Administration for use in the food industry [18]. Further research and development of this product are necessary to enable its application.

As seen thus far, ethylene traps, while effective at "capturing" ethylene, are not efficient at preventing its action. Some of the main drawbacks of these technologies are as follows:

1. Ethylene is only adsorbed on the surface or absorbed into the interior of these materials but cannot be decomposed.
2. Desorption phenomena (the opposite process of adsorption/absorption) may occur, whereby a substance is released from or through a surface.
3. Over time, the effectiveness of adsorption/absorption tends to decrease as these materials easily become saturated and require replacement.

5.3. Ethylene Removal by Catalytic Oxidation

This method of ethylene elimination relies on chemical oxidation–reduction reactions in which ethylene irreversibly dissociates into CO_2 and water. The effectiveness of this method depends on the oxidizing capacity of the various compounds that can be used for this purpose.

5.3.1. Potassium Permanganate ($KMnO_4$)

Potassium permanganate acts on the double bond of the ethylene molecule to oxidize it in the presence of water. Ethylene, when oxidized by potassium permanganate, initially forms acetaldehyde, which is subsequently oxidized to acetic acid and then to carbon dioxide and water.

The redox reaction caused by potassium permanganate results in a colour change in the permanganate itself, shifting from purple (MnO_4 ions) to brown (MnO_2). The studies conducted using potassium permanganate as an ethylene remover are quite diverse. Its effects have been tested in various applications involving apples [211], apricots [120], blueberries [212], tomatoes [41,139,213], pears [114,214], and peaches [42,115], among others. Many studies conducted with this molecule aim to enhance ethylene removal by increasing the contact surface of potassium permanganate with this phytohormone. This is achieved by mixing it with inert materials such as zeolites or activated carbon mentioned earlier, compounds that also aid in the absorption and adsorption of ethylene (see Table 2).

Table 2. Some studies on the application of potassium permanganate, in different formats, in the preservation of fruits.

Format	Concentration	Fruit	Conditions	Significative results	Reference
LDPE Sachet	Saturated and dried $KMnO_4$ solution (50 mL) together with natural clays and activated carbon	'Pollock' Avocado	12 °C, 94% RH, 21 days	Reduction in CO_2 and C_2H_4 concentrations in the packaging. Maintained good visual quality. Inhibited disease incidence and reduced chilling damage. Delayed weight and firmness losses. Soluble solid content and flesh colour were not significantly affected.	[215]
LDPE Sachet	5 g $KMnO_4$ and vermiculite mixture at doses of 1.5 and 1% *w*/*w*	'Musa' Banana	18 °C, 70–80% RH, 16 days	Delayed yellowing of the skin. Slower increase in CSS and decrease in AT. Reduced loss in firmness and weight. Minimal increase in the SSC/TA ratio.	[216]
Sachet	Mixture of natural clays and KMnO4 (9 g per kg of fruit)	'Hayward' Kiwi	0 °C, 85–95% RH, 200 days	Delayed firmness. Reduction in the increase in SSC and decrease in TA. Reduction in total chlorophyll and ascorbic acid degradation.	[217]
LDPE Sachet	20 g $KMnO_4$ together with aluminium oxide	'Nijisseiki' Pear	0 °C, 90% RH, 252 days	Delayed yellowing. Decrease in C_2H_4 levels and respiration rate. Reduced severity of disorders (browning of flesh and heart).	[218]
Sachet	20% $KMnO_4$ in bentonite zeolite (2:1 *w*/*w*)	'Valouro' Tomato	7 °C, 90% RH, 35 days	Reduction in C_2H_4. Delayed softening of the fruit. Minimization of the rate of decrease in phenol content and antioxidant activity. Decreased ascorbic content and increased lycopene content. Delayed water weight loss and reduced severity of rotting.	[41]

This molecule has significant industrial applications due to its low production cost, ease of incorporation into storage and transportation systems, and high effectiveness in ethylene removal. The main drawback lies in its rapid consumption and potential toxicity upon contact with food. However, it is worth mentioning that this issue has been addressed by presenting it (in most of its commercial forms) in easily replaceable sachets when they become saturated.

5.3.2. UV Radiation

Short-wave UV radiation between 100 and 280 nm (UV-C) is also capable of acting on the double bond of ethylene, generating oxidizing radicals that ultimately produce CO_2 and water in the reaction process, a phenomenon known as photocatalysis [219,220]. Additionally, this radiation can produce small amounts of ozone at these wavelengths, a molecule that also acts on ethylene, as will be discussed in the following points.

Bu and colleagues [221] applied UV-C radiation at 254 nm to tomatoes of the 'Zhenzhu' variety. Despite the direct impact of UV radiation on the fruit potentially causing adverse effects, these researchers observed a significant 20% reduction in the ethylene production rate. Mabusela and colleagues [222] in their review article studied ultraviolet radiation at 254 nm and 185 nm in a vacuum environment. At both wavelengths, ethylene elimination was observed through two pathways: firstly, the direct action of UV-C radiation on the double bond of ethylene and, secondly, by producing ROS by reacting with water and oxygen from the environment (considering that the relative humidity of fruit storage environments shows values around 90% and 21% for oxygen). These ROS then react with ethylene, producing CO_2 and water (Figure 11).

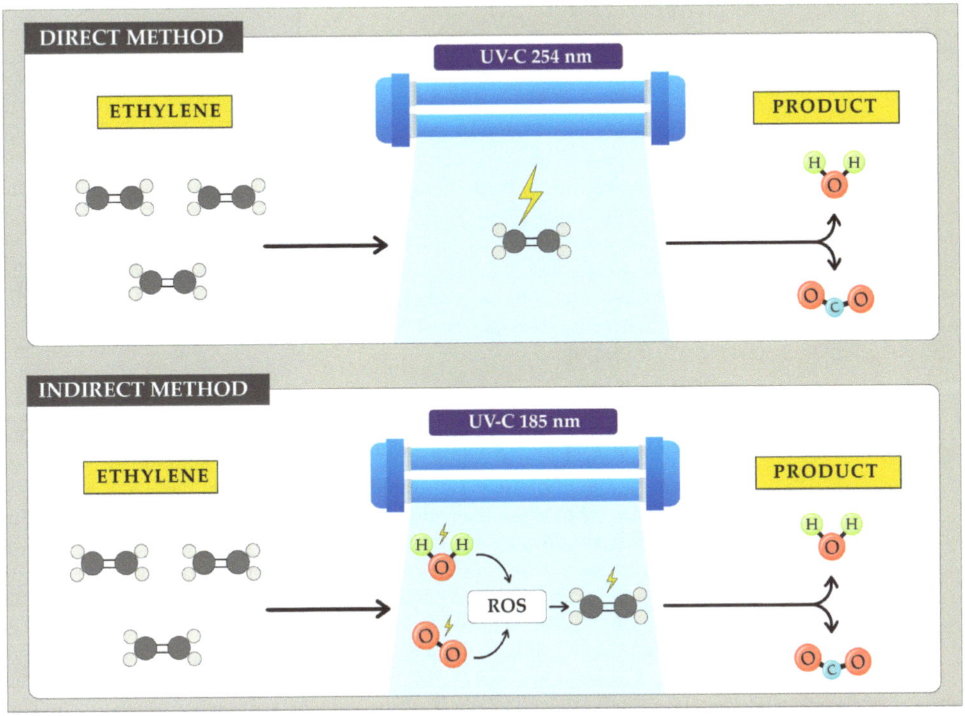

Figure 11. How two different wavelengths of UV-C radiation affect ethylene removal. The first one, at 254 nm, produces an indirect ethylene elimination by producing free radicals that act on ethylene. The second wavelength works directly on the ethylene molecules (own adaptation from [222]).

The effectiveness of the treatments studied by Mabusela and colleagues [222] depends on numerous factors that greatly influence the ethylene removal rate. One example is relative humidity. Water is the main source of ROS when attacked by radiation; therefore, low relative humidity significantly affects ethylene removal using this method. Currently, there is no consensus on the performance of this ethylene oxidation technique.

Additionally, UV-C radiation has been extensively studied for its ability to eliminate spores of microorganisms present in fruit storage environments [223]. This is an effect that must be considered, as fungal and bacterial diseases, among many other effects, increase the ethylene production rate in infected fruits [222].

This method has numerous advantages, such as ease of implementation, low energy consumption, effective ethylene removal under optimal conditions, and the elimination of spores, minimizing fungal and bacterial infections. However, it has several disadvantages that, if not considered, would reduce its effectiveness:

- Relative humidity: As mentioned earlier, environmental water is the main source of certain ROS, crucial for the efficacy of this method. The lower the relative humidity, the lower the ethylene removal achievable with UV-C radiation [224].
- Initial ethylene concentration: Increasing the initial concentration of ethylene, while keeping the photon energy constant, reduces the energy received by a greater number of ethylene molecules, thus reducing ethylene removal [157,225].
- Oxygen concentration: Oxygen is the precursor of certain ROS and ozone that facilitate this process. A low concentration of this gas will hinder ethylene removal [225].
- Direct incidence on fruits: When UV-C radiation is directly targeted at fruits, it causes structural changes that negatively affect their quality [222].

5.3.3. Titanium Dioxide (TiO_2)

Photocatalysis is another approach for eliminating ethylene [16,219,226]. Titanium dioxide (TiO_2) is a semiconductor material and one of the most frequently employed photocatalysts for ethylene degradation, a capacity primarily attributed to its unique photochemical reactivity and physical properties, which include high brightness (due to high refractive index) and resistance to discolouration [29].

The complete mechanism of organic compound oxidation by UV light radiation on titanium dioxide is described by Pathak and colleagues [225]. In summary, the reaction is based on the irradiation with UV wavelengths (hv) (around 240–380 nm) which causes titanium dioxide to generate electrons (e-) and other effects (reaction 1). These, when acting on the water in the environment, produce highly reactive hydroxyl radicals (·OH) (reaction 2). These radicals then react with organic compounds, resulting in CO_2 and water (reaction 3) [219,220].

- Reaction 1: TiO_2 + hν (UV energy/radiation) → TiO_2 + e− + h+
- Reaction 2: H_2O + h+ → ·OH + H+
- Reaction 3: ·OH + Organic compound (ethylene) → CO_2 + H_2O

Although the mechanism of photocatalytic oxidation has been described [225], the exact mechanism of reaction with ethylene remains a topic of debate due to the presence of various reaction intermediates that have not been fully elucidated [157].

Numerous studies have been conducted on the application of this ethylene elimination technology. Alves and colleagues [227] applied it to cherry tomatoes stored at 18 °C with a relative humidity of 85%. The treatment involved passing a constant airflow through the titanium dioxide and UV light system. During the storage period, cherry tomatoes treated with titanium dioxide irradiated with UV light showed a lower concentration of ethylene, lower respiration rate, lower total soluble solids content, and higher concentrations of lycopene and titratable acidity compared to the control, demonstrating that fruits treated with photocatalysis did not reach full ripeness. Li and colleagues [228] obtained similar results using the same tomato variety for their study.

In recent years, many studies on the effect of titanium dioxide–UV on ethylene have focused on its application as a component of films or fruit packaging. Among them, Fonseca and colleagues [229] evaluated the efficacy of polyethylene foam nets coated with a photocatalytic nanocomposite composed of gelatine and titanium dioxide for degrading the ethylene produced by papayas (*Carica papaya* L.) of the 'Golden' variety. The fruits treated with gelatine–titanium dioxide and irradiated with UV light showed a 60% lower ethylene accumulation than the fruit in the control group after four days of storage, as well as higher firmness and lower ripening index (soluble solids/titratable acidity ratio). Ghosh and colleagues [230] applied photoactivated titanium dioxide and chitosan to form a film to cover peppers of the 'Tejaswani' variety stored at room temperature (25 °C). The treated peppers maintained CSS and had higher firmness, cell integrity, hydration, and colour. De Chiara and colleagues [231] applied a powder composed of titanium dioxide and silicon dioxide in different proportions for the preservation of 'Camone' tomatoes. In tomatoes treated with an 80–20 ratio (TiO_2–SiO_2), ethylene completely degraded after 14 days of study. However, the fruits in this treatment did not reach the red colouration stage.

Titanium dioxide is categorized as a food additive by the U.S. Food and Drug Administration under code 21CFR73.575 and is inexpensive. This makes it cost-effective for application in fruit storage and transportation. However, there are some drawbacks to consider for its implementation:

- It requires very controlled storage conditions as the effectiveness of titanium dioxide depends on numerous variables such as humidity, temperature, or UV light intensity [157,193,206].
- It is not a selective method, as seen in the summary of the reaction triggered by titanium dioxide and UV radiation. In addition to ethylene, it acts on other organic

compounds such as aromatic compounds, which can affect the organoleptic quality of the fruits [232].

- It can increase the storage temperature, which may lead to fruit damage [233].
- If UV radiation, essential for this method to be effective, directly impacts the fruits, it can lead to negative consequences such as loss in aroma, rupture of cell membranes, and degradation of structures [234].

5.3.4. Ozone (O_3)

Ozone, typically produced with UV lamps, is a strongly oxidizing molecule that can also be used as an ethylene eliminator [219]. Ozone, being a highly reactive gas, acts on the double bond of ethylene, breaking it and dissociating this phytohormone into CO_2 and water according to the following reaction: $O_3 + C_2H_4 \rightarrow CO_2 + H_2O$. The application of this method is straightforward. It involves passing a flow of air from fruit storage facilities (rich in ethylene) through an ozone generator and returning it to the warehouse free of ethylene [235]. Liang and colleagues [236] analysed ethylene production, firmness, vitamin C content, malondialdehyde content, levels of ascorbate peroxidase, peroxidase, and aromatic compound contents of 'Hard Pink' tomatoes stored for 25 days at 10 °C and treated with 17.14 mg $(m^3)^{-1}$ of ozone for 1 h at the start of the study. The results showed that ozone treatment delayed the loss in firmness, bioactive compounds, and aromatic compounds and reduced the ethylene production rate by 50% during the first 15 days. Toti and colleagues [237] treated 'Cantaloupe' melons with 0.3 ppm of ozone at 6 °C for 13 days. Their results indicated that ozone treatment reduced the activity of cell wall enzymes (α-arabinopyranosidase, β-galactopyranosidase, and polygalacturonase) responsible for their degradation and hence the softening of melons. Ethylene concentration was also reduced. With all this, the quality of the treated melons was maintained for a longer period. More recently, Gao and colleagues [238] treated 'Muscat Hamburg' grapes, despite being non-climacteric fruits, with 14.98 mg $(m^3)^{-1}$ of ozone. The results revealed a reduction in ethylene production close to 30% in the fruits subjected to the treatment. Additionally, there was maintenance of antioxidant capacity with higher concentrations of enzymes such as catalase (CAT), peroxidase (POD), superoxide dismutase (SOD), and polyphenol oxidase (PPO), capable of removing reactive oxygen species (ROS) from plant tissues. Furthermore, a lower incidence of fungal diseases was observed in the treated fruits. Although considered safe by the FDA (U.S. Food and Drug Administration) and not complicated for industrial implementation, ozone presents several drawbacks:

- Handling is very challenging due to its easy decomposition into oxygen [159].
- If certain levels are exceeded, ozone shifts from offering significant improvements to being harmful as it can destroy tissues, leading to wounds that can promote ethylene production due to stress [219].
- Even though it is safe at low concentrations, ozone begins to have harmful effects on human health from 5 ppm onwards, including vision problems, sensation of asphyxia, headaches, pulmonary oedema, and coma when concentrations reach 50 ppm [239].

5.3.5. Palladium

Palladium is a scarce metal with numerous applications, including its use as a catalyst in chemical reactions and in medicine for surgical material production. It also has the ability to absorb hydrogen into its metallic structure and purify hydrocarbons in the automotive industry. This latter application prompted its use as a potential ethylene eliminator through catalysis. Similar to other ethylene oxidants, palladium irreversibly decomposes ethylene into CO_2 and water.

Bailén and colleagues [206] employed a mixture of granulated (20–60 mesh) activated carbon with 1% palladium on three tomato varieties, 'Beef', 'Mendez', and 'Raf'. Within the first 50 h, nearly 60% ethylene elimination was achieved in all three varieties, along with better preservation of firmness and colour, both internally and externally. Mok and colleagues [240] applied a mixture of ZSM-5 zeolite and 0.36% palladium in the conservation of 'Fuji' apples, introduced into quartz tubes through which air from the storage chamber passed. The results showed nearly 100% ethylene elimination after 40 days of storage. Tzeng and colleagues [241] observed a 99% ethylene removal in banana preservation using a similar mixture and application method.

Mansourbahmani and colleagues [41] employed palladium with nano zeolites in various concentrations (0%, 1%, 2.5%, and 5%) and compared this ethylene removal method with other ethylene eliminators or inhibitors such as potassium permanganate, 1-MCP, and UV-C, concluding that the palladium method was the most effective.

Although this method does not have many drawbacks, its industrial application proves to be very complicated due to palladium being a very scarce and difficult-to-extract metal and consequently extremely expensive. Additionally, the country with the highest production of this element worldwide is the Russian Federation, which is currently subject to numerous economic and commercial sanctions (due to the invasion of Ukraine), further increasing its price.

5.3.6. Cold Plasma and Other Technologies

Cold plasma technology has emerged as a promising method for ethylene scavenging, offering a non-thermal and eco-friendly approach to extend the shelf life and maintain the quality of fresh produce. This promising and novel method stands out (1) for being a non-thermal technology, which means it does not rely on heat to achieve its effects, and is particularly beneficial for ethylene scavenging as it avoids thermal damage to the produce, maintaining its nutritional and sensory properties; (2) as favouring microbial decontamination where cold plasma has shown effectiveness in microbial decontamination, enzyme denaturation, and pesticide degradation, which are all crucial for preserving the quality and safety of food products; and (3) for its versatility and applications where cold plasma technology allows for its application in various forms, such as direct treatment, plasma-treated water, and functional coatings. This flexibility makes it suitable for a wide range of uses in the food industry [242]. In addition, cold plasma is considered an eco-friendly technology due to its low energy requirements and minimal environmental impact. It represents a sustainable option for ethylene scavenging and food preservation [243]. The potential of cold plasma technology in food processing is vast, with ongoing research exploring its scalability and commercial viability. Future advancements are expected to further enhance its effectiveness and applicability in the food industry [242].

Other novel advances and developments in ethylene scavenging beyond cold plasma technology are as follows: (1) The first is nanotechnology, where the incorporation of nanoparticles into polymer matrices has been shown to play a major role in reducing the permeability of gases and the absorption of ethylene. This approach enhances the effectiveness of packaging materials in controlling ethylene levels around fresh produce [105]. For this reason, nanotechnology is at the forefront of advancing ethylene scavenging techniques to maintain the quality and extend the shelf life of fruits and vegetables. For example, nanosilica has been isolated from rice straw and used in a polyvinyl alcohol (PVA) formulation to coat paperboard, which showed improved barrier, mechanical, and surface properties, along with higher ethylene scavenging activity compared to uncoated paperboard [244]. Also, palladium encapsulated nanofibres has been developed for ethylene scavenging. Encapsulation of 1–2% PdCl2 in nanomats has increased the ethylene scavenging capacity significantly, proving to be effective in the presence of fruits like sapota [245]. (2) The second is mechanochemical synthesis, where recent studies have investigated the use of mechanochemically synthesized absorbents for ethylene removal. These materials have shown promising ethylene scavenging activity and offer a safer, innovative, and eco-friendlier approach to active packaging [246]. This process involves the use of mechanical force to induce chemical reactions, which can be used to synthesize metal–organic frameworks (MOFs) that are effective in ethylene scavenging [247].

6. Practical Implications and Future Directions

6.1. Energy Consumption

Controlled atmosphere storage or refrigeration require significant energy inputs to maintain optimal storage conditions. High energy consumption associated with refrigeration systems contributes to greenhouse gas emissions and environmental footprint, highlighting the need for energy-efficient technologies and alternative cooling methods [248–250].

6.2. Food, Microbiological, and Chemical Safety

Ensuring the safety of ethylene inhibition strategies involves addressing potential risks to human health, worker safety, and food quality: On one hand, workers involved in handling ethylene inhibition chemicals or applying postharvest treatments may be exposed to potential health hazards, such as respiratory irritation or skin sensitization. Proper training, personal protective equipment (PPE), and adherence to safety protocols are essential for minimizing chemical exposure risks and ensuring worker safety. On the other hand, ethylene inhibition treatments and packaging materials must comply with regulatory standards and food safety guidelines to prevent contamination and ensure consumer protection. Residual chemicals or contaminants from postharvest treatments or packaging materials should be carefully monitored and controlled to maintain food safety and quality throughout the supply chain. In addition, improper handling or storage of fruits treated with ethylene inhibition strategies may increase the risk of microbial contamination and foodborne illness. Effective sanitation practices, temperature control, and hygienic handling procedures are essential for minimizing microbial risks and ensuring food safety from farm to fork [160,251–254].

6.3. Exploring Sustainable and Eco-Friendly Alternatives

Packaging plays a critical role in maintaining fruit quality and extending shelf life by creating protective barriers against external factors, including ethylene exposure, physical damage, and microbial contamination. Increased adoption of single-use plastics in packaging can exacerbate plastic pollution and waste management issues, necessitating the adoption of recyclable or biodegradable packaging alternatives and promoting circular economy initiatives. Utilizing biodegradable or compostable packaging materials derived from renewable sources, such as bioplastics or cellulose-based films, can help to reduce plastic waste and promote circular economy principles, in addition to exploring innovative packaging designs and technologies that enhance gas barrier properties, extend shelf life, and minimize resource consumption throughout the packaging lifecycle. Innovation of smart packaging solutions equipped with sensors, indicators, and active components for real-time monitoring and control of ethylene levels, fruit quality, and storage conditions enhances supply chain transparency, traceability, and product integrity, ensuring end-to-end quality assurance and consumer confidence [255–263].

6.4. Biocontrol Agents

Biocontrol agents offer sustainable alternatives to chemical treatments, providing effective control of postharvest pathogens and decay while promoting environmental sustainability and food safety. Implementing integrated pest management (IPM) strategies that incorporate biocontrol agents can reduce reliance on synthetic pesticides and minimize environmental contamination risks [264,265].

6.5. Commercial and Industrial Applications

The implementation of ethylene inhibition strategies in the fruit industry encompasses various postharvest handling and storage practices aimed at extending shelf life, preserving quality, and ensuring market competitiveness.

Pretreatments and harvest treatments may be employed to mitigate ethylene production and enhance fruit quality. Techniques such as preharvest stress conditioning, including deficit irrigation or exposure to plant growth regulators, can modulate ethylene biosynthesis and improve fruit firmness, colour development, and flavour. Additionally, preharvest application of natural or bio-based ethylene inhibitors may offer sustainable alternatives to synthetic chemicals, reducing environmental impacts and chemical residues. Harvesting fruits at the appropriate stage of maturity, using sharp tools to prevent mechanical damage, and handling fruits gently to avoid bruising or injury are essential considerations. Prompt removal of harvested fruits from the field and rapid cooling to reduce respiration rates and ethylene production are key steps in preserving postharvest quality and extending shelf life [266–269].

Efficient transportation and distribution networks are essential for minimizing postharvest losses and ensuring timely delivery of fresh fruits to markets. Temperature-controlled transport vehicles, such as refrigerated trucks or containers, maintain optimal storage conditions throughout transit, preventing temperature fluctuations and ethylene build-up. Monitoring devices and data loggers enable real-time tracking of environmental parameters, ensuring compliance with quality standards and regulatory requirements [270,271].

Proper handling and display practices at retail outlets are crucial for preserving fruit quality and enhancing consumer appeal. Temperature-controlled display cases, ethylene filters, and airflow management systems help minimize ethylene exposure and maintain freshness. Training retail staff on proper handling techniques, including gentle handling, regular rotation of stock, and removal of damaged or overripe fruits, ensures optimal product presentation and reduces waste [272].

Finally, as for consumer education and behaviour change, it is possible to promote eco-friendly alternatives, sustainable consumption behaviours, and waste reduction strategies through education campaigns, labelling initiatives, and consumer engagement programmes [273,274].

7. Conclusions

In conclusion, this review provides a comprehensive examination of ethylene-mediated ripening in climacteric fruits and the strategies employed to delay this process, thereby enhancing postharvest quality. Beginning with an introduction to climacteric fruits and the pivotal role of ethylene in their ripening, this review delineated the scope and structure of the discussion. Ethylene, as a plant hormone, plays a fundamental role in initiating and coordinating fruit ripening, orchestrating a cascade of physiological and biochemical changes. Various ethylene inhibition strategies were explored, encompassing chemical inhibitors, temperature control, modified atmospheres, and controlled atmosphere storage. These strategies act by reducing ethylene levels or its effects, thereby extending shelf life and preserving fruit quality. Understanding ethylene production and receptor mechanisms is crucial for effective inhibition strategies. Insights into ethylene biosynthesis, enzyme regulation, and receptor targeting were discussed to elucidate the underlying processes.

This review highlighted the significant impact of ethylene inhibition on shelf life extension and postharvest quality. Examples and case studies underscored the efficacy of these strategies in maintaining firmness, colour, sugar content, and nutritional value while reducing decay and pathogen development. Sensory quality, including flavour, aroma, and texture, emerged as a key consideration in ethylene inhibition. Examination of sensory attributes and consumer preferences emphasized the importance of treated fruits in meeting market demands. Furthermore, alterations in volatile compounds due to ethylene inhibition were investigated, shedding light on aroma profiles and the potential for off-flavours. Molecular and genetic approaches offered promising avenues for reducing ethylene responsiveness and developing genetically modified varieties with improved shelf life. Integrated approaches, combining multiple inhibition strategies, demonstrated enhanced efficacy and synergy between postharvest treatments and packaging techniques. Commercial and industrial applications showcased success stories and economic benefits, illustrating the real-world impact of ethylene inhibition strategies. Looking ahead, future directions in ethylene inhibition encompass emerging research areas and technological advancements, including nanomaterials and precision agriculture. Leveraging these innovations holds the promise of further enhancing efficacy, sustainability, and market competitiveness in the fruit industry. Finally, environmental and safety considerations underscored the importance of exploring sustainable and eco-friendly alternatives, ensuring responsible stewardship of resources and minimizing adverse impacts on the environment and human health. In essence, this review underscores the critical role of ethylene inhibition in preserving postharvest quality, driving innovation, and shaping the future of the fruit industry. By embracing emerging technologies, advancing scientific knowledge, and fostering collaborative efforts, stakeholders can continue to unlock new opportunities and address challenges in ethylene-mediated ripening, ultimately benefiting producers, consumers, and the environment.

Author Contributions: Conceptualization, R.A.-S., J.R.A.-M., S.L.-M. and A.J.P.-L.; methodology, R.A.-S., J.R.A.-M., S.L.-M. and A.J.P.-L.; software, R.A.-S. and J.R.A.-M.; validation, J.R.A.-M., S.L.-M. and A.J.P.-L.; formal analysis, R.A.-S., J.R.A.-M., S.L.-M. and A.J.P.-L.; investigation, R.A.-S., J.R.A.-M., S.L.-M. and A.J.P.-L.; resources, J.R.A.-M. and S.L.-M.; data curation, R.A.-S., J.R.A.-M. and A.J.P.-L.; writing—original draft preparation, R.A.-S., J.R.A.-M., S.L.-M. and A.J.P.-L.; writing—review and editing, R.A.-S., J.R.A.-M., S.L.-M. and A.J.P.-L.; visualization, A.J.P.-L., J.R.A.-M. and S.L.-M.; supervision, J.R.A.-M. and S.L.-M.; project administration, J.R.A.-M. and S.L.-M.; funding acquisition, J.R.A.-M. and S.L.-M. All authors have read and agreed to the published version of the manuscript.

Funding: Nuevas Tecnologías Agroalimentarias (KEEPCOOL): CFE/KE/76-19; "Cátedra Emprendimiento en el Ámbito Agroalimentario" from Universidad Católica San Antonio de Murcia (UCAM) and Banco Santander: CFE/KE/76-19.

Data Availability Statement: No new data were created or analyzed in this study. Data sharing is not applicable to this article.

Acknowledgments: The authors gratefully acknowledge Antonio Cerdá (Cátedra UCAM-Banco Santander de Emprendimiento en el Ámbito Agroalimentario director), FECOAM ("Federación de Cooperativas Agrarias de Murcia"), especially Antonio Sanz, Lola Mondéjar, and Miriam Monje for their help with figures.

Conflicts of Interest: The doctoral thesis (Ph.D.) of Ramiro Alonso Salinas, co-author of this work, is being co-financed by the company Nuevas Tecnologías Agroalimentarias (KEEPCOOL) within the UCAM Universidad Católica de Murcia Industrial Doctorate Program. The rest of the authors of this article declare that their contribution to this research was carried out in the absence of commercial or financial relationships that could be constructed as possible conflicts of interest.

References

1. Ministerio de Agricultura, Alimentación y Medio Ambiente. Estrategia Más Alimento, Menos Desperdicio. 2013, 60. Aprovecha los alimentos (alimentosdespana.es). Available online: https://www.alimentosdespana.es/es/estrategia-alimentos-espana/aprovecha-los-alimentos/#:~:text=Esta%20iniciativa%20del%20Ministerio%20de%20Agricultura,%20Pesca%20y,contribuir%20a%20la%20reducci%C3%B3n%20del%20desperdicio%20de%20alimentos (accessed on 25 June 2024).
2. De Laurentiis, V.; Corrado, S.; Sala, S. Quantifying Household Waste of Fresh Fruit and Vegetables in the EU. *Waste Manag.* **2018**, *77*, 238–251. [CrossRef] [PubMed]
3. Informe Semanal RTVE. Comer Bien, Tirar Menos. 2023. Available online: https://www.rtve.es/play/videos/informe-semanal/comer-bien-tirar-menos/6832236/ (accessed on 25 June 2024).
4. Yang, S.F.; Hoffman, N.E. Ethylene Biosynthesis and Its Regulation in Higher Plants. *Annu. Rev. Plant Physiol.* **1984**, *35*, 155–189. [CrossRef]
5. Kader, A. *Tecnología Postcosecha de Cultivos Hortofrutícolas*; UCANR Publications: Irvine, CA, USA, 2011; Volume 311, ISBN 1601077440.
6. Neljubow, D. Uber Die Horizontale Nutation Der Stengel von Pisum Sativum Und Einiger Anderen Planzen. *Bot. Centralbl. Beih.* **1901**, *10*, 128–139.
7. Cousins, H.H., III. Agricultural Experiments: Citrus. *Annu. Rep. Dep. Agric. Jam.* **1910**, 1–7.
8. Doubt, S.L. The Response of Plants to Illuminating Gas. *Bot. Gaz.* **1917**, *63*, 209–224. [CrossRef]
9. Denny, F.E. Effect of Ethylene Upon Respiration of Lemons. *Bot. Gaz.* **1924**, *77*, 322–329. [CrossRef]
10. Gane, R. Production of Ethylene by Some Ripening Fruits. *Nature* **1934**, *134*, 1008. [CrossRef]
11. Gane, R. The Formation of Ethylene by Plant Tissues, and Its Significance in the Ripening of Fruits. *J. Pomol. Hortic. Sci.* **1935**, *13*, 351–358. [CrossRef]

12. Crocker, W.; Hitchcock, A.E.; Zimmerman, P.W. Similarities in the Effects of Ethylene and the Plant Auxins. *Contrib. Boyce Thompson Inst.* **1935**, *7*, 231–248.
13. Forsyth, F.R.; Eaves, C.A.; Lightfoot, H.J. Storage Quality of McIntosh Appels as Affected Nby Removal of Ethylene from the Storage Atmosphere. *Can. J. Plant Sci.* **1969**, *49*, 567–572. [CrossRef]
14. Southwick, F.W.; Smock, R.M. Lengthening the Storage Life of Apples By Removal of Volatile Materials From the Storage Atmosphere. *Plant Physiol.* **1943**, *18*, 716–717. [CrossRef] [PubMed]
15. Scott, K.J.; Wills, R.B.H.; Patterson, B.D. Removal by Ultra-Violet Lamp of Ethylene and Other Hydrocarbons Produced by Bananas. *J. Sci. Food Agric.* **1971**, *22*, 496–497. [CrossRef]
16. Keller, N.; Ducamp, M.-N.; Robert, D.; Keller, V. Ethylene Removal and Fresh Product Storage: A Challenge at the Frontiers of Chemistry. Toward an Approach by Photocatalytic Oxidation. *Chem. Rev.* **2013**, *113*, 5029–5070. [CrossRef] [PubMed]
17. Álvarez-Hernández, M.H.; Martínez-Hernández, G.B.; Avalos-Belmontes, F.; Castillo-Campohermoso, M.A.; Contreras-Esquivel, J.C.; Artés-Hernández, F. Potassium Permanganate-Based Ethylene Scavengers for Fresh Horticultural Produce as an Active Packaging. *Food Eng. Rev.* **2019**, *11*, 159–183. [CrossRef]
18. Awalgaonkar, G.; Beaudry, R.; Almenar, E. Ethylene-Removing Packaging: Basis for Development and Latest Advances. *Compr. Rev. Food Sci. Food Saf.* **2020**, *19*, 3980–4007. [CrossRef] [PubMed]
19. Iqbal, N.; Nazar, R.; Khan, M.I.R.; Khan, N.A. Variation in Photosynthesis and Growth of Mustard Cultivars: Role of Ethylene Sensitivity. *Sci. Hortic.* **2012**, *135*, 1–6. [CrossRef]
20. Giovannoni, J.J. Genetic Regulation of Fruit Development and Ripening. *Plant Cell* **2004**, *16*, S170–S180. [CrossRef] [PubMed]
21. Carrari, F.; Fernie, A.R. Metabolic Regulation Underlying Tomato Fruit Development. *J. Exp. Bot.* **2006**, *57*, 1883–1897. [CrossRef] [PubMed]
22. Cherian, S.; Figueroa, C.R.; Nair, H. "Movers and Shakers" in the Regulation of Fruit Ripening: A Cross-Dissection of Climacteric versus Non-Climacteric Fruit. *J. Exp. Bot.* **2014**, *65*, 4705–4722. [CrossRef] [PubMed]
23. Liu, M.; Pirrello, J.; Chervin, C.; Roustan, J.P.; Bouzayen, M. Ethylene Control of Fruit Ripening: Revisiting the Complex Network of Transcriptional Regulation. *Plant Physiol.* **2015**, *169*, 2380–2390. [CrossRef] [PubMed]
24. Leng, P.; Yuan, B.; Guo, Y. The Role of Abscisic Acid in Fruit Ripening and Responses to Abiotic Stress. *J. Exp. Bot.* **2014**, *65*, 4577–4588. [CrossRef] [PubMed]
25. Serrano, M.; Martínez-Madrid, M.C.; Romojaro, F. Ethylene Biosynthesis and Polyamine and ABA Levels in Cut Carnations Treated with Aminotriazole. *J. Am. Soc. Hortic. Sci. JASHS* **1999**, *124*, 81–85. [CrossRef]
26. Symons, G.M.; Chua, Y.-J.; Ross, J.J.; Quittenden, L.J.; Davies, N.W.; Reid, J.B. Hormonal Changes during Non-Climacteric Ripening in Strawberry. *J. Exp. Bot.* **2012**, *63*, 4741–4750. [CrossRef] [PubMed]
27. Liao, X.; Li, M.; Liu, B.; Yan, M.; Yu, X.; Zi, H.; Liu, R.; Yamamuro, C. Interlinked Regulatory Loops of ABA Catabolism and Biosynthesis Coordinate Fruit Growth and Ripening in Woodland Strawberry. *Proc. Natl. Acad. Sci. USA* **2018**, *115*, E11542–E11550. [CrossRef] [PubMed]
28. Forlani, S.; Masiero, S.; Mizzotti, C. Fruit Ripening: The Role of Hormones, Cell Wall Modifications, and Their Relationship with Pathogens. *J. Exp. Bot.* **2019**, *70*, 2993–3006. [CrossRef] [PubMed]
29. Wei, H.; Seidi, F.; Zhang, T.; Jin, Y.; Xiao, H. Ethylene Scavengers for the Preservation of Fruits and Vegetables: A Review. *Food Chem.* **2021**, *337*, 127750. [CrossRef] [PubMed]
30. Lara, I.; Vendrell, M. Cold-Induced Ethylene Biosynthesis Is Differentially Regulated in Peel and Pulp Tissues of 'Granny Smith' Apple Fruit. *Postharvest Biol. Technol.* **2003**, *29*, 109–119. [CrossRef]
31. Sdiri, S.; Navarro, P.; Monterde, A.; Benabda, J.; Salvador, A. New Degreening Treatments to Improve the Quality of Citrus Fruit Combining Different Periods with and without Ethylene Exposure. *Postharvest Biol. Technol.* **2012**, *63*, 25–32. [CrossRef]
32. Mathooko, F.M.; Kubo, Y.; Inaba, A.; Nakamura, R. Characterization of the Regulation of Ethylene Biosynthesis in Tomato Fruit by Carbon Dioxide and Diazocyclopentadiene. *Postharvest Biol. Technol.* **1995**, *5*, 221–233. [CrossRef]
33. Alberto, O.; Alzate, T. Hallazgos De La Biosíntesis Del Etileno En Frutas Climatéricas Y De Los Factores Que Afectan La Ruta Metabólica. *Rev. Aliment. Hoy* **2014**, *22*, 46–63.
34. Asif, M.H.; Pathak, N.; Solomos, T.; Trivedi, P.K. Effect of Low Oxygen, Temperature and 1-Methylcyclopropene on the Expression of Genes Regulating Ethylene Biosynthesis and Perception during Ripening in Apple. *S. Afr. J. Bot.* **2009**, *75*, 137–144. [CrossRef]
35. Yang, Y.; Zheng, Y.; Liu, C.; Chen, L.; Ma, J.; Sheng, J.; Shen, L. Inhibition of Nitric Oxide Synthesis Delayed Mature-Green Tomato Fruits Ripening Induced by Inhibition of Ethylene. *Sci. Hortic.* **2016**, *211*, 95–101. [CrossRef]
36. Li, Y.; Golding, J.B.; Arcot, J.; Wills, R.B.H. Continuous Exposure to Ethylene in the Storage Environment Adversely Affects "Afourer" Mandarin Fruit Quality. *Food Chem.* **2018**, *242*, 585–590. [CrossRef] [PubMed]
37. Li, S.-J.; Xie, X.-L.; Liu, S.-C.; Chen, K.-S.; Yin, X.-R. Auto- and Mutual-Regulation between Two CitERFs Contribute to Ethylene-Induced Citrus Fruit Degreening. *Food Chem.* **2019**, *299*, 125163. [CrossRef] [PubMed]
38. Wills, R.B.H.; Golding, J.B. Reduction of Energy Usage in Postharvest Horticulture through Management of Ethylene. *J. Sci. Food Agric.* **2015**, *95*, 1379–1384. [CrossRef] [PubMed]
39. Vilela, C.; Kurek, M.; Hayouka, Z.; Röcker, B.; Yildirim, S.; Antunes, M.D.C.; Nilsen-Nygaard, J.; Pettersen, M.K.; Freire, C.S.R. A Concise Guide to Active Agents for Active Food Packaging. *Trends Food Sci. Technol.* **2018**, *80*, 212–222. [CrossRef]
40. Cara, B.; Giovannoni, J.J. Molecular Biology of Ethylene during Tomato Fruit Development and Maturation. *Plant Sci.* **2008**, *175*, 106–113. [CrossRef]

41. Mansourbahmani, S.; Ghareyazie, B.; Zarinnia, V.; Kalatejari, S.; Mohammadi, R.S. Study on the Efficiency of Ethylene Scavengers on the Maintenance of Postharvest Quality of Tomato Fruit. *J. Food Meas. Charact.* **2018**, *12*, 691–701. [CrossRef]
42. Emadpour, M.; Ghareyazie, B.; Kalaj, Y.R.; Entesari, M.; Bouzari, N. Effect of the Potassium Permanganate Coated Zeolite Nanoparticles on the Quality Characteristic and Shelf Life of Peach and Nectarine. *Int. J. Agric. Technol.* **2015**, *11*, 1411–1421.
43. Gavin, C.; Barzallo, D.; Vera, H.; Lazo, R. Revisión Bibliográfica: Etileno En Poscosecha, Tecnologías Para Su Manejo y Control. *Ecuadorian Sci. J.* **2021**, *5*, 163–178. [CrossRef]
44. Li, L.; Lichter, A.; Chalupowicz, D.; Gamrasni, D.; Goldberg, T.; Nerya, O.; Ben-Arie, R.; Porat, R. Effects of the Ethylene-Action Inhibitor 1-Methylcyclopropene on Postharvest Quality of Non-Climacteric Fruit Crops. *Postharvest Biol. Technol.* **2016**, *111*, 322–329. [CrossRef]
45. Zhang, W.Q.; Hu, Y.X.; Liu, J.; Wang, H.; Wei, J.H.; Sun, P.D.; Wu, L.F.; Zheng, H.J. Progress of Ethylene Action Mechanism and Its Application on Plant Type Formation in Crops. *Saudi J. Biol. Sci.* **2020**, *27*, 1667–1673. [CrossRef] [PubMed]
46. Serna-Escolano, V.; Gimenez, M.J.; Garcia-Pastor, M.E.; Dobon-Suarez, A.; Pardo-Pina, S.; Zapata, P.J. Effects of Degreening Treatment on Quality and Shelf-Life of Organic Lemons. *Agronomy* **2022**, *12*, 270. [CrossRef]
47. Li, L.; Kaplunov, T.; Zutahy, Y.; Daus, A.; Porat, R.; Lichter, A. The Effects of 1-Methylcyclopropane and Ethylene on Postharvest Rachis Browning in Table Grapes. *Postharvest Biol. Technol.* **2015**, *107*, 16–22. [CrossRef]
48. Özdemir, A.; Çandır, E. Impact of Hot Water and Modified Atmosphere Packaging Treatments on the Postharvest Quality of Pomegranate Fruit Cv. Hicaznar. *Tarım Bilim. Derg.* **2021**, *27*, 304–311. [CrossRef]
49. Zhang, L.-H.; Zhang, Y.-H.; Li, L.-L.; Li, Y.-X. Effects of 1-MCP on peel browning of pomegranates. In Proceedings of the Acta Horticulturae, International Society for Horticultural Science (ISHS), Leuven, Belgium, 31 October 2008; pp. 275–282.
50. Flores, F.; El Yahyaoui, F.; de Billerbeck, G.; Romojaro, F.; Latché, A.; Bouzayen, M.; Pech, J.; Ambid, C. Role of Ethylene in the Biosynthetic Pathway of Aliphatic Ester Aroma Volatiles in Charentais Cantaloupe Melons. *J. Exp. Bot.* **2002**, *53*, 201–206. [CrossRef] [PubMed]
51. Saltveit, M.E. Effect of Ethylene on Quality of Fresh Fruits and Vegetables. *Postharvest Biol. Technol.* **1999**, *15*, 279–292. [CrossRef]
52. Abeles, F.B.; Morgan, P.W.; Saltveit, M.E. *Chapter 3—The Biosynthesis of Ethylene*, 2nd ed.; Abeles, F.B., Morgan, P.W., Saltveit, M.E.B.T.-E., Eds.; Academic Press: New York, NY, USA, 1992; pp. 26–55. ISBN 978-0-08-091628-6.
53. Bleecker, A.B.; Kende, H. Ethylene: A Gaseous Signal Molecule in Plants. *Annu. Rev. Cell Dev. Biol.* **2000**, *16*, 1–18. [CrossRef] [PubMed]
54. Sadeghi, K.; Lee, Y.; Seo, J. Ethylene Scavenging Systems in Packaging of Fresh Produce: A Review. *Food Rev. Int.* **2021**, *37*, 155–176. [CrossRef]
55. Bleecker, A.B.; Estelle, M.A.; Somerville, C.; Kende, H. Insensitivity to Ethylene Conferred by a Dominant Mutation in Arabidopsis Thaliana. *Science* **1988**, *241*, 1086–1089. [CrossRef] [PubMed]
56. Lieberman, M.; Kunishi, A. Stimulation of Ethylene Production in Apple Tissue Slices by Methionine. *Plant Physiol.* **1966**, *41*, 376–382. [CrossRef]
57. Bradford, K.J. Shang Fa Yang: Pioneer in Plant Ethylene Biochemistry. *Plant Sci.* **2008**, *175*, 2–7. [CrossRef]
58. Zhang, T.C. Ethylene Biosynthesis and Signal Pathway Model. *Chin. Bull. Bot.* **2006**, *23*, 519–530.
59. Buchanan, B.B.; Gruissem, W.; Jones, R.L. *Biochemistry and Molecular Biology of Plants*; American Society of Plant Biologists: Chichester, UK, 2000.
60. Li, J.-F.; Qu, L.-H.; Li, N. Tyr152 Plays a Central Role in the Catalysis of 1-Aminocyclopropane-1-Carboxylate Synthase. *J. Exp. Bot.* **2005**, *56*, 2203–2210. [CrossRef]
61. Ravanel, S.; Gakière, B.; Job, D.; Douce, R. The Specific Features of Methionine Biosynthesis and Metabolism in Plants. *Proc. Natl. Acad. Sci. USA* **1998**, *95*, 7805–7812. [CrossRef] [PubMed]
62. Pretel, M.; Serrano, M.; Amoros, A.; Romojaro, F. Ripening and Ethylene Biosynthesis in Controlled Atmosphere Stored Apricots. *Eur. Food Res. Technol.* **1999**, *209*, 130–134. [CrossRef]
63. Yamagami, T.; Tsuchisaka, A.; Yamada, K.; Haddon, W.F.; Harden, L.A.; Theologis, A. Biochemical Diversity among the 1-Amino-Cyclopropane-1-Carboxylate Synthase Isozymes Encoded by the Arabidopsis Gene Family. *J. Biol. Chem.* **2003**, *278*, 49102–49112. [CrossRef] [PubMed]
64. De Paepe, A.; Van Der Straeten, D. Ethylene Biosynthesis and Signaling: An Overview. In *Plant Hormones*; Litwack, G., Ed.; Vitamins & Hormones; Academic Press: Cambridge, MA, USA, 2005; Volume 72, pp. 399–430.
65. Chung, M.-C.; Chou, S.-J.; Kuang, L.-Y.; Charng, Y.-Y.; Yang, S.F. Subcellular Localization of 1-Aminocyclopropane-1-Carboxylic Acid Oxidase in Apple Fruit. *Plant Cell Physiol.* **2002**, *43*, 549–554. [CrossRef] [PubMed]
66. Kawai, Y.; Ono, E.; Mizutani, M. Evolution and Diversity of the 2-Oxoglutarate-Dependent Dioxygenase Superfamily in Plants. *Plant J. Mol. Biol.* **2014**, *78*, 328–343. [CrossRef] [PubMed]
67. Van de Poel, B.; Van Der Straeten, D. 1-Aminocyclopropane-1-Carboxylic Acid (ACC) in Plants: More than Just the Precursor of Ethylene! *Front. Plant Sci.* **2014**, *5*, 640. [CrossRef]
68. Hudgins, J.W.; Ralph, S.G.; Franceschi, V.R.; Bohlmann, J. Ethylene in Induced Conifer Defense: CDNA Cloning, Protein Expression, and Cellular and Subcellular Localization of 1-Aminocyclopropane-1-Carboxylate Oxidase in Resin Duct and Phenolic Parenchyma Cells. *Planta* **2006**, *224*, 865–877. [CrossRef] [PubMed]
69. Pattyn, J.; Vaughan-Hirsch, J.; van de Poel, B. The Regulation of Ethylene Biosynthesis: A Complex Multilevel Control Circuitry. *New Phytol.* **2021**, *229*, 770–782. [CrossRef] [PubMed]

70. Argueso, C.T.; Hansen, M.; Kieber, J.J. Regulation of Ethylene Biosynthesis. *J. Plant Growth Regul.* **2007**, *26*, 92–105. [CrossRef]
71. Binder, B.M. Ethylene Signaling in Plants. *J. Biol. Chem.* **2020**, *295*, 7710–7725. [CrossRef] [PubMed]
72. Javid, E.; Corpas Iguarán, E.; Alberto, O.; Alzate, T. Hallazgos de la biosíntesis del etileno en frutas climatéricas y de los factores que afectan la ruta metabólica. *Aliment. Hoy* **2014**, *22*, 46–63.
73. Stearns, J.C.; Glick, B.R. Transgenic Plants with Altered Ethylene Biosynthesis or Perception. *Biotechnol. Adv.* **2003**, *21*, 193–210. [CrossRef] [PubMed]
74. Ju, C.; Yoon, G.M.; Shemansky, J.M.; Lin, D.Y.; Ying, Z.I.; Chang, J.; Garrett, W.M.; Kessenbrock, M.; Groth, G.; Tucker, M.L.; et al. CTR1 Phosphorylates the Central Regulator EIN2 to Control Ethylene Hormone Signaling from the ER Membrane to the Nucleus in Arabidopsis. *Proc. Natl. Acad. Sci. USA* **2012**, *109*, 19486–19491. [CrossRef]
75. Qiao, H.; Shen, Z.; Huang, S.C.; Schmitz, R.J.; Urich, M.A.; Briggs, S.P.; Ecker, J.R. Processing and Subcellular Trafficking of ER-Tethered EIN2 Control Response to Ethylene Gas. *Science* **2012**, *338*, 390–393. [CrossRef]
76. Merchante, C.; Brumos, J.; Yun, J.; Hu, Q.; Spencer, K.R.; Enríquez, P.; Binder, B.M.; Heber, S.; Stepanova, A.N.; Alonso, J.M. Gene-Specific Translation Regulation Mediated by the Hormone-Signaling Molecule EIN2. *Cell* **2015**, *163*, 684–697. [CrossRef] [PubMed]
77. Binder, B.M.; Mortimore, L.A.; Stepanova, A.N.; Ecker, J.R.; Bleecker, A.B. Short-Term Growth Responses to Ethylene in Arabidopsis Seedlings Are EIN3/EIL1 Independent. *Plant Physiol.* **2004**, *136*, 2921–2927. [CrossRef] [PubMed]
78. Chang, K.N.; Zhong, S.; Weirauch, M.T.; Hon, G.; Pelizzola, M.; Li, H.; Huang, S.-S.C.; Schmitz, R.J.; Urich, M.A.; Kuo, D.; et al. Temporal Transcriptional Response to Ethylene Gas Drives Growth Hormone Cross-Regulation in Arabidopsis. *eLife* **2013**, *2*, e00675. [CrossRef] [PubMed]
79. Gao, R.; Stock, A.M. Biological Insights from Structures of Two-Component Proteins. *Annu. Rev. Microbiol.* **2009**, *63*, 133–154. [CrossRef] [PubMed]
80. Kehoe, D.M.; Grossman, A.R. Similarity of a Chromatic Adaptation Sensor to Phytochrome and Ethylene Receptors. *Science* **1996**, *273*, 1409–1412. [CrossRef] [PubMed]
81. Martin, W.; Rujan, T.; Richly, E.; Hansen, A.; Cornelsen, S.; Lins, T.; Leister, D.; Stoebe, B.; Hasegawa, M.; Penny, D. Evolutionary Analysis of Arabidopsis, Cyanobacterial, and Chloroplast Genomes Reveals Plastid Phylogeny and Thousands of Cyanobacterial Genes in the Nucleus. *Proc. Natl. Acad. Sci. USA* **2002**, *99*, 12246–12251. [CrossRef] [PubMed]
82. Mount, S.M.; Chang, C. Evidence for a Plastid Origin of Plant Ethylene Receptor Genes. *Plant Physiol.* **2002**, *130*, 10–14. [CrossRef] [PubMed]
83. Timmis, J.N.; Ayliffe, M.A.; Huang, C.Y.; Martin, W. Endosymbiotic Gene Transfer: Organelle Genomes Forge Eukaryotic Chromosomes. *Nat. Rev. Genet.* **2004**, *5*, 123–135. [CrossRef] [PubMed]
84. Schaller, G.E.; Shiu, S.-H.; Armitage, J.P. Two-Component Systems and Their Co-Option for Eukaryotic Signal Transduction. *Curr. Biol.* **2011**, *21*, R320–R330. [CrossRef] [PubMed]
85. Hérivaux, A.; Dugé de Bernonville, T.; Roux, C.; Clastre, M.; Courdavault, V.; Gastebois, A.; Bouchara, J.-P.; James, T.Y.; Latgé, J.-P.; Martin, F.; et al. The Identification of Phytohormone Receptor Homologs in Early Diverging Fungi Suggests a Role for Plant Sensing in Land Colonization by Fungi. *mBio* **2017**, *8*, e01739-16. [CrossRef] [PubMed]
86. Rivarola, M.; McClellan, C.A.; Resnick, J.S.; Chang, C. ETR1-Specific Mutations Distinguish ETR1 from Other Arabidopsis Ethylene Receptors as Revealed by Genetic Interaction with RTE1. *Plant Physiol.* **2009**, *150*, 547–551. [CrossRef] [PubMed]
87. Rodríguez, F.I.; Esch, J.J.; Hall, A.E.; Binder, B.M.; Schaller, G.E.; Bleecker, A.B. A Copper Cofactor for the Ethylene Receptor ETR1 from Arabidopsis. *Science* **1999**, *283*, 996–998. [CrossRef] [PubMed]
88. Lacey, R.F.; Binder, B.M. Ethylene Regulates the Physiology of the Cyanobacterium Synechocystis Sp. PCC 6803 via an Ethylene Receptor. *Plant Physiol.* **2016**, *171*, 2798–2809. [CrossRef] [PubMed]
89. Allen, C.J.; Lacey, R.F.; Binder Bickford, A.B.; Beshears, C.P.; Gilmartin, C.J.; Binder, B.M. Cyanobacteria Respond to Low Levels of Ethylene. *Front. Plant Sci.* **2019**, *10*, 950. [CrossRef] [PubMed]
90. Evans, D.E.; Bengochea, T.; Cairns, A.J.; Dodds, J.H.; Hall, M.A. Studies on Ethylene Binding by Cell-Free Preparations from Cotyledons of Phaseolus Vulgaris L.: Subcellular Localization. *Plant Cell Environ.* **1982**, *5*, 101–107. [CrossRef]
91. Evans, D.E.; Dodds, J.H.; Lloyd, P.C.; Apgwynn, I.; Hall, M.A. A Study of the Subcellular Localisation of an Ethylene Binding Site in Developing Cotyledons of *Phaseolus vulgaris* L. by High Resolution Autoradiography. *Planta* **1982**, *154*, 48–52. [CrossRef]
92. Chen, Y.-F.; Randlett, M.D.; Findell, J.L.; Schaller, G.E. Localization of the Ethylene Receptor ETR1 to the Endoplasmic Reticulum of Arabidopsis. *J. Biol. Chem.* **2002**, *277*, 19861–19866. [CrossRef] [PubMed]
93. Chen, Y.-F.; Shakeel, S.N.; Bowers, J.; Zhao, X.-C.; Etheridge, N.; Schaller, G.E. Ligand-Induced Degradation of the Ethylene Receptor ETR2 through a Proteasome-Dependent Pathway in Arabidopsis. *J. Biol. Chem.* **2007**, *282*, 24752–24758. [CrossRef] [PubMed]
94. Ma, B.; Cui, M.-L.; Sun, H.-J.; Takada, K.; Mori, H.; Kamada, H.; Ezura, H. Subcellular Localization and Membrane Topology of the Melon Ethylene Receptor CmERS1. *Plant Physiol.* **2006**, *141*, 587–597. [CrossRef] [PubMed]
95. Zhong, S.; Lin, Z.; Grierson, D. Tomato Ethylene Receptor-CTR Interactions: Visualization of NEVER-RIPE Interactions with Multiple CTRs at the Endoplasmic Reticulum. *J. Exp. Bot.* **2008**, *59*, 965–972. [CrossRef] [PubMed]
96. Dong, C.-H.; Rivarola, M.; Resnick, J.S.; Maggin, B.D.; Chang, C. Subcellular Co-Localization of Arabidopsis RTE1 and ETR1 Supports a Regulatory Role for RTE1 in ETR1 Ethylene Signaling. *Plant J. Cell Mol. Biol.* **2008**, *53*, 275–286. [CrossRef]

97. Grefen, C.; Städele, K.; Růzicka, K.; Obrdlik, P.; Harter, K.; Horák, J. Subcellular Localization and in Vivo Interactions of the Arabidopsis Thaliana Ethylene Receptor Family Members. *Mol. Plant* **2008**, *1*, 308–320. [CrossRef] [PubMed]
98. Chang, C.; Kwok, S.F.; Bleecker, A.B.; Meyerowitz, E.M. Arabidopsis Ethylene-Response Gene ETR1: Similarity of Product to Two-Component Regulators. *Science* **1993**, *262*, 539–544. [CrossRef] [PubMed]
99. Hua, J.; Chang, C.; Sun, Q.; Meyerowitz, E.M. Ethylene Insensitivity Conferred by Arabidopsis ERS Gene. *Science* **1995**, *269*, 1712–1714. [CrossRef] [PubMed]
100. Hua, J.; Sakai, H.; Nourizadeh, S.; Chen, Q.G.; Bleecker, A.B.; Ecker, J.R.; Meyerowitz, E.M. EIN4 and ERS2 Are Members of the Putative Ethylene Receptor Gene Family in Arabidopsis. *Plant Cell* **1998**, *10*, 1321–1332. [CrossRef] [PubMed]
101. Schaller, G.E.; Bleecker, A.B. Ethylene-Binding Sites Generated in Yeast Expressing the Arabidopsis ETR1 Gene. *Science* **1995**, *270*, 1809–1811. [CrossRef] [PubMed]
102. Alexander, L.; Grierson, D. Ethylene Biosynthesis and Action in Tomato: A Model for Climacteric Fruit Ripening. *J. Exp. Bot.* **2002**, *53*, 2039–2055. [CrossRef] [PubMed]
103. Kays, S.J.; Paull, R.E. *Postharvest Biology*; Exon Press: Athens, Greece, 2004; ISBN 1888186542/9781888186543.
104. Tucker, G.; Yin, X.; Zhang, A.; Wang, M.; Zhu, Q.; Liu, X.; Xie, X.; Chen, K.; Grierson, D. Ethylene† and Fruit Softening. *Food Qual. Saf.* **2017**, *1*, 253–267. [CrossRef]
105. Ebrahimi, A.; Khajavi, M.Z.; Ahmadi, S.; Mortazavian, A.M.; Abdolshahi, A.; Rafiee, S.; Farhoodi, M. Novel Strategies to Control Ethylene in Fruit and Vegetables for Extending Their Shelf Life: A Review. *Int. J. Environ. Sci. Technol.* **2021**, *19*, 4599–4610. [CrossRef]
106. Mope, C.; Adegoroye, A.; Oluwalade, T.A.; Adeyelu, A.A. Ethylene Management in Fresh Produce Transport. *Asian J. Food Res. Nutr.* **2024**, *3*, 60–71.
107. Dong, T.; Zheng, T.; Fu, W.; Guan, L.; Jia, H.; Fang, J. The Effect of Ethylene on the Color Change and Resistance to Botrytis Cinerea Infection in 'Kyoho' Grape Fruits. *Foods* **2020**, *9*, 892. [CrossRef] [PubMed]
108. Yang, X.; Chen, L.; Yang, Y.; Guo, X.; Chen, G.; Xiong, X.; Dong, D.; Li, G. Transcriptome Analysis Reveals That Exogenous Ethylene Activates Immune and Defense Responses in a High Late Blight Resistant Potato Genotype. *Sci. Rep.* **2020**, *10*, 21294. [CrossRef] [PubMed]
109. Ma, N.; Sun, P.; Li, Z.-Y.; Zhang, F.-J.; Wang, X.-F.; You, C.-X.; Zhang, C.-L.; Zhang, Z. Plant Disease Resistance Outputs Regulated by AP2/ERF Transcription Factor Family. *Stress Biol.* **2024**, *4*, 2. [CrossRef] [PubMed]
110. Thirugnanasambantham, K.; Durairaj, S.; Saravanan, S.; Karikalan, K.; Muralidaran, S.; Islam, V.I.H. Role of Ethylene Response Transcription Factor (ERF) and Its Regulation in Response to Stress Encountered by Plants. *Plant Mol. Biol. Report.* **2015**, *33*, 347–357. [CrossRef]
111. Chen, Y.-F.; Etheridge, N.; Schaller, G.E. Ethylene Signal Transduction. *Ann. Bot.* **2005**, *95*, 901–915. [CrossRef] [PubMed]
112. Álvarez-Hernández, M.H.; Artés-Hernández, F.; Ávalos-Belmontes, F.; Castillo-Campohermoso, M.A.; Contreras-Esquivel, J.C.; Ventura-Sobrevilla, J.M.; Martínez-Hernández, G.B. Current Scenario of Adsorbent Materials Used in Ethylene Scavenging Systems to Extend Fruit and Vegetable Postharvest Life. *Food Bioprocess Technol.* **2018**, *11*, 511–525. [CrossRef]
113. Wu, B.; Guo, Q.; Wang, G.; Peng, X.; Wang, J.; Che, F. Effects of Different Postharvest Treatments on the Physiology and Quality of 'Xiaobai' Apricots at Room Temperature. *J. Food Sci. Technol.* **2015**, *52*, 2247–2255. [CrossRef] [PubMed]
114. Alonso-Salinas, R.; Acosta-Motos, J.R.; Pérez-López, A.J.; Noguera-Artiaga, L.; Núñez-Delicado, E.; Burló, F.; López-Miranda, S. Effect of Combination of $KMnO_4$ Oxidation and UV-C Radiation on Postharvest Quality of Refrigerated Pears Cv. "Ercolini". *Horticulturae* **2022**, *8*, 1078. [CrossRef]
115. Alonso-Salinas, R.; Acosta-Motos, J.R.; Núñez-Delicado, E.; Gabaldón, J.A.; López-Miranda, S. Combined Effect of Potassium Permanganate and Ultraviolet Light as Ethylene Scavengers on Post-Harvest Quality of Peach at Optimal and Stressful Temperatures. *Agronomy* **2022**, *12*, 616. [CrossRef]
116. Zhou, H.-W.; Dong, L.i.; Ben-Arie, R.; Lurie, S. The Role of Ethylene in the Prevention of Chilling Injury in Nectarines. *J. Plant Physiol.* **2001**, *158*, 55–61. [CrossRef]
117. Fan, X.; Xi, Y.; Zhao, H.; Liu, B.; Cao, J.; Jiang, W. Improving Fresh Apricot (*Prunus armeniaca* L.) Quality and Antioxidant Capacity by Storage at near Freezing Temperature. *Sci. Hortic.* **2018**, *231*, 1–10. [CrossRef]
118. Hayama, H.; Shimada, T.; Fujii, H.; Ito, A.; Kashimura, Y. Ethylene-Regulation of Fruit Softening and Softening-Related Genes in Peach. *J. Exp. Bot.* **2006**, *57*, 4071–4077. [CrossRef] [PubMed]
119. Palou, L.; Crisosto, C.H.; Garner, D.; Basinal, L.M. Effect of Continuous Exposure to Exogenous Ethylene during Cold Storage on Postharvest Decay Development and Quality Attributes of Stone Fruits and Table Grapes. *Postharvest Biol. Technol.* **2003**, *27*, 243–254. [CrossRef]
120. Álvarez-Hernández, M.H.; Martínez-Hernández, G.B.; Avalos-Belmontes, F.; Miranda-Molina, F.D.; Artés-Hernández, F. Postharvest Quality Retention of Apricots by Using a Novel Sepiolite–Loaded Potassium Permanganate Ethylene Scavenger. *Postharvest Biol. Technol.* **2020**, *160*, 111061. [CrossRef]
121. Pech, J.-C.; Purgatto, E.; Bouzayen, M.; Latché, A. Ethylene and Fruit Ripening. In *Annual Plant Reviews Volume 44: The Plant Hormone Ethylene*; Wiley: Hoboken, NJ, USA, 2018; pp. 275–304. ISBN 9781119312994.
122. Pott, D.; Vallarino, J.; Osorio, S. Metabolite Changes during Postharvest Storage: Effects on Fruit Quality Traits. *Metabolites* **2020**, *10*, 187. [CrossRef] [PubMed]

123. Yahia, E.; Serrano, M.; Valero, D.; Aguilar, G. Influence of Postharvest Technologies and Handling Practices on Phytochemicals in Fruits and Vegetables: Chemistry and Human Health. In *Fruit and Vegetable Phytochemicals: Chemistry and Human Health*, 2nd ed.; John Wiley & Sons: Hoboken, NJ, USA, 2017; pp. 609–628. ISBN 9781119157946.
124. Zhang, B.; Peng, B.; Zhang, C.; Song, Z.; Ma, R. Determination of Fruit Maturity and Its Prediction Model Based on the Pericarp Index of Absorbance Difference (IAD) for Peaches. *PLoS ONE* **2017**, *12*, e0177511. [CrossRef] [PubMed]
125. Zhang, P.; Shao, X.; Wei, Y.; Xu, F.; Wang, H. At-Harvest Fruit Maturity Affects Sucrose Metabolism during Cold Storage and Is Related to Chilling Injury in Peach. *J. Food Sci. Technol.* **2020**, *57*, 2000–2009. [CrossRef] [PubMed]
126. Beckles, D.M. Factors Affecting the Postharvest Soluble Solids and Sugar Content of Tomato (*Solanum lycopersicum* L.) Fruit. *Postharvest Biol. Technol.* **2012**, *63*, 129–140. [CrossRef]
127. Brecht, J.; Yahia, E. Postharvest Physiology. In *The Mango, 2nd Edition: Botany, Production and Uses*; CABI: Wallingford, UK, 2009; pp. 484–528. ISBN 978-1-84593-489-7.
128. Brizzolara, S.; Manganaris, G.; Fotopoulos, V.; Watkins, C.; Tonutti, P. Primary Metabolism in Fresh Fruits During Storage. *Front. Plant Sci.* **2020**, *11*, 80. [CrossRef] [PubMed]
129. McBey, D.; Nadathur, S. Flavors, Taste Preferences, and the Consumer: Taste Modulation and Influencing Change in Dietary Patterns for a Sustainable Earth. In *Sustainable Protein Sources*; Academic Press: Cambridge, MA, USA, 2024; pp. 629–647. ISBN 9780323916523.
130. Hossain, M.A.; Karim, M.; Juthee, S. Postharvest Physiological and Biochemical Alterations in Fruits: A Review. *Fundam. Appl. Agric.* **2020**, *5*, 453–469. [CrossRef]
131. Prasad, K.; Jacob, S.; Siddiqui, M. Fruit Maturity, Harvesting, and Quality Standards. In *Preharvest Modulation of Postharvest Fruit and Vegetable Quality*; Academic Press: Cambridge, MA, USA, 2017; pp. 41–69.
132. Singh, D.; Sharma, R.; Kesharwani, A. Postharvest Losses of Horticultural Produce. In *Postharvest Handling and Diseases of Horticultural Produce*; CRC Press: Boca Raton, FL, USA, 2021; pp. 1–24. ISBN 9781003045502.
133. Mohammed, M.; Khalid Alqahtani, N.; Munir, M. Artificial Ripening Technologies for Dates. In *New Discoveries in the Ripening Processes*; BoD–Books on Demand: Norderstedt, Germany, 2023; ISBN 978-0-85014-126-9.
134. Madani, B.; Mirshekari, A.; Yahia, E.; Golding, J. Sapota (*Manilkara achras* Forb.): Factors Influencing Fresh and Processed Fruit Quality. *Hortic. Rev.* **2018**, *45*, 105–142.
135. Arabia, A.; Munné-Bosch, S.; Muñoz, P. Ascorbic Acid as a Master Redox Regulator of Fruit Ripening. *Postharvest Biol. Technol.* **2024**, *207*, 112614. [CrossRef]
136. Ludwig-Ohm, S.; Dirksmeyer, W.; Klockgether, K. Approaches to Reduce Food Losses in German Fruit and Vegetable Production. *Sustainability* **2019**, *11*, 6576. [CrossRef]
137. Gallagher, M.S.; Mahajan, P.V. The stability and shelf life of fruit and vegetables. In *Food and Beverage Stability and Shelf-Life*; Kilcast, D., Subramaniam, P., Eds.; Woodhead Publishing Ltd.: Cambridge, UK, 2011; Chapter 22; pp. 641–656.
138. Debebe, S. Post-Harvest Losses of Crops and Its Determinants in Ethiopia: Tobit Model Analysis. *Agric. Food Secur.* **2022**, *11*, 13. [CrossRef]
139. Alonso-Salinas, R.; López-Miranda, S.; Pérez-López, A.J.; Noguera-Artiaga, L.; Carbonell-Barrachina, Á.A.; Núñez-Delicado, E.; Acosta-Motos, J.R. Novel Combination of Ethylene Oxidisers to Delay Losses on Postharvest Quality, Volatile Compounds and Sensorial Analysis of Tomato Fruit. *LWT* **2022**, *170*, 114054. [CrossRef]
140. Gaikwad, K.K.; Singh, S.; Negi, Y.S. Ethylene Scavengers for Active Packaging of Fresh Food Produce. *Environ. Chem. Lett.* **2020**, *18*, 269–284. [CrossRef]
141. Zerbini, P. Role of Maturity for Improved Flavour. *Fruit Veg. Flavour Recent Adv. Future Prospect.* **2008**, 180–199. [CrossRef]
142. Cruz-Hernández, A.; Paredes-López, O. Fruit Quality: New Insights for Biotechnology. *Crit. Rev. Food Sci. Nutr.* **2012**, *52*, 272–289. [CrossRef] [PubMed]
143. Klee, H.J.; Giovannoni, J.J. Genetics and Control of Tomato Fruit Ripening and Quality Attributes. *Annu. Rev. Genet.* **2011**, *45*, 41–59. [CrossRef] [PubMed]
144. Zhang, A.; Zhang, Q.; Li, J.; Gong, H.; Fan, X.; Yang, Y.; Liu, X.; Yin, X. Transcriptome Co-Expression Network Analysis Identifies Key Genes and Regulators of Ripening Kiwifruit Ester Biosynthesis. *BMC Plant Biol.* **2020**, *20*, 103. [CrossRef] [PubMed]
145. Jia, H.; Xu, Y.; Deng, Y.; Xie, Y.; Gao, Z.; Lang, Z.; Niu, Q. Key Transcription Factors Regulate Fruit Ripening and Metabolite Accumulation in Tomato. *Plant Physiol.* **2024**, *195*, 2256–2273. [CrossRef] [PubMed]
146. Wang, Q.; Gao, F.; Chen, X.; Wu, W.; Wang, L.; Shi, J.; Huang, Y.; Shen, Y.; Wu, G.; Guo, J. Characterization of Key Aroma Compounds and Regulation Mechanism of Aroma Formation in Local Binzi (*Malus pumila* × *Malus asiatica*) Fruit. *BMC Plant Biol.* **2022**, *22*, 532. [CrossRef] [PubMed]
147. Lu, L.; Delrot, S.; Liang, Z. From Acidity to Sweetness: A Comprehensive Review of Carbon Accumulation in Grape Berries. *Mol. Hortic.* **2024**, *4*, 22. [CrossRef] [PubMed]
148. Jiang, Y.; Liu, Z.; Peydayesh, M.; Zhang, B.; Jia, X.; Huang, Q. Ethylene Control in Fruit Quality Assurance: A Material Science Perspective. *Aggregate* **2024**, e565. [CrossRef]
149. Datta, H.S. Physiological Approaches for Regulation of Fruit Ripening: A Review. *IJCS* **2019**, *7*, 4587–4597.
150. Abbas, F.; Zhou, Y.; O'Neill Rothenberg, D.; Alam, I.; Ke, Y.; Wang, H.-C. Aroma Components in Horticultural Crops: Chemical Diversity and Usage of Metabolic Engineering for Industrial Applications. *Plants* **2023**, *12*, 1748. [CrossRef] [PubMed]

151. Chai, J.; Liao, B.; Li, R.; Liu, Z. Changes in Taste and Volatile Compounds and Ethylene Production Determined the Eating Window of 'Xuxiang' and 'Cuixiang' Kiwifruit Cultivars. *Postharvest Biol. Technol.* **2022**, *194*, 112093. [CrossRef]
152. Wendt, L.; Soldateli, F.; Both, V.; Thewes, F.; Ludwig, V.; Berghetti, M.; Ferrão, T.; Wagner, R.; Brackmann, A. Effects of Ethylene Control and Dynamic Controlled Atmosphere Storage on 'Galaxy' Apple Quality. *J. Plant Growth Regul.* **2022**, *42*, 701–718. [CrossRef]
153. Alonso-Salinas, R.; López-Miranda, S.; González-Báidez, A.; Pérez-López, A.J.; Noguera-Artiaga, L.; Núñez-Delicado, E.; Carbonell-Barrachina, Á.; Acosta-Motos, J.R. Effect of Potassium Permanganate, Ultraviolet Radiation and Titanium Oxide as Ethylene Scavengers on Preservation of Postharvest Quality and Sensory Attributes of Broccoli Stored with Tomatoes. *Foods* **2023**, *12*, 2418. [CrossRef] [PubMed]
154. Büchele, F.; Khera, K.; Thewes, F.; Kittemann, D.; Neuwald, D. Dynamic Control of Atmosphere and Temperature Based on Fruit CO_2 Production: Practical Application in Apple Storage and Effects on Metabolism, Quality, and Volatile Profiles. *Food Bioprocess Technol.* **2023**, *16*, 2497–2510. [CrossRef]
155. Zaitoon, A.; Luo, X.; Lim, L.-T. Triggered and Controlled Release of Active Gaseous/Volatile Compounds for Active Packaging Applications of Agri-Food Products: A Review. *Compr. Rev. Food Sci. Food Saf.* **2022**, *21*, 541–579. [CrossRef] [PubMed]
156. Öztürk, M.; Ayhan, Z. Combined Effects of Ethylene Scavenging-Active Packaging System and Modified Atmosphere to Reduce Postharvest Losses of Ethylene Sensitive Produce: Banana and Kiwifruit. *Packag. Technol. Sci.* **2023**, *36*, 951–967. [CrossRef]
157. Pathak, N.; Caleb, O.J.; Geyer, M.; Herppich, W.B.; Rauh, C.; Mahajan, P. V Photocatalytic and Photochemical Oxidation of Ethylene: Potential for Storage of Fresh Produce—A Review. *Food Bioprocess Technol.* **2017**, *10*, 982–1001. [CrossRef]
158. Pathak, N. Photocatalysis and Vacuum Ultraviolet Light Photolysis as Ethylene Removal Techniques for Potential Application in Fruit Storage. Ph.D. Thesis, Technische Universität Berlin, Berlin, Germany, 2019.
159. Janjarasskul, T.; Suppakul, P. Active and Intelligent Packaging: The Indication of Quality and Safety. *Crit. Rev. Food Sci. Nutr.* **2018**, *58*, 808–831. [CrossRef] [PubMed]
160. Baswal, A.K.; Ramezanian, A. 1-Methylcyclopropene Potentials in Maintaining the Postharvest Quality of Fruits, Vegetables, and Ornamentals: A Review. *J. Food Process. Preserv.* **2021**, *45*, 15129. [CrossRef]
161. Watkins, C. 1-Methylcyclopropene (1-MCP) Based Technologies for Storage and Shelf Life Extension. *Int. J. Postharvest Technol. Innov.* **2006**, *1*, 62–68. [CrossRef]
162. Tsantili, E.; Gapper, N.E.; Arquiza, J.M.R.A.; Whitaker, B.D.; Watkins, C.B. Ethylene and α-Farnesene Metabolism in Green and Red Skin of Three Apple Cultivars in Response to 1-Methylcyclopropene (1-MCP) Treatment. *J. Agric. Food Chem.* **2007**, *55*, 5267–5276. [CrossRef] [PubMed]
163. Yarılgaç, T.; Kadim, H.; Ozturk, B. Role of Maturity Stages and Modified-Atmosphere Packaging on the Quality Attributes of Cornelian Cherry Fruits (*Cornus mas* L.) throughout Shelf Life. *J. Sci. Food Agric.* **2019**, *99*, 421–428. [CrossRef] [PubMed]
164. Awalgaonkar, G.S. Development of active packaging trays with ethylene removing capacity. Master's Thesis, Michigan State University, East Lansing, MI, USA, 2018.
165. Guillén, F.; Castillo, S.; Zapata, P.J.; Martínez-Romero, D.; Serrano, M.; Valero, D. Efficacy of 1-MCP Treatment in Tomato Fruit. 1. Duration and Concentration of 1-MCP Treatment to Gain an Effective Delay of Postharvest Ripening. *Postharvest Biol. Technol.* **2007**, *43*, 23–27. [CrossRef]
166. Charoenchongsuk, N.; Matsumoto, D.; Itai, A.; Murayama, H. Ripening Characteristics and Pigment Changes in Russeted Pear Fruit in Response to Ethylene and 1-MCP. *Horticulturae* **2018**, *4*, 22. [CrossRef]
167. Taye, A.M.; Tilahun, S.; Seo, M.H.; Park, D.S.; Jeong, C.S. Effects of 1-MCP on Quality and Storability of Cherry Tomato (*Solanum lycopersicum* L.). *Horticulturae* **2019**, *5*, 29. [CrossRef]
168. Xu, F.; Liu, S.; Liu, Y.; Xu, J.; Liu, T.; Dong, S. Effectiveness of Lysozyme Coatings and 1-MCP Treatments on Storage and Preservation of Kiwifruit. *Food Chem.* **2019**, *288*, 201–207. [CrossRef] [PubMed]
169. Pongprasert, N.; Srilaong, V.; Kaewsukseang, S. 1-MCP Micro-Bubbles Delaying Postharvest Ripening of "Khai" Banana. In Proceedings of the III Asia Pacific Symposium on Postharvest Research, Education and Extension: APS2014, Hochiminh City, Vietnam, 5 October 2018; Volume 1213, pp. 245–250.
170. Manenoi, A.; Bayogan, E.R.V.; Thumdee, S.; Paull, R.E. Utility of 1-Methylcyclopropene as a Papaya Postharvest Treatment. *Postharvest Biol. Technol.* **2007**, *44*, 55–62. [CrossRef]
171. Wang, J.; Allan, A.C.; Wang, W.Q.; Yin, X.R. The Effects of Salicylic Acid on Quality Control of Horticultural Commodities. *New Zealand J. Crop Hortic. Sci.* **2022**, *50*, 99–117. [CrossRef]
172. Sayyari, M.; Babalar, M.; Kalantari, S.; Serrano, M.; Valero, D. Effect of Salicylic Acid Treatment on Reducing Chilling Injury in Stored Pomegranates. *Postharvest Biol. Technol.* **2009**, *53*, 152–154. [CrossRef]
173. Babalar, M.; Asghari, M.; Talaei, A.; Khosroshahi, A. Effect of Pre- and Postharvest Salicylic Acid Treatment on Ethylene Production, Fungal Decay and Overall Quality of Selva Strawberry Fruit. *Food Chem.* **2007**, *105*, 449–453. [CrossRef]
174. Asghari, M.; Aghdam, M.S. Impact of Salicylic Acid on Post-Harvest Physiology of Horticultural Crops. *Trends Food Sci. Technol.* **2010**, *21*, 502–509. [CrossRef]
175. Leslie, C.A.; Romani, R.J. Inhibition of Ethylene Biosynthesis by Salicylic Acid. *Plant Physiol.* **1988**, *88*, 833–837. [CrossRef] [PubMed]
176. Srivastava, M.K.; Dwivedi, U.N. Delayed Ripening of Banana Fruit by Salicylic Acid. *Plant Sci.* **2000**, *158*, 87–96. [CrossRef] [PubMed]

177. Elmenofy, H.M.; Okba, S.K.; Salama, A.M.; Alam-Eldein, S.M. Yield, Fruit Quality, and Storability of "Canino" Apricot in Response to Aminoethoxyvinylglycine, Salicylic Acid, and Chitosan. *Plants* **2021**, *10*, 1838. [CrossRef] [PubMed]
178. Gong, D.; Bi, Y.; Li, Y.; Wang, Y.; Prusky, D.; Alkan, N. Preharvest Elicitors Spray Improves Antioxidant Activity, Alleviates Chilling Injury, and Maintains Quality in Harvested Fruit. *Horticulturae* **2022**, *8*, 1208. [CrossRef]
179. Johnson, D.S.; Colgan, R. Low Ethylene Controlled Atmosphere Induces Adverse Effects on the Quality of 'Cox's Orange Pippin' Apples Treated with Aminoethoxyvinylglycine during Fruit Development. *Postharvest Biol. Technol.* **2003**, *27*, 59–68. [CrossRef]
180. Balaguera-López, H.; Salamanca-Gutiérrez, F.; García, J.; Herrera, A. Ethylene and Maturation Retardants in the Postharvest of Perishable Horticultural Products. A Review. *Rev. Colomb. Cienc. Hortícolas* **2014**, *8*, 302–313. [CrossRef]
181. Capitani, G.; Tschopp, M.; Eliot, A.C.; Kirsch, J.F.; Grütter, M.G. Structure of ACC Synthase Inactivated by the Mechanism-Based Inhibitor l-Vinylglycine. *FEBS Lett.* **2005**, *579*, 2458–2462. [CrossRef] [PubMed]
182. Saltveit, M.E. Aminoethoxyvinylglycine (AVG) Reduces Ethylene and Protein Biosynthesis in Excised Discs of Mature-Green Tomato Pericarp Tissue. *Postharvest Biol. Technol.* **2005**, *35*, 183–190. [CrossRef]
183. Ozturk, B.; Kucuker, E.; Yıldız, K.; Celik, S.M. Changes of Bioactive Compounds and Ethylene Production of Japanese Plums Treated with Pre-Harvest Aminoethoxyvinylglycine. *Int. J. Food Prop.* **2015**, *18*, 2165–2186. [CrossRef]
184. Ozturk, B.; Uzun, S.; Karakaya, O. Combined Effects of Aminoethoxyvinylglycine and MAP on the Fruit Quality of Kiwifruit during Cold Storage and Shelf Life. *Sci. Hortic.* **2019**, *251*, 209–214. [CrossRef]
185. Yuan, R.; Carbaugh, D.H. Effects of NAA, AVG, and 1-MCP on Ethylene Biosynthesis, Preharvest Fruit Drop, Fruit Maturity, and Quality of 'Golden Supreme' and 'Golden Delicious' Apples. *HortScience Horts* **2007**, *42*, 101–105. [CrossRef]
186. Muñoz-Robredo, P.; Rubio, P.; Infante, R.; Campos-Vargas, R.; Manríquez, D.; González-Agüero, M.; Defilippi, B.G. Ethylene Biosynthesis in Apricot: Identification of a Ripening-Related 1-Aminocyclopropane-1-Carboxylic Acid Synthase (ACS) Gene. *Postharvest Biol. Technol.* **2012**, *63*, 85–90. [CrossRef]
187. Xie, X.; Einhorn, T.; Wang, Y. Inhibition of Ethylene Biosynthesis and Associated Gene Expression by Aminoethoxyvinylglycine and 1-Methylcyclopropene and Their Consequences on Eating Quality and Internal Browning of 'Starkrimson' Pears. *J. Am. Soc. Hortic. Sci. J. Amer. Soc. Hort. Sci.* **2015**, *140*, 587–596. [CrossRef]
188. Kim, Y.T.; Ha, S.T.T.; Chun, I.; In, B.C. Inhibition of Ethylene Binding and Biosynthesis Maintains Fruit Quality of "Formosa" Plums during Postharvest Storage. *Hortic. Sci. Technol.* **2021**, *39*, 368–378. [CrossRef]
189. Win, N.M.; Yoo, J.; Lwin, H.P.; Lee, E.J.; Kang, I.K.; Lee, J. Effects of 1-Methylcyclopropene and Aminoethoxyvinylglycine Treatments on Fruit Quality and Antioxidant Metabolites in Cold-Stored "Sangjudungsi" Persimmons. *Hortic. Environ. Biotechnol.* **2021**, *62*, 891–905. [CrossRef]
190. Salazar, N.A.; Molina-Corral, F.J.; Aguilar, G.; Otero, A.; Sepulveda, D.R.; Olivas, G. Volatile Production by 'Golden Delicious' Apples Is Affected by Preharvest Application of Aminoethoxyvinylglycine. *Sci. Hortic.* **2011**, *130*, 436–444. [CrossRef]
191. Romani, R.; Labavitch, J.; Yamashita, T.; Hess, B.; Rae, H. Preharvest AVG Treatment of 'Bartlett' Pear Fruits: Effects on Ripening, Color Change, and Volatiles. *J. Am. Soc. Hortic. Sci.* **1983**, *108*, 1046–1049. [CrossRef]
192. López, P.; Neisa, D.P.; Bacca, C.; Flórez, V.J. Evaluación de Preservantes Florales En La Poscosecha de Tres Variedades de Clavel Estándar. *Agron. Colomb.* **2008**, *26*, 116–126.
193. Martínez-Romero, D.; Bailén, G.; Serrano, M.; Guillén, F.; Valverde, J.M.; Zapata, P.; Castillo, S.; Valero, D. Tools to Maintain Postharvest Fruit and Vegetable Quality through the Inhibition of Ethylene Action: A Review. *Crit. Rev. Food Sci. Nutr.* **2007**, *47*, 543–560. [CrossRef] [PubMed]
194. Stabyl, G.; Basel, R.; Reid, M.; Dodge, L. Efficacies of Commercial Anti Ethylene Products for Fresh Cut Flowers. *HortTechnology* **1993**, *3*, 199–202. [CrossRef]
195. Bulantseva, E.A.; Nguyen, T.T.; Ruzhitsky, A.O.; Protsenko, M.A.; Korableva, N.P. The Effect of Ethylene Biosynthesis Regulators on Metabolic Processes in the Banana Fruits in Various Physiological States. *Appl. Biochem. Microbiol.* **2009**, *45*, 93–96. [CrossRef]
196. Lima, P.C.C.; Santos, M.N.D.; Guimaraes, M.E.D.; de Araujo, N.O.; Krause, M.R.; Finger, F.L. Ethylene and Its Inhibitors Affect the Quality of Processed Sweet Potatoes. *Food Sci. Technol.* **2021**, *41*, 825–832. [CrossRef]
197. Kovaleva, L.V.; Zakharova, E.V.; Timofeeva, G.V.; Andreev, I.M.; Golivanov, Y.Y.; Bogoutdinova, L.R.; Baranova, E.N.; Khaliluev, M.R. Aminooxyacetic Acid (AOA), Inhibitor of 1-Aminocyclopropane-1-Carboxilic Acid (ACC) Synthesis, Suppresses Self-Incompatibility-Induced Programmed Cell Death in Self-Incompatible *Petunia hybrida* L. Pollen Tubes. *Protoplasma* **2020**, *257*, 213–227. [CrossRef] [PubMed]
198. Instituto Latinoamericano de la Comunicación Educativa ¿Que Es Una Zeolita? Available online: http://bibliotecadigital.ilce.edu.mx/sites/ciencia/volumen1/ciencia2/55/htm/sec_3.html (accessed on 24 May 2024).
199. Patdhanagul, N.; Rangsriwatananon, K.; Siriwong, K.; Hengrasmee, S. Combined Modification of Zeolite NaY by Phenyl Trimethyl Ammonium Bromide and Potassium for Ethylene Gas Adsorption. *Microporous Mesoporous Mater.* **2012**, *153*, 30–34. [CrossRef]
200. Coloma, A.; Rodríguez, F.; Bruna, J.; Guarda, A.; Galotto, M. Development of an Active Film with Natural Zeolite as Ethylene Scavenger. *J. Chil. Chem. Soc.* **2014**, *59*, 2409–2414. [CrossRef]
201. Erdoğan, B.; Sakızcı, M.; Yörükoğulları, E. Characterization and Ethylene Adsorption of Natural and Modified Clinoptilolites. *Appl. Surf. Sci.* **2008**, *254*, 2450–2457. [CrossRef]
202. Patdhanagul, N.; Srithanratana, T.; Rangsriwatananon, K.; Hengrasmee, S. Ethylene Adsorption on Cationic Surfactant Modified Zeolite NaY. *Microporous Mesoporous Mater.* **2010**, *131*, 97–102. [CrossRef]

203. de Bruijn, J.; Gomez, A.; Loyola, C.; Melin, P.; Solar, V.; Abreu, N.; Azzolina-Jury, F.; Valdes, H. Use of a Copper- and Zinc-Modified Natural Zeolite to Improve Ethylene Removal and Postharvest Quality of Tomato Fruit. *Crystals* **2020**, *10*, 471. [CrossRef]
204. Mariah, M.A.; Vonnie, J.M.; Erna, K.H.; Nur'Aqilah, N.M.; Huda, N.; Abdul Wahab, R.; Rovina, K. The Emergence and Impact of Ethylene Scavengers Techniques in Delaying the Ripening of Fruits and Vegetables. *Membranes* **2022**, *12*, 117. [CrossRef] [PubMed]
205. Carbotecnia ¿Qué Es El Carbón Activado y Para Qué Sirve? Available online: https://www.carbotecnia.info/aprendizaje/carbon-activado/que-es-carbon-activado/ (accessed on 15 February 2024).
206. Bailén, G.; Guillén, F.; Castillo, S.; Zapata, P.J.; Serrano, M.; Valero, D.; Martínez-Romero, D. Use of a Palladium Catalyst to Improve the Capacity of Activated Carbon to Absorb Ethylene, and Its Effect on Tomato Ripening. *Span. J. Agric. Res.* **2007**, *5*, 579–586. [CrossRef]
207. Jaimun, R.; Sangsuwan, J. Efficacy of Chitosan-Coated Paper Incorporated with Vanillin and Ethylene Adsorbents on the Control of Anthracnose and the Quality of Nam Dok Mai Mango Fruit. *Packag. Technol. Sci.* **2019**, *32*, 383–394. [CrossRef]
208. Kuppler, R.J.; Timmons, D.J.; Fang, Q.-R.; Li, J.-R.; Makal, T.A.; Young, M.D.; Yuan, D.; Zhao, D.; Zhuang, W.; Zhou, H.-C. Potential Applications of Metal-Organic Frameworks. *Coord. Chem. Rev.* **2009**, *253*, 3042–3066. [CrossRef]
209. Li, J.-R.; Sculley, J.; Zhou, H.-C. Metal–Organic Frameworks for Separations. *Chem. Rev.* **2012**, *112*, 869–932. [CrossRef] [PubMed]
210. Chopra, S.; Dhumal, S.; Abeli, P.; Beaudry, R.; Almenar, E. Metal-Organic Frameworks Have Utility in Adsorption and Release of Ethylene and 1-Methylcyclopropene in Fresh Produce Packaging. *Postharvest Biol. Technol.* **2017**, *130*, 48–55. [CrossRef]
211. Shorter, A.J.; Scott, K.J.; Ward, G.; Best, D.J. Effect of Ethylene Absorption on the Storage of Granny Smith Apples Held in Polyethylene Bags. *Postharvest Biol. Technol.* **1992**, *1*, 189–194. [CrossRef]
212. Álvarez-Hernández, M.H.; Martínez-Hernández, G.B.; Avalos-Belmontes, F.; Rodríguez-Hernández, A.M.; Castillo-Campohermoso, M.A.; Artés-Hernández, F. An Innovative Ethylene Scrubber Made of Potassium Permanganate Loaded on a Protonated Montmorillonite: A Case Study on Blueberries. *Food Bioprocess Technol.* **2019**, *12*, 524–538. [CrossRef]
213. Mujtaba, A.; Masud, T.; Butt, S.J.; Qazalbash, M.; Fareed, W.; Shahid, A. Potential Role of Calcium Chloride, Potassium Permanganate and Boric Acid on Quality Maintenance of Tomato Cv. Rio Grandi at Ambient Temperature. *Int. J. Biosci.* **2014**, *5*, 9–20.
214. Scott, K.J.; Wills, R.B.H. Reduction of Brown Heart in Pears by Absorption of Ethylene from the Storage Atmosphere. *Aust. J. Exp. Agric.* **1974**, *14*, 266–268. [CrossRef]
215. Illeperuma, C.K.; Nikapitiya, C. Extension of the Postharvest Life of 'Pollock' Avocado Using Modified Atmosphere Packaging. *Fruits* **2002**, *57*, 287–295. [CrossRef]
216. García, J.; Herrera, A.; García, J. Conservación Del Fruto de Banano Bocadillo (Musa AA Simmonds) Con La Aplicación de Permanganato de Potasio ($KMnO_4$). *Rev. Colomb. De Cienc. Hortícolas* **2012**, *6*, 161–171. [CrossRef]
217. Çelik, S.; Bal, E. The Effects of Postharvest Treatments of Salicylic Acid and Potassium Permanganate on the Storage of Kiwifruit. *Bulg. J. Agric. Sci.* **2010**, *16*, 576–584.
218. Giraldo, E.; Szczerbanik, M.J.; Scottpez, K.J.; Paton, J.E.; Best, D.J. Effects of Polyethylene Bags, Ethylene Absorbent and 1-Methylcyclopropene on the Storage of Japanese Pears. *J. Hortic. Sci. Biotechnol.* **2005**, *80*, 162–166. [CrossRef]
219. Kim, S.; Jeong, G.H.; Kim, S.-W. Ethylene Gas Decomposition Using ZSM-5/WO3-Pt-Nanorod Composites for Fruit Freshness. *ACS Sustain. Chem. Eng.* **2019**, *7*, 11250–11257. [CrossRef]
220. Kaewklin, P.; Siripatrawan, U.; Suwanagul, A.; Lee, Y.S. Active Packaging from Chitosan-Titanium Dioxide Nanocomposite Film for Prolonging Storage Life of Tomato Fruit. *Int. J. Biol. Macromol.* **2018**, *112*, 523–529. [CrossRef] [PubMed]
221. Bu, J.; Yu, Y.; Aisikaer, G.; Ying, T. Postharvest UV-C Irradiation Inhibits the Production of Ethylene and the Activity of Cell Wall-Degrading Enzymes during Softening of Tomato (*Lycopersicon esculentum* L.) Fruit. *Postharvest Biol. Technol.* **2013**, *86*, 337–345. [CrossRef]
222. Mabusela, B.P.; Belay, Z.A.; Godongwana, B.; Pathak, N.; Mahajan, P.V.; Caleb, O.J. Advances in Vacuum Ultraviolet Photolysis in the Postharvest Management of Fruit and Vegetables Along the Value Chains: A Review. *Food Bioprocess Technol.* **2022**, *15*, 28–46. [CrossRef]
223. Collazo Cordero, C.; Charles, F.; Aguiló-Aguayo, I.; Marín-Sáez, J.; Lafarga, T.; Abadias, M.; Vinas, I. Decontamination of Listeria Innocua from Fresh-Cut Broccoli Using UV-C Applied in Water or Peroxyacetic Acid, and Dry-Pulsed Light. *Innov. Food Sci. Emerg. Technol.* **2019**, *52*, 438–449. [CrossRef]
224. Orhewere, A.; Ajayi, O.; Ajayi, A. Advances in the Development of a Tomato Postharvest Storage System: Towards Eradicating Postharvest Losses. *J. Phys. Conf. Ser.* **2019**, *1378*, 22064. [CrossRef]
225. Pathak, N.; Caleb, O.J.; Rauh, C.; Mahajan, P. V Efficacy of Photocatalysis and Photolysis Systems for the Removal of Ethylene under Different Storage Conditions. *Postharvest Biol. Technol.* **2019**, *147*, 68–77. [CrossRef]
226. Ibhadon, A.; Fitzpatrick, P. Heterogeneous Photocatalysis: Recent Advances and Applications. *Catalysts* **2013**, *3*, 189–218. [CrossRef]
227. Alves, M.J.D.; Nobias, M.C.; Soares, L.S.; Coelho, D.S.; Maraschin, M.; Basso, A.; Moreira, R.; Jose, H.J.; Monteiro, A.R. Physiological Changes in Green and Red Cherry Tomatoes after Photocatalytic Ethylene Degradation Using Continuous Air Flux. *Food Sci. Technol. Int.* **2023**, *29*, 3–12. [CrossRef]
228. Li, W.H.; Liu, Z.Y.; Li, X.J.; Li, X.H. Quality Maintenance of 1-Methylcyclopropene Combined with Titanium Dioxide Photocatalytic Reaction on Postharvest Cherry Tomatoes. *J. Food Process. Preserv.* **2022**, *46*, e16500. [CrossRef]

229. Fonseca, J.D.; Pabon, N.Y.L.; Nandi, L.G.; Valencia, G.A.; Moreira, R.; Monteiro, A.R. Gelatin-TiO_2-Coated Expanded Polyethylene Foam Nets as Ethylene Scavengers for Fruit Postharvest Application. *Postharvest Biol. Technol.* **2021**, *180*, 111602. [CrossRef]
230. Ghosh, A.; Saha, I.; Fujita, M.; Debnath, S.C.; Hazra, A.K.; Adak, M.K.; Hasanuzzaman, M. Photoactivated TiO_2 Nanocomposite Delays the Postharvest Ripening Phenomenon through Ethylene Metabolism and Related Physiological Changes in Capsicum Fruit. *Plants* **2022**, *11*, 513. [CrossRef] [PubMed]
231. de Chiara, M.L.V.; Pal, S.; Licciulli, A.; Amodio, M.L.; Colelli, G. Photocatalytic Degradation of Ethylene on Mesoporous TiO_2/SiO_2 Nanocomposites: Effects on the Ripening of Mature Green Tomatoes. *Biosyst. Eng.* **2015**, *132*, 61–70. [CrossRef]
232. Chawengkijwanich, C.; Hayata, Y. Efficiency of TiO_2 Photocatalytic Reaction on Delay of Fruit Ripening and Removal of Off-Flavors from the Fruit Storage Atmosphere. *Trans. ASABE* **2006**, *49*, 833–837. [CrossRef]
233. Fonseca, J.D.; Alves, M.J.D.; Soares, L.S.; Moreira, R.; Valencia, G.A.; Monteiro, A.R. A Review on TiO_2-Based Photocatalytic Systems Applied in Fruit Postharvest: Set-Ups and Perspectives. *Food Res. Int.* **2021**, *144*, 110378. [CrossRef] [PubMed]
234. Yildirim, S.; Röcker, B.; Pettersen, M.K.; Nilsen-Nygaard, J.; Ayhan, Z.; Rutkaite, R.; Radusin, T.; Suminska, P.; Marcos, B.; Coma, V. Active Packaging Applications for Food. *Compr. Rev. Food Sci. Food Saf.* **2018**, *17*, 165–199. [CrossRef] [PubMed]
235. Wang, Z.Q.; Wang, B.; Jin, W.F. Experimental Investigation on Decomposition of Ethylene by Ozone: Harmful Product, Food Safety, and Control Strategy. *J. Food Process. Preserv.* **2021**, *45*, e15861. [CrossRef]
236. Liang, Y.Z.; Ji, L.L.; Chen, C.K.; Dong, C.H.; Wang, C.R. Effects of Ozone Treatment on the Storage Quality of Post-Harvest Tomato. *Int. J. Food Eng.* **2018**, *14*, 20180012. [CrossRef]
237. Toti, M.; Carboni, C.; Botondi, R. Postharvest Gaseous Ozone Treatment Enhances Quality Parameters and Delays Softening in Cantaloupe Melon during Storage at 6 °C. *J. Sci. Food Agric.* **2018**, *98*, 487–494. [CrossRef] [PubMed]
238. Gao, C.C.; Lin, Q.; Dong, C.H.; Ji, H.P.; Yu, J.Z.; Chen, C.K.; Zhu, Z.Q.; Ban, Z.J.; Zhang, N.; Bao, Y.Y. Effects of Ozone Concentration on the Postharvest Quality and Microbial Diversity of Muscat Hamburg Grapes. *RSC Adv.* **2020**, *10*, 9037–9045. [CrossRef] [PubMed]
239. Yan, Y.; Wang, X.S.; Huang, Z.C.; Qu, K.; Shi, W.B.; Peng, Z.M.; Zeng, L.M.; Xie, S.D.; Zhang, Y.H. Impacts of Synoptic Circulation on Surface Ozone Pollution in a Coastal Eco-City in Southeastern China during 2014–2019. *J. Environ. Sci.* **2023**, *127*, 143–157. [CrossRef] [PubMed]
240. Mok, Y.S.; Kim, S.G.; Han, J.; Nguyen, D.B.; Lee, H.W.; Jeon, H.; Kim, S.B. Removal of Dilute Ethylene Using Repetitive Cycles of Adsorption and Plasma-Catalytic Oxidation over Pd/ZSM-5 Catalyst. *J. Phys. D Appl. Phys.* **2020**, *53*, 334002. [CrossRef]
241. Tzeng, J.H.; Weng, C.H.; Huang, J.W.; Shiesh, C.C.; Lin, Y.H.; Lin, Y.T. Application of Palladium-Modified Zeolite for Prolonging Post-Harvest Shelf Life of Banana. *J. Sci. Food Agric.* **2019**, *99*, 3467–3474. [CrossRef] [PubMed]
242. Mehta, D.; Yadav, S.K. Recent Advances in Cold Plasma Technology for Food Processing. *Food Eng. Rev.* **2022**, *14*, 555–578. [CrossRef]
243. Gururani, P.; Bhatnagar, P.; Bisht, B.; Kumar, V.; Joshi, N.C.; Tomar, M.S.; Pathak, B. Cold Plasma Technology: Advanced and Sustainable Approach for Wastewater Treatment. *Environ. Sci. Pollut. Res.* **2021**, *28*, 65062–65082. [CrossRef] [PubMed]
244. Saini, S.; Kardam, S.K.; Kadam, A.A.; Singh, S.P.; Kumar, V.; Gaikwad, K.K. Ethylene Gas Scavenging Properties of Nano-Silica Isolated from Rice Straw and of Its PVA-Based Formulation Coated on Duplex Paperboard Made from Softwood and Rice Straw. *Biomass Convers. Biorefinery* **2023**, 1–10. [CrossRef]
245. Gundewadi, G.; Rudra, S.G.; Prasanna, R.; Banerjee, T.; Singh, S.K.; Dhakate, S.R.; Gupta, A.; Anand, A. Palladium Encapsulated Nanofibres for Scavenging Ethylene from Sapota Fruits. *Front. Nutr.* **2022**, *9*, 994813. [CrossRef] [PubMed]
246. Rupérez, D.; Gracia-Vallés, N.; Clavero, E.; Silva, F.; Nerín, C. Mechanochemically Scaled-Up Alpha Cyclodextrin Nanosponges: Their Safety and Effectiveness as Ethylene Scavenger. *Nanomaterials* **2022**, *12*, 2900. [CrossRef] [PubMed]
247. Sultana, A.; Kathuria, A.; Gaikwad, K.K. Metal–Organic Frameworks for Active Food Packaging. A Review. *Environ. Chem. Lett.* **2022**, *20*, 1479–1495. [CrossRef] [PubMed]
248. Büchele, F.; Hivare, K.; Khera, K.; Thewes, F.R.; Argenta, L.C.; Hoffmann, T.G.; Mahajan, P.V.; Prange, R.K.; Pareek, S.; Neuwald, D.A. Novel Energy-Saving Strategies in Apple Storage: A Review. *Sustainability* **2024**, *16*, 1052. [CrossRef]
249. Abedrabboh, O.; Koç, M.; Biçer, Y. Sustainability Performance of Space-Cooling Technologies and Approaches. *Energy Sources Part A Recovery Util. Environ. Eff.* **2022**, *44*, 9017–9042. [CrossRef]
250. Yenare, R.; Sonawane, C.; Sur, A.; Singh, B.; Panchal, H.; Kumar, A.; Sadasivuni, K.K.; Siddiqui, M.; Bhalerao, Y. A Comprehensive Review of Portable Cold Storage: Technologies, Applications, and Future Trends. *Alex. Eng. J.* **2024**, *94*, 23–33. [CrossRef]
251. Marimuthu, S.; Saikumar, A.; Badwaik, L. Food Losses and Wastage within Food Supply Chain: A Critical Review of Its Generation, Impact, and Conversion Techniques. *Waste Dispos. Sustain. Energy* **2024**, 1–16. [CrossRef]
252. Onyeaka, H.; Ghosh, S.; Obileke, K.; Miri, T.; Odeyemi, O.; Nwaiwu, O.; Tamasiga, P. Preventing Chemical Contaminants in Food: Challenges and Prospects for Safe and Sustainable Food Production. *Food Control* **2023**, *155*, 110040. [CrossRef]
253. Chen, Y.; Wan, G.; Song, J.; Dai, J.; Shi, W.; Wang, L. Food Safety Practices of Food Handlers in China and Their Correlation with Self-Reported Foodborne Illness. *J. Food Prot.* **2024**, *87*, 100202. [CrossRef] [PubMed]
254. Usman, A. Influence of Agricultural Practices on Food Safety in Nigeria. *Int. J. Food Sci.* **2024**, *7*, 43–54. [CrossRef]
255. Kumar, N.; Devgan, K.; Kumar, A.; Kaur, P.; Mahajan, P. Active and Passive Modified Atmosphere Packaging. In *Nonthermal Food Engineering Operations*; John Wiley & Sons: Hoboken, NJ, USA, 2024; pp. 225–259. ISBN 9781119776468.

256. Nian, L.; Wang, M.; Sun, X.; Zeng, Y.; Xie, Y.; Cheng, S.; Cao, C. Biodegradable Active Packaging: Components, Preparation, and Applications in the Preservation of Postharvest Perishable Fruits and Vegetables. *Crit. Rev. Food Sci. Nutr.* **2024**, *64*, 2304–2339. [CrossRef] [PubMed]
257. Ozal, G.; Ilyasova, C.; Ilgiz, V. Post-Harvest Storage and Processing Technology in Russia: Reducing Yield Loss. *Techno Agric. Stud. Res.* **2024**, *1*, 28–40.
258. Mavai, S.; Bains, A.; Sridhar, K.; Rashid, S.; Elossaily, G.M.; Ali, N.; Chawla, P.; Sharma, M. Formulation and Application of Poly Lactic Acid, Gum, and Cellulose-Based Ternary Bioplastic for Smart Food Packaging: A Review. *Int. J. Biol. Macromol.* **2024**, *268*, 131687. [CrossRef] [PubMed]
259. Verma, S.; Prasad, A.; Sonika; Katiyar, V. State of Art Review on Sustainable Biodegradable Polymers with a Market Overview for Sustainability Packaging. *Mater. Today Sustain.* **2024**, *26*, 100776. [CrossRef]
260. Singh, A.; Itkor, P.; Lee, M.; Shin, J.; Lee, Y.S. Promoting Sustainable Packaging Applications in the Circular Economy by Exploring and Advancing Molded Pulp Materials for Food Products: A Review. *Crit. Rev. Food Sci. Nutr.* **2022**, *63*, 11010–11025. [CrossRef] [PubMed]
261. Hasanin, M.; Abdelkhalek, A. Packaging of the Future: Smart Technologies and Food Quality and Safety. In *Intelligent Packaging*; Academic Press: Cambridge, MA, USA, 2024; pp. 1–30. ISBN 9780443153884.
262. Arijeniwa, F.; Akinsemolu, A.; Chukwugozie, D.; Onawo, G.; Ochulor, C.; Uju, N.; Kawino, D.; Onyeaka, H. Closing the Loop: A Framework for Tackling Single-Use Plastic Waste in the Food and Beverage Industry through Circular Economy—A Review. *J. Environ. Manag.* **2024**, *359*, 120816. [CrossRef] [PubMed]
263. Wong, E.; Ting, H.; Atanda, A. Enhancing Supply Chain Traceability through Blockchain and IoT Integration: A Comprehensive Review. *Green Intell. Syst. Appl.* **2024**, *4*, 11–28. [CrossRef]
264. Ninama, N.; Gangal, L.; Khayum, A.; Harshitha, S.B.; Divya, C.; Sowmya, H.M.; Singh, A. Anushi Post-Harvest Biotechnology or Genetic Engineering Solutions: Extending Shelf Life and Reducing Food Waste. *J. Adv. Biol. Biotechnol.* **2024**, *27*, 1–26. [CrossRef]
265. Shi, C.; Xiang, L.; Jiahu, G. Exploring the Frontier of Fruit Diseases Management: Advances in Nano-Based and Biocontrol Strategies and Underlying Action Mechanism. *S. Afr. J. Bot.* **2024**, *166*, 612–623. [CrossRef]
266. Doerflinger, F.; Al Shoffe, Y.; Sutano, G.; Nock, J.F.; Watkins, C. Preharvest 1-Methylcyclopropene (1-MCP) Treatment Effects on Quality of Spot and Strip Picked 'Gala' Apples at Harvest and after Storage as Affected by Postharvest 1-MCP and Temperature Conditioning Treatments. *Sci. Hortic.* **2023**, *325*, 112682. [CrossRef]
267. Zhu, M.; Li, J.; Liu, Y.; Wang, Q.; Fan, Z.; Zeng, J.; Yu, J. Preharvest Nano-Calcium Reduces the Table Grape Berry Abscission by Regulating Ethylene Production During Storage. *J. Plant Growth Regul.* **2023**, *43*, 1400–1409. [CrossRef]
268. Opara, U.; Pathare, P. Mechanical Damage of Fresh Produce: An Overview. In *Mechanical Damage in Fresh Horticultural Produce: Measurement, Analysis and Control*; Springer: Berlin/Heidelberg, Germany, 2024; pp. 1–19. ISBN 978-981-99-7095-7.
269. Baloyi, R.G.; Mafeo, T.P.; Mathaba, N. Effect of Pre- and Post-Harvest Factors on 'Benny' Valencia Fruit Rind Phenolics on Mitigation of Chilling and Non-Chilling Disorders during Cold Storage. *J. Hortic. Postharvest Res.* **2023**, *6*, 299–316. [CrossRef]
270. Rattanakaran, J.; Saengrayap, R.; Prahsarn, C.; Kitazawa, H.; Chaiwong, S. Application of Room Cooling and Thermal Insulation Materials to Maintain Quality of Okra during Storage and Transportation. *Horticulturae* **2021**, *7*, 188. [CrossRef]
271. Waldhans, C.; Albrecht, A.; Ibald, R.; Wollenweber, D.; Sy, S.-J.; Kreyenschmidt, J. Temperature Control and Data Exchange in Food Supply Chains: Current Situation and the Applicability of a Digitalized System of Time–Temperature-Indicators to Optimize Temperature Monitoring in Different Cold Chains. *J. Packag. Technol. Res.* **2024**, *8*, 79–93. [CrossRef]
272. Bisht, A.; Singh, S. Postharvest Losses and Management of Horticultural Produce: A Review. *J. Sci. Res. Rep.* **2024**, *30*, 305–320. [CrossRef]
273. Gupta, N.; Daruwalla, F. Examining the Effectiveness of Green Marketing Communication on Consumer Behavior towards Sustainable Purchases. *Educ. Adm. Theory Pract.* **2024**, *30*, 6861–6868. [CrossRef]
274. Uda, S.; Basrowi, B. Environmental Education Using SARITHA-Apps to Enhance Environmentally Friendly Supply Chain Efficiency and Foster Environmental Knowledge towards Sustainability. *Uncertain Supply Chain Manag.* **2024**, *12*, 359–372. [CrossRef]

horticulturae

MDPI

Review

Emerging Technologies for Prolonging Fresh-Cut Fruits' Quality and Safety during Storage

Rey David Iturralde-García, Francisco Javier Cinco-Moroyoqui, Oliviert Martínez-Cruz, Saúl Ruiz-Cruz, Francisco Javier Wong-Corral, Jesús Borboa-Flores, Yaeel Isbeth Cornejo-Ramírez, Ariadna Thalia Bernal-Mercado * and Carmen Lizette Del-Toro-Sánchez *

Departamento de Investigación y Posgrado en Alimentos, Universidad de Sonora, Blvd. Luis Encinas y Rosales S/N, Col. Centro, Hermosillo 83000, Mexico
* Correspondence: thalia.bernal@unison.mx (A.T.B.-M.); carmen.deltoro@unison.mx (C.L.D.-T.-S.)

Abstract: Fresh-cut fruits have been in great demand by consumers owing to the convenience of buying them in shopping centers as ready-to-eat products, and various advantages, such as the fact that they are healthy and fresh products. However, their shelf lives are brief due to their physiological changes and maturation. Therefore, this review includes information from the physicochemical, microbiological, nutritional, and sensory points of view on the deterioration mechanisms of fresh-cut fruits. In addition, updated information is presented on the different emerging technologies, such as active packaging (edible films, coatings, and modified atmospheres), natural preservatives (antioxidants and antimicrobials), and physical treatments (high hydrostatic pressure, UV-C radiation, and ozone). The benefits and disadvantages of each of these technologies and the ease of their applications are discussed. Having alternatives to preserve fresh-cut fruit is essential both for the consumer and the merchant, since the consumer could then obtain a high-quality product maintaining all its properties without causing any damage, and the merchant would receive economic benefits by having more time to sell the product.

Keywords: microbial quality; sensory quality; edible coating and films; natural antioxidants; natural antimicrobials; modified atmospheres; UV-C; ozone; high hydrostatic pressure

Citation: Iturralde-García, R.D.; Cinco-Moroyoqui, F.J.; Martínez-Cruz, O.; Ruiz-Cruz, S.; Wong-Corral, F.J.; Borboa-Flores, J.; Cornejo-Ramírez, Y.I.; Bernal-Mercado, A.T.; Del-Toro-Sánchez, C.L. Emerging Technologies for Prolonging Fresh-Cut Fruits' Quality and Safety during Storage. *Horticulturae* **2022**, *8*, 731. https://doi.org/10.3390/horticulturae8080731

Academic Editors: Paola Sánchez-Bravo and Luis Noguera-Artiaga

Received: 1 July 2022
Accepted: 12 August 2022
Published: 14 August 2022

1. Introduction

In recent years, fruit consumption has increased in a large portion of the population due to the current concern for a healthy lifestyle [1]. Fruits are a rich source of vitamins, minerals, and dietary fiber, which are essential for the human diet. Due to a population that is very busy with their different activities, and which consequently has less time to prepare its food, there has been a tendency to demand fresh, nutritious, convenient, and quickly accessible products, such as fresh-cut produce. Therefore, fresh-cut fruits constitute one of the fastest-growing food industry segments [2]. The Food and Drug Administration (FDA) defines fresh-cut fruit as any fruit that has been changed physically from its natural state by minimal processing, such as chopping, dicing, peeling, shredding, slicing, spiralizing, or tearing [3]. Fresh-cut fruits remain fresh without additional treatment, such as blanching, freezing, canning, cooking, or adding juice, syrup, or dressing. These products can be considered ready-to-eat but may or may not be washed before being packed for use by the customer or a retail food store. Fresh-cut fruits can include single or mixed fruits in the same packaging, providing great convenience, nutritional value, taste, and freshness [3].

The maintenance of the quality of fresh-cut fruits is attributed to the physiological and biological mechanisms of each fruit and environmental conditions such as storage conditions, temperature, and humidity [4]. The main disadvantage of fresh-cut fruits is their short shelf-lives, often less than two weeks [5]. When fresh fruits are cut or minimally processed, they are susceptible to chemical, physical, microbiological, and sensory changes.

This affects the product's marketability and reduces its nutritional value, and several food-borne outbreaks are linked to fresh-cut produce [1]. Additionally, it must be considered that many other factors can impact the final product, such as the type of packaging, postharvest processing, and type of cultivar, among others [6]. Whole fruits have a natural barrier (peel) that protects them from spoilage, microorganisms, or environmental conditions; however, this barrier is typically removed during the processing of fresh-cut fruits, making them more susceptible to decay (Figure 1) [7]. It is even more challenging to maintain the quality of fresh-cut fruits than vegetables due to their complicated physiology [8].

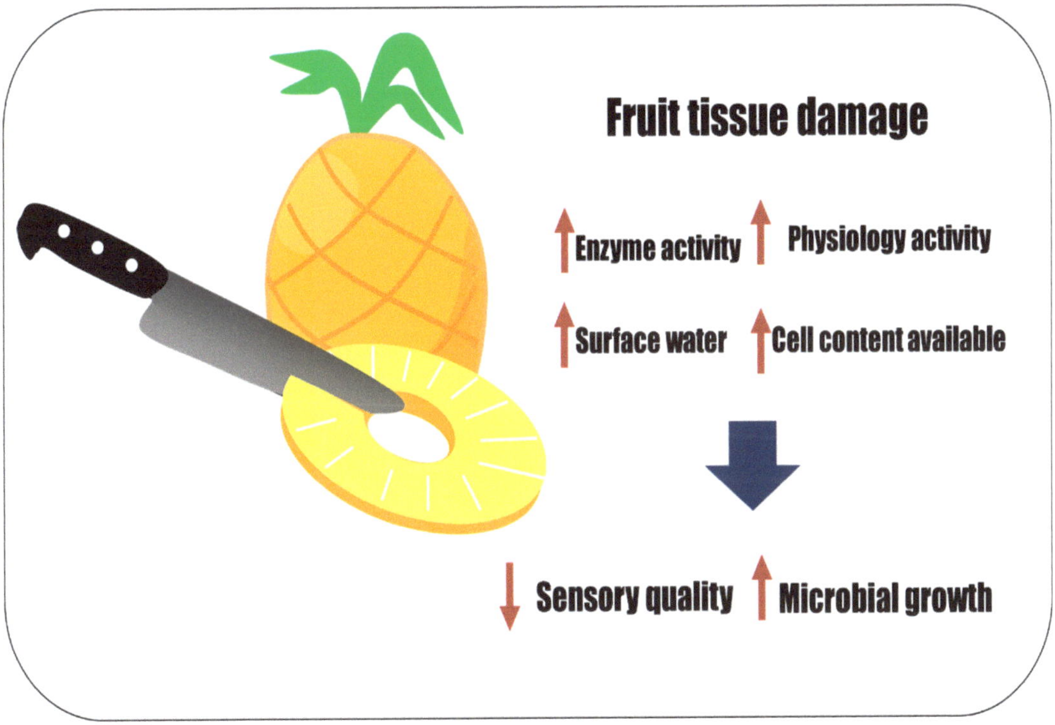

Figure 1. Minimal processing affects sensory quality and promotes microbial growth. The red up arrow indicates increase and the red down arrow indicates decrease.

Several methods could be employed to preserve fruits; however, as fresh-cut fruits are consumed fresh, thermal or freeze techniques are unsuitable because they impair the sensory, nutritional, and physicochemical quality. As a result, finding effective strategies to prevent microorganism growth while maintaining the quality of fresh-cut fruits is of great interest to food industries. The conventional way of retaining the fresh-cut products' microbiological quality is through washing and disinfection with chlorine due to its low cost and ease of use. However, it has been hypothesized that chlorination's disinfection byproducts may pose a health risk to humans [9]. Moreover, consumers demand more natural and fresh products without synthetic additives, making them nutritionally healthier. Therefore, technologies such as active packing, natural additives, ultraviolet light, high hydrostatic pressure, and ozone treatments are emerging to improve fresh-cut fruits' microbiology and sensory quality. This review summarizes some recent findings concerning the use of emerging technologies to improve the quality of fresh-cut fruit.

Our article summarizes recent studies about the main alternatives with which to maintain the sensory physicochemical quality and ensure the microbiological safety of fresh-cut fruits. In the last year, several reviews have addressed the issue of fresh-cut fruits' preservation with some differences. For example, Yousuf et al. [10] conducted a critical study to summarize the impacts of utilizing essential-oil edible coatings on fresh/fresh-cut fruits and vegetables. The primary emphasis was on the inclusion of essential oils in edible coatings, their benefits and drawbacks, methods of extracting essential oils, and the results obtained from their use in fruits, fresh cut fruits, and vegetables to extend their shelf lives. The key difference between our study and theirs is that they only focused on a single technology, whereas we summarize multiple conservation strategies. Giannakourou and Tsironi [1] outlined the hurdles of technology applied in the preservation and shelf-life extension of fresh-cut fruits and vegetables. This review highlights the combinations of traditional methods, such as temperature control and chemical agents, and a number of emerging technologies, including high hydrostatic pressure, UV-C irradiation, pulse light, ozone, ultrasound, and some packaging methods. The primary goal of this study was to emphasize the combination of technologies to boost efficacy; it also reviews studies on entire fruits and vegetables in comparison to our study.

Other reviews individually addressed different technologies; for example, Botondi, Barone, and Grasso [2] in their review aimed to increase understanding of environmentally beneficial technologies, such as ozone, which prolongs shelf life and preserves the quality of fresh-cut fruits and vegetables without emitting toxic chemicals that can harm plant material and the environment. Zhang et al. [11] highlighted the most recently published studies based on the use of plant extracts for browning suppression in fresh-cut fruits and vegetables. They described the types of plant extracts, their anti-browning capabilities, the main extraction techniques, and application procedures for fresh-cut fruits and vegetables. Kocira et al. [12] shows the application, trends and perspectives of polysaccharide coatings and edible films, and their influences on the quality of fruits and vegetables, while demonstrating their main functions and advantages. However, this review does not consider fresh-cut fruits.

2. Fresh-Cut Fruit Processing Impacts Physicochemical, Sensory, and Microbial Quality

When fresh fruits are cut or minimally processed (peeling, cutting, or slicing), they are susceptible to chemical, physical, sensory, and microbiological changes. Wounding during processing results in an increase in off-flavor compounds, loss of firmness and respiration, reduced fresh-cut shelf life, and senescence processes [4]. One of the degradation factors of these products is the color change. For the consumer, the product's first appearance is significant, since each product's characteristic color indicates its freshness and quality [13]. Polyphenol oxidase (PPO) and peroxidase (POD) are the main enzymes that cause color degradation and degradation of other sensory properties, such as taste and flavor, in fresh-cut fruits [14]. In particular, the PPO enzyme catalyzes the oxidation of polyphenols to o-quinones in the presence of oxygen, producing undesirable pigments (brown coloration of many fruits and vegetables) during ripening, storage, processing, and handling. At the same time, the POD enzyme (considered antioxidant) catalyzes the conversion of hydrogen peroxide to water using various substrates, such as polyphenols, lipids, or other compounds [15]. There are several studies focused on the inhibition of these enzymes using different inhibitory compounds, such as volatile compounds, cysteine, and ascorbic acid [16,17]; or by other processes, such as microwaving [18], ultrasonication [19], high isostatic pressure, and thermal conditions [20,21].

Flavor and texture are other attributes that must be considered in consumer satisfaction. The flavor is commonly related to aroma (odor) and taste (salty, sour, bitter, sweet, and umami). Umami is described by salts of amino acids and nucleotides [22]. The flavor in fresh-cut fruits can be affected by the increases in sugar contents, decrease in organic acids, and changes in aroma (volatile compounds) [13]. Enzymes such as lipoxygenase or peroxidase lead to the development of undesirable flavors—rancid, cardboard, or oxidized off-flavors (Figure 2) [23].

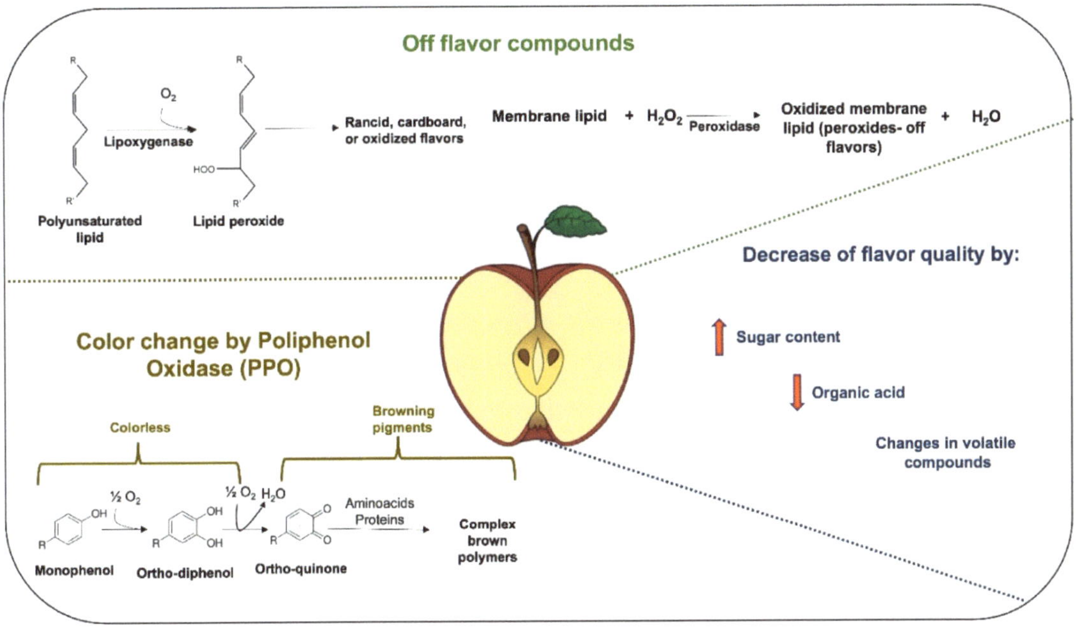

Figure 2. Enzymatic reactions and other factors that produce browning and off-flavors of fresh-cut fruits. The red up arrow indicates increase and the red down arrow indicates decrease.

On the other hand, the textural properties of fresh-cut fruits are related to the deformation, disintegration, and flow of the food under a force. The texture is related to the cell wall components (pectin, hemicellulose, and cellulose), generating the softening of fresh-cut fruits during storage by enzymatic or non-enzymatic mechanisms. This is due to the loss of cohesiveness, which decreases intermolecular bonding between cell wall polymers. Consequently, an increase in the solubility of the wall constituents is observed, principally pectin, generating the softening. Other factors, such as pH, salt concentration, and water, deficiency could influence texture loss [24].

Salinas-Hernández et al. [25] predicted fresh-cut mango's shelf life, indicating that physicochemical variables related to sensory changes can be identified and used as quality indicators. Therefore, physicochemical variables (weight, volume, pH, titratable acidity, soluble solids content, volatiles) with high correlations with the sensory attributes (color, texture, and flavor) can be used as indicators of the sensory changes, since they are easy to measure, objective, and relatively inexpensive. Additionally, the identification of these variables was carried out by regression analysis. There are several studies relating physicochemical and sensory characteristics of papaya [26], persimmon [27], plums [28], mango-based fruit bar [29], jackfruit [30], apples [31], and oranges [32], among other fresh-cut fruits.

Fresh-cut fruits are considered perishable due to their intrinsic characteristics and minimal processing that favor microbial growth, which can cause changes in safety and sensory and physicochemical properties. Fresh-cut fruit cells naturally deteriorate with time and are influenced by conditions after harvest, processing, and storage. Peeling and processing cause physiological damage and cell disruption, resulting in the release of their components, such as carbohydrates and proteins, which serve as a source of nutrients for native or exogenous microorganisms [33]. Minimal processing destroys living plants' protective membranes and barriers, allowing microbial pathogens to enter and contaminate them. The chances of food-borne illness due to pathogens or spoilage organisms growing in these products are very high. In addition, fresh-cut fruits have water activity levels between 0.8 and 0.99 and pH values between 3.0 and 7.5, suitable conditions for the growth of microorganisms [2]. The types and growth rates of the microorganisms will be significantly influenced by the product's temperature, the relative humidity, the atmosphere, and intrinsic factors such as pH, water content, and nutrients [34].

Fresh-cut fruits are prone to the invasion and colonization of microorganisms such as mesophiles and psychrophiles, molds, and yeasts [2]. Two types of microorganisms can contaminate these products: deteriorative and pathogenic. Deteriorative microorganisms cause damage to fresh-cut fruits and make them sensorily unacceptable. For example, this type of microorganism causes the production of lactic acid, acetic acid, hydrogen gases, and carbon dioxide, which results in sour odors and puffing up of packages. Other products include thiols, esters, amines, and peroxides that cause off-flavors, odors, and color changes. In addition, spoilage microorganisms can produce enzymes such as proteases, lipases, and amylases that cause structural changes in tissue and flavor [34]. The incidence of pathogenic microorganisms represents a food safety risk. Fresh-cut fruits when consumed raw are vehicles for the transmission of pathogens. Among the most common pathogenic microorganisms found in fresh-cut fruits are bacteria such as *Listeria monocytogenes*, *Escherichia coli*, *Salmonella enterica*, *Campylobacter* spp., and *Staphylococcus aureus*; fungi such as *Alternaria* spp., *Penicillium* spp., *Botrytis* spp., *Rhizopus* spp., and *Colletotrichum* spp.; and some viruses, such as norovirus and hepatitis A [2,35–37]. These microorganisms are responsible for many outbreaks worldwide [37]. The microbiological safety of fresh-cut fruits is crucial to maintaining their commercial value. Microbial contamination has an economic impact on food loss because it reduces the product's shelf life through deterioration and risks the public's health by causing foodborne illnesses [33,38]. As these products are consumed without any thermal treatment, it is essential to keep the microbial loads of fresh-cut fruits as low as possible to avoid foodborne diseases.

One of the key elements influencing the quality of fresh-cut fruits is exposure to cold temperatures immediately after harvest to reduce the effects of cutting stress. Sometimes a temperature just above that which would cause chilling injury gives the optimal condition for quality. Due to the perishable nature of fresh-cut fruits, on some occasions, it is preferable to store them under refrigeration at a temperature that could cause slight chilling injury as opposed to one that would promote quick natural deterioration [39]. Chilling damage to fresh-cut fruits can have various symptoms, some of which are noted despite minor visual manifestations; for example, poor flavor retention associated with the inhibition of volatile aroma production, increased respiration rates in fresh-cuts relative to the corresponding whole fruit, and tissue transparency and juice leakage because membrane damage, particularly in fresh-cut tropical and subtropical fruit. The signs of chilling injury in the entire fruit can also be generalized for fresh-cut products, such as softening or other textural alterations, pigment loss, and increased CO_2 production [40]. In fresh-cut fruits, chilling injury symptoms are caused mainly by the physical shifting of the membrane from a liquid–crystalline to a solid–gel phase during chilling. This process is highly reliant on the degree of membrane lipid saturation. The fluidity-dependent membranes start solidifying at cool temperatures, causing membrane integrity/leakiness problems, solute diffusion, tissue water loss, and membrane-bound protein agglomeration. Due to disruptions in the membranes connected to the electron transport chain, suboptimal chilling temperatures may also hasten oxidatively induced senescence and increase the accumulation of active oxygen species [39].

Cold storage temperatures have an impact on final product quality. For example, fresh-cut mango, a tropical fruit, is susceptible to chilling injury when stored at low temperatures, compromising its overall sensory quality. The ideal storage temperature for fresh-cut fruits is never more than 5 °C, a chilling temperature for freezing delicate tropical fruits such as mangoes. Despite the possibility of chilling injury, Dea, Brecht, Nunes, and Baldwin [40] showed that fresh-cut mango slices had a longer shelf-life when stored at 5 °C than at 12 °C because the negative changes in appearance and aroma that occurred at the higher temperature were more unpleasant than the moderate negative changes that appeared at a lower temperature. Marrero and Kader [41] reported that temperature was the primary factor affecting the quality of fresh-cut pineapple. The pulp pieces' post-cutting life was 4 days at 10 °C, yet more than 2 weeks at 0 °C. This increased longevity at levels below the chilling injury limit is also the case for whole fruits. At all temperatures, a rapid increase in respiration followed by an increase in ethylene production marked the end of commercial life. Beyond this point, storage was continued, which resulted in the development of off-tastes, smells, and microbiological deterioration. Even after being maintained at 0 °C for two weeks, the pulp fragments exhibited no symptoms of chilling damage. Since some of the symptoms may not appear for several days after being transferred to non-chilling temperatures, this may be because of the short time (approximately 3 h) between removal from refrigerated storage and quality evaluation. The impacts of temperature on fresh-cut fruits' physicochemical, sensory, and microbiological quality still require more research.

3. Emerging Technologies to Preserve the Shelf Lives of Fresh-Cut Fruits

3.1. *Active Packaging*

3.1.1. Edible Films and Coatings

Some technologies have been developed to preserve fresh-cut fruits, such as edible packaging [42–44]. This includes any edible material used to wrap food to prolong its shelf life that can be safely consumed with it. Edible films and coatings present exciting features, including biocompatibility, edibility, and a wide range of applications, making them excellent alternatives for fruit preservation [45]. The terms "film" and "coating" are frequently used interchangeably to describe a relatively thin layer of edible material covering a product's surface. While a coating is placed and formed directly on the food's surface, a film is sometimes distinguished from a coating because it is a stand-alone wrapping material. Edible coatings are typically used for liquid applications, whereas edible films are used as solid sheets and subsequently applied to food products [46].

Edible films and coating materials must be safe for human consumption, since they are not removed before consuming the product and must not alter the original product's taste, texture, smell, or appearance [47]. These emerging technologies can address customers' demands for more natural, nutritious, ready-to-eat, and minimally processed products without generating waste [48]. In addition, edible films and coatings may replace, to some extent, plastic packaging with natural and biodegradable substances. Therefore, their use could significantly reduce packaging requirements and waste disposal problems [49].

Edible films are generally good moisture barriers, thereby restricting moisture exchange between fresh-cut fruits and the atmosphere, hence reducing microbial development, weight loss, texture changes, and undesired chemical and enzymatic reactions. Fresh-cut fruits experience less respiration and senescence due to the changing environment caused by edible components' good oxygen and gas barriers [50]. Furthermore, coatings may improve the visual quality by providing gloss to the coated commodities [51]. Edible films and coatings can also be used as carriers of antioxidants, flavoring agents, coloring agents, growth regulators, and antimicrobials that will improve food quality and safety [52–54].

Coating materials include carbohydrates, proteins, lipids, and combinations [55]. Polysaccharides such as alginate, pectin, cellulose, starch, chitosan, carrageenan, arabic gum, and xanthan gum are polysaccharides that constitute the fundamental structure of the polymeric matrix [56–58]. These polymers are also effective gas barriers at low and intermediate relative humidity levels because they are hydrophilic, but due to their high water vapor permeability, they are poor moisture barriers [59]. Paraffin, carnauba, beeswax, shellac resin, and certain oils are the most popular lipids used for edible coatings because they have strong barrier qualities [60]. Proteins such as milk casein, milk zein, maize, and whey are beneficial as gas barriers (O_2 and CO_2) and antimicrobial carriers in coating treatments; however, they have limited water barrier capabilities [61,62]. Plasticizers, stabilizers, and emulsifiers can also be used to improve edible coatings' physical and chemical characteristics [60]. The functionality of covers and films is commonly evaluated through mechanical properties, such as strength, elasticity, and rheological properties [46].

Edible coatings are applied using various methods, such as immersion (dipping), spraying, and brushing (Figure 3) [46]. The dipping process submerges the product in the coating solution for a period and then lets it dry for a few minutes. For fresh-cut fruits, the immersion approach is the most popular [63]. The spraying direction is appropriate when the coating solution has a low viscosity and can be sprayed over the product. Brushing involves applying the coating solution directly to the product's surface with a brush; nevertheless, some variables, such as the amount left in the brush, are difficult to control and may influence this process [64].

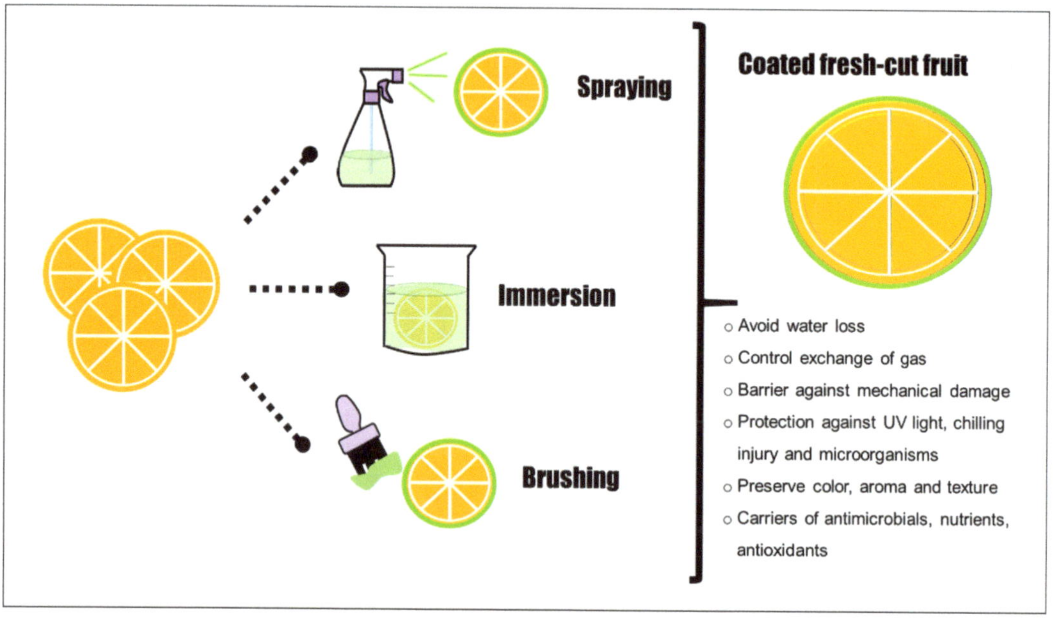

Figure 3. Main goals of edible coatings and how they are applied to fresh-cut fruits.

The loss of quality of fresh-cut fruits is due to physiological changes in the fruits, such as an altered respiration rate, loss of water, changes in texture, a decrease in organic acids, an increase in soluble solids, and starch breakdown, among others. Numerous studies have been conducted to study edible coatings in various fruits to avoid physiological changes and prolong their shelf lives (Table 1), considering the appropriate coating materials [4]. During postharvest deterioration, fresh fruits are particularly prone to weight loss, which contributes to product wilting and loss of textural qualities, such as softening and crispness, resulting in low market value and poor consumer acceptance [4]. In this context, adding ascorbic acid (1%) to a chitosan-based coating on strawberry fruits during cold storage for 15 days reduced weight loss. It also suppressed fruit softening by reducing cell wall degrading enzymes, such as polygalacturonase, cellulase, and pectin methyl esterase activities [65]. Furthermore, this treatment reduced the percentage of decayed fruits while maintaining soluble solids, titratable acidity, and antioxidant activity without affecting the fruits' sensory quality (color, taste, glossiness, and overall acceptability).

Respiration rate is one element that contributes to postharvest loss and the shelf lives of fresh-cut fruits [10]. Edible coatings can reduce respiration by establishing a modified internal atmosphere that acts as a barrier to oxygen and carbon dioxide. Afifah et al. [66] studied how the respiration rate of fresh-cut apple slices coated with chitosan and stevia combinations changed after three days of storage in modified atmosphere packing (polypropylene-PP, 30 μm) at +1 °C. The edible coatings reduced the respiration rate, which was explained by the diffusion of oxygen from the packaging atmosphere to the fruit tissue, due to the high O_2 resistance generated by the edible coating, which slowed plant metabolism.

Table 1. Edible coatings and films used to improve fresh-cut fruits' quality.

Fresh-Cut Fruit	Edible Film/Coating Material	Treatment Application	Storage Condition	Results	Reference
Apple	Sodium alginate + Tween-80 + glycerol + thymol-ethanol solution	The container was covered	5 days at 4 °C	• Increase antioxidant status • Inhibition of *S. aureus* and *E. coli* growth • Reduction of weight loss • Retain nutrition and surface color	[67]
Apple	Whey protein concentrate + apple pomace extract	Immersion	12 days at 5 °C	• Decrease in the weight loss • Reduction of browning index • Reduction of microbial loads • Slight effect in sensory evaluation	[68]
Apple	Sodium alginate + carboxymethyl cellulose + glycerol + calcium chloride + citric acid + shallot waste extracts	Wrapped films	12 days at 4 °C	• Prevention of browning index • Maintain the overall quality	[69]
Apple	Chitosan + ascorbic acid	Immersion	14 days at 5 °C	• Suppress browning • Retain flesh firmness • Maintain phenolic compounds throughout • Delay the microbial growth	[70]
Apple	Pectin + whey protein + sweet orange essential oil or lemon essential oil	Immersion	7 days at 4 °C	• Reduce weight loss • Reduce changes in color and texture • Reduce microbial counts • No change in flavor	[71]
Apple	Chitosan + gelatin + tannic acid	Cover for polyethylene terephthalate packages box	10 days at 4 °C	• Decrease weight loss • Delay browning degree • Inhibit the lipid oxidase activity • Decrease the malondialdehyde content	[72]
Kiwifruit	Aloe vera gel + hydroxypropyl methylcellulose + lemon essential oil	Spraying	10 days at 4 °C	• Act as a barrier to gas exchange • Reduce microbial load • Present Herbaceous and lemon taste • Preserve the quality of fresh-cut kiwifruit	[73]

Table 1. *Cont.*

Fresh-Cut Fruit	Edible Film/Coating Material	Treatment Application	Storage Condition	Results	Reference
Mango	Carrageenan + beeswax	Immersion	6 days at 6 °C	• Reduce weight loss • Delay an increase in total soluble solids • Maintain pH, total acidity, and growth of microorganisms. • No difference in sensory perception	[66]
Mango	Citric, ascorbic + potassium sorbate acid + aloe vera	Immersion	6 days at 7 °C	• Resist the loss of water • More weight shrinkage • Retain the brightness • Inhibit the development of redness • Maintain the yellow color • Maintain vitamin C levels	[74]
Melon	Citral nanoemulsions + chitosan or carboxymethyl cellulose	Immersion	7 °C for 14 days	• Reduce microbial counts antimicrobial protection • Extent product's storability	[75]
Orange	Sodium alginate + cocoa	Immersion	9 days at 6 °C	• Maintain texture quality • Maintain the microbiological properties (yeast and mesophilic aerobic bacteria) • Maintain the sensory properties	[32]
Papaya	Alginate + oregano essential oil	Immersion	12 days at 4 °C	• Reduce water loss • Increase sensory scores with low oil concentrations	[76]
Papaya	Starch + stearic acid + aloe vera	Immersion	12 days at 10 °C	• Reduce weight loss • Improve firmness • Retain color • Slight decrease in ascorbic acid • Extent microbial quality	[77]
Pear	Whey protein	Immersion	4 °C for 28 days	• Reduce browning • Maintain firmness • Maintain taste, smell, color, and hardness properties	[43]
Pineapple	Sodium alginate + citral nanoemulsion	Immersion	4 °C for 12 days	• Improve color retention (higher L* and b* values) • Reduce respiration rate • Reduce microbial growth (*S. enterica* and *L. monocytogenes*) • Higher concentrations of citral cause a decrease in texture and sensory acceptance.	[78]

Table 1. *Cont.*

Fresh-Cut Fruit	Edible Film/Coating Material	Treatment Application	Storage Condition	Results	Reference
Strawberry	Alginate + calcium chloride	Immersion	15 days at 4 °C	• Reduce respiration and transpiration rates • Delay the increase in the pH and soluble solid content • Delay surface mold growth • Preserve the sensory properties (color and texture)	[79]

L* = brightness; b* = yellow/blue coordinates.

Citric acid, malic acid, and glutamic acid are the primary sources of titrable acidity in fresh-cut fruits and serve as substrates for respiration; therefore, titrable acidity drops with fruit ripening or maturity. By reducing respiration, edible coatings effectively postpone the loss of organic acids. For example, Alharaty and Ramaswamy [79] coated fresh-cut strawberries, and they showed higher citric acid (%) than the uncoated samples. The acidity decreased slightly in the coated samples during the 15 days of storage compared to uncoated fruits. On the other hand, the breakdown of starch into soluble sugars or the hydrolysis of cell walls increases the total soluble solids of fruits during storage. By limiting the respiration rate, the edible coating reduces the breakdown of complex sugars into simple sugars. In this approach, fresh-cut Fuji apples were treated with three edible coatings based on Aloe vera gel and lemon essential oils applied by the spraying method [80]. The results showed that soluble solids content was reduced in uncoated samples, and the treatment was able to limit the loss of soluble solids content from the first three days of storage by maintaining this behavior until the last day of evaluation. In addition, this edible coating reduced weight loss and color changes and delayed the browning process during cold storage without affecting apples' taste, aroma, or flavor. The total soluble solids allow for the measurement of sugar content, and when combined with titratable acidity, provide helpful input on consumer satisfaction with the fruit.

The potential for developing edible antimicrobial coatings as a cost-effective way to extend the shelf lives of fresh-cut fruit has been examined. Antimicrobial chemicals can be delivered through edible films and coatings to prevent food spoiling caused by bacteria, extending product shelf life by many folds [75]. Antimicrobial compounds can be found naturally in the raw material used to make the packaging, or they can be added to it specifically. Some antimicrobial chemicals have been isolated from plants and tested in various polymeric matrixes for antibacterial activity [81]. Zhang et al. [82] produced sodium alginate/pullulan composite films with different capsaicin concentrations. Their physicochemical characteristics were studied thoroughly, and the film's transmittance, elongation at break, and moisture content dropped as the capsaicin concentration increased. In contrast, the tensile strength, water vapor permeability, and surface contact angle increased. Furthermore, the composite films showed good antibacterial activity against *E. coli* and *S. aureus* in liquid culture tests. Different composite films covered fresh mature apple cubes. As the capsaicin content increased in the film, the apples stayed fresh longer, which demonstrated that the edible films could prolong the shelf life of apples and inhibit the growth of bacteria.

Organic acids (acetic, benzoic, lactic, propionic, sorbic), fatty acid esters (glyceryl monolaurate), polypeptides (lysozyme, peroxidase, lactoferrin, nisin), plant essential oils (EOs) (cinnamon, oregano, lemongrass), nitrites, and sulfites are among the antimicrobials that can be incorporated into edible coatings [46]. In a recent study, Marquez et al. [83] coated fresh-cut apples with a blended whey protein/pectin film made in transglutaminase, and the coated and untreated fruits samples were compared throughout 10 days of storage. The treatment-preserved phenolic content prevented microbial growth, and reduced hardness and chewiness without affecting sensory acceptability. All these studies prove the potential of edible coatings to keep fresh-cut fruits' physicochemical, microbiological, and sensory quality. However, there is still a lack of studies that handle more samples and are applied in real conditions.

Even though various edible coatings and films have been successfully deposited on fresh-cut fruits, this technique may have some limitations and negatively impact the final product's quality. For example, a thick coating on the surface of fruit creates an unfavorable barrier between the interior and exterior atmosphere and prevents the exchange of respiratory gases (CO_2 and O_2), lowering the fruit's quality [84]. Therefore, it is essential to modify the coating's thickness for the variety, storage conditions, and marketing temperatures. Additionally, for edible coatings, the type of material should be considered based on whether the fruit is climacteric or not to cover the fruit's needs, delay ripening, and reduce weight loss. Incorporating antimicrobials, antioxidants, nutraceuticals, or other substances into the coatings could give fresh-cut fruits an undesirable odor, flavor, or color, mainly if essential oils are employed [85]. Additionally, the incorporation of bioactive compounds can affect the mechanical properties of the edible coatings [86]. This must be contemplated before its application to assure sensory quality and consumer acceptance. Another restriction may be increased cost, since more materials and operations were added to the fresh-cut fruits [85]. Therefore, it is crucial to search for low-cost materials and employ highly effective manufacturing and application techniques. Another drawback found in edible coatings is that they must be regulated and declared on the labels; and it must be ensured that the material used and added ingredients are non-toxic, food grade, and meet the highest hygiene standards [86].

3.1.2. Modified Atmosphere Packaging

Modified atmosphere packaging (MAP) is an inexpensive preservation technique used to extend the shelf life of fruit postharvest by slowing its respiration rate and senescence, and inhibiting microbial growth [87,88]. This technology reduces food waste and constitutes 12.3 of the United Nations' 2030 Sustainable Development Goals [89,90].

Active MAP is a method that involves the removal of air from inside the package (by flushing or evacuation-backflushing) and its replacement by a gas or gas mixture (N_2, CO_2, and O_2) supplied from pressurized cylinders or otherwise [87,91]. Therefore, applying MAP requires packing which is airtight to maintain gas concentrations during the necessary exposure time [92]. Another advantage of this package type is the biogeneration of a MA: the commodity of producing the MA by reducing the O_2 and increasing the CO_2 levels through respiratory metabolism [88]. Packaging permeability is another factor that influences gaseous composition. Gaseous composition can be modified by microperforations in the film (commonly 60–120 μm), and the size, shape, and method of hole production are crucial for MAP's effectiveness [88].

Oxygen and carbon dioxide are the most important gasses, because these two gases are crucial parts of the respiration processes in fruits. These gases inside the package decrease the respiration rate affecting all the fresh-cut fruits' metabolic pathways. Elevating the carbon dioxide concentration at the expense of oxygen decreases the respiration rate [91].

The reduction of the respiration rate with MAP support prevents oxidation and reduces ethylene biosynthesis. Oxidation is a factor that induces the browning in fruits and leads to the loss of the nutritional value of fruits due to the destruction of many nutrients, such as vitamins and proteins [93]. Fresh-cut fruits contain phenolic compounds that prevent staining or discoloration when processed into fresh-cut products. This, combined with MAP, allows enzyme-mediated browning to decrease, and MAP with low oxygen and high carbon dioxide content increases the shelf-life by reducing ethylene biosynthesis and perception [94].

Some authors reported that MAP increases the fruit's bioactive compounds during storage (Table 2). Phenolic compounds are valuable for human health as antioxidants and free-radical inhibitors, and affect the products' organoleptic properties [95]. In addition, ethylene synthesis is interrupted when oxygen is restricted and carbon dioxide increases, reducing the respiration rate [93]. When the respiration rate is reduced, the amount of carbohydrate consumed is also reduced, and the accumulated carbohydrates are used to produce phenolic compounds [96]. This increase in phenol content is due to genetic potential and environmental factors during growing and postharvest, such as a high CO_2 concentration that induces abiotic stress [93]. Some authors have also reported that high CO_2 content (5–7 kPa) prevents the oxidation of the main antioxidant compounds in fresh-cut strawberries [97]. On the other hand, carotenoid biosynthesis continues after fruit harvest and increases when exposed to a CO_2-rich modified atmosphere [93].

The carotenoids are related to vitamin A, and researchers believe that carotenoids absorb light wavelengths that cause chlorophyll optical oxidation, preserving chlorophyll and reducing damage [98]. Carotenoid oxidative degradation depends on environmental factors (oxygen level, temperature, and light) and the non-enzymatic oxidation of carotenoids by reactive oxygen [98,99]. Therefore, the increasing carotenoid content in MAP is probably due to the prevention of oxidative stress in the presence of high CO_2.

Table 2. Bioactive compounds in the fresh-cut fruits after the use of MAP during storage.

Bioactive Compounds	MAP	Exposure Time	Result	Fresh-Cut Fruit	References
Vitamin C	20% CO_2, in air	10 days	124 mg/L	Apple (braeburn)	[91]
Vitamin C	7 Kpa CO_2	28 days	5.9 mg/100 g	Apple (golden delicious)	[100]
Hydroxybenzoic acid	2.5 O_2 + 7 Kpa CO_2	21 days	10.1 mg/kg	Strawberry	[101]
p-Coumaric acid	7 Kpa CO_2	21 days	7.8 mg/kg	Strawberry	[101]
Ellagic acid	7 Kpa CO_2	21 days	73.8 mg/kg	Strawberry	[101]
Myricetin	7 Kpa CO_2	5 days	5.2 mg/kg	Strawberry	[101]
Quercetin	7 Kpa CO_2	21 days	33.5 mg/kg	Strawberry	[101]
Kaempferol	7 Kpa CO_2	21 days	4.0 mg/kg	Strawberry	[101]
Vitamin C	7 Kpa CO_2	21 days	400 mg/kg	Strawberry	[101]

Several studies have shown physicochemical qualities such as a firmness to increase in MAP with high CO_2 levels of 10% to 20% during storage at 4 °C (e.g., for apple slices) [91]. Fruit softening occurs in the middle lamellar pectin degradation of cell walls due to cell wall hydrolase activity [102]. Furthermore, MAP induces the loss of tissue strength, depending on O_2 availability [91]. On the order hand, MAP treatment did not affect soluble solid content or total acidity during storage [91,103–105].

Postharvest handling of fresh-cut fruits such as apples and pears has been of concern due to contamination with different microorganisms. For example, fungal and total aerobic mesophilic bacteria counts on fresh-cut apples stored in MAP flushed with CO_2 (20%) decreased to 5 and 3.5 log CFU/g (control samples) until below 3.5 and 2.5 log CFU/g (MAP samples), respectively, on day 10 at 4 °C [91]. Furthermore, for fresh-cut pears, fungal and bacterial counts decreased from 5.4 and 6.5 CFU/g (control samples) to 4.0 and 6.0 CFU/g (21% CO_2), respectively, by 21 days at 4 °C [105].

The main pathogens found in strawberries are *Botrytis cinerea*, *Rhizopus* spp., *Mucor* spp., *Colletotrichum* spp., and *Penicillium* spp. [106]. The technique of MAP with a CO_2-rich atmosphere has been shown to effectively reduce the development of *B. cinerea*, *Rhizopus stolonifera*, and *Mucor* species in strawberry fruit [107,108]. For this reason, using MAP technology is one of the best methods and most studied for controlling fresh-cut fruits.

3.2. Natural Preservatives

3.2.1. Antioxidants

Besides peeling, cutting, or slicing, the loss of membrane integrity of the fruits is related to several stress factors that can generate reactive oxygen species (ROS). ROS can act as secondary messengers in various critical physiological phenomena; however, ROS also induce oxidative damage under stress conditions [109]. The ROS mainly comprise singlet oxygen ($^1O_2$), hydrogen peroxide (H_2O_2), superoxide radicals ($O^{\bullet -}{}_2$), and hydroxyl radicals (OH•). Degradation of pigments, lipids, proteins, and other biomolecules indicates cellular damage by ROS, which leads to cell death. To prevent or attenuate this phenomenon, exogenous antioxidants could be a helpful solution for fresh-cut fruits. These compounds inactivate ROS, PPO, and POX to reduce the formation of brown pigments [110]. Synthetic antioxidants have been used for this due to their high efficacy; however, consumers demand more natural and sustainable compounds as food ingredients.

Plant phenolic compounds are potent antioxidants due to their redox properties, reducing agents, hydrogen donators, and singlet oxygen quenchers [111]. Flavonoids are the main phenolic compounds of their entire class and are divided into flavonols, flavones, isoflavones, and anthocyanins. They have a role in scavenging principally $^1O_2$ [109]. Phenolic extracts from plants or fruits and their antioxidant potential to preserve fresh-cut fruits have been extensively studied (Table 3). For example, yerba mate, pomegranate, kiwifruit, elderberry flower, strawberry tree (leaves and branches), vine (leaves and branches), olive (leaves and branches), pear (pulp, peel, pomace), apple (peel and pomace), bitter melon, and mango seed extracts are used to reduce browning and increase the antioxidant capacities of fresh-cut mangoes, apples, and pears [110,112–115]. Some examples of phenolics from plant or fruit extracts are rosmarinic acid, p-coumaric acid, trans-cinnamic acid, hydroxyphenyl lactic acid, caffeic acid, gallic acid, vanillin, quercetin, resveratrol, and eugenol [116–118].

On the other hand, ascorbic acid is the most abundant and extensively studied antioxidant compound. Its mechanism of action is through electron donation to a wide range of enzymatic and non-enzymatic reactions. Ascorbic acid participates in the first defense against ROS attacks, reacting with H_2O_2, OH•, and $O^{\bullet -}{}_2$; and it regenerates α-tocopherol from tocopheroxyl radicals, protecting membranes from oxidative stress damage [109]. Table 3 shows some studies on reducing the browning of fresh-cut fruits, principally apples using ascorbic acid.

Another antioxidant commonly used in fresh-cut fruits is citric acid. This compound is often considered safe. It can prevent browning and fruit disease by reducing the respiration of postharvest fruits. Citric acid treatment can slow down the decreases in the soluble sugars and titratable acidity and is beneficial to maintaining fruit quality during storage [124]. Usually, this antioxidant compound affects PPO, and with its inhibition, produces an antibrowning effect on fresh-cut fruits, as is the case with apples [14,113,114].

Melatonin is a hormone from fruits or vegetables that can protect cell structure, reduce peroxide levels by removing free radicals, reduce lipid peroxidation, enhance oxidation resistance, and prevent DNA damage [120]. According to Dan et al. [125], this molecule also inhibits browning during the cultivation of plant tissue; therefore, it can inhibit the oxidation of phenolic compounds, being a powerful antioxidant. Studies in recent years have promoted fruit ripeness and improved its storage quality, as is the case for tomatoes [126] and strawberries [127]. They have also reduced surface browning on fresh-cut apples and pears [120]. Antioxidants such as natural preservatives are essential in the food industry because they allow foods to conserve their nutritional properties and quality levels, preventing contact between oxidative enzymes and phenols, and indirectly inhibiting enzymatic browning. However, limitations have been observed when applying antioxidants: for example, they could modify the natural flavor of fresh-cut fruits, which generally depends on the concentration—that is, the higher the concentration, the greater the susceptibility of the fresh-cut fruits to having their flavor modified [113,114]. This has the consequence that consumers reject fresh-cut fruits for consumption, as they do not perceive their natural flavor. Therefore, most studies focus on determining the optimal concentrations of antioxidants that preserve the nutritional and sensory quality of fresh-cut fruits.

Table 3. Effects of pure or extracted antioxidant compounds on the quality of fresh-cut fruits.

Antioxidant/ Source of Antioxidant	Fresh-Cut Fruit	Treatment Application	Principal Results	References
Rosmarinic acid, p-Coumaric acid, Trans-Cinnamic acid, Hydroxyphenyllactic acid, caffeic acid, ascorbic acid, gallic acid, citric acid and BHA	Apple	Immersion (500 μg/mL of each antioxidant)	Reduced the browning, maintained the acidic pH and restricted growth of *L. monocytogenes* even after 10 days of treatment.	[116]
Calcium ascorbate; vanillin or cinnamic acid	Nectarine	Immersion (6% calcium L-ascorbate)	Reduced browning	[117]
Mango seed extract	Mango	Immersion (6.25 g/L of the extract)	Preserved fresh-cut fruits, increasing polyphenols, flavonoids and antioxidant capacity.	[112]
Apple polyphenols	Red pitaya	Spraying (5 g/L apple polyphenols)	Maintained sensory (retention of color, delay of the softening) and nutritional attributes of fresh-cut red pitaya fruit.	[119]
Yerba mate (*Ilex paraguariensis*) Citric acid Ascorbic acid	Apple	Infusion (1.2% yerba mate + 0.9% citric acid + 1.0% ascorbic acid)	Increased antioxidant capacity and decreased browning. The color, flavor and texture of the apples were kept.	[113,114]
Phenolics from juice or extract of pomegranate and kiwifruit	Pear	Immersion (0.3% of antioxidants)	Improved antioxidant capacity and prevented enzymatic browning.	[115]
Melatonin	Pear	Soaked with 0, 0.05, 0.1 and 0.5 mM melatonin	Reduced the surface browning, maintained the titratable acidity, enhanced total phenolic content and antioxidant capacity, and delayed the reduction of ascorbic acid.	[120]
Extracts of: Elderberry flower (*Sambucus* L.) Vine (*Vitis vinifera* L.) leaves and branches Pear (*Pyrus communis* L. "Rocha") pulp, peel and pomace Olive (*Oleo europaea* L.) leaves and branches Apple (*Malus domestica* L.) peel and pomace Acorn (*Quercus* L.) bark Bitter Melon (*Momordica charantia* L.) whole plant Strawberry tree (*Arbutus Unedo* L.) leaves and branches Potato plant (*Solanum tuberosum* L.) leaves	Pear	Sprayed (9.5 mg/mL, 5 mg/mL and 16 mg/mL)	Delaying fresh-cut pear browning expansion. Strawberry leaves and branches were the best antioxidant extracts.	[110]
Coconut liquid endosperm	Apple	Immersion (100% into the coconut liquid)	Coconut liquid endosperms are feasible natural agent inhibiting browning incidence of fresh-cut fruits during storage.	[121]
Melatonin	Apple and Pear	Immersion (0.05, 0.1, and 0.2 mM melatonin)	Reduced surface browning in fresh-cut foods.	[122]
Citric acid	Apple	Immersion (5% citric acid)	Inactivation of *Salmonella* and polyphenol oxidase.	[123]
Eugenol	Water chestnut	Immersion (0.4% and 1.5% eugenol)	Eugenol exhibited inhibitory effect on fresh-cut water chestnuts browning. Eugenol could also enhance the enzymatic/non-enzymatic antioxidant capacity and alleviate the ROS damage to membrane	[118]

3.2.2. Antimicrobials

As was stated before, fresh-cut fruits are more susceptible to microbial spoilage than raw produce. Traditional antimicrobials have been widely used for many years to preserve food products. However, currently, there is great interest in using natural antimicrobial agents to prevent microbial deterioration, assure safe consumption, and retain the physicochemical and sensory quality of food due to the demand for healthier, fresher products with lower amounts of synthetic preservatives [128]. Emerging technologies involve using natural compounds with bioactive properties to enhance the shelf lives of fresh-cut fruits. They have been used against food spoilage and pathogenic microorganisms [129].

Several antimicrobials from various sources, such as animals (chitosan), plants (phenolic extracts, essential oils), and microbial (nisin), have been effectively used for the preservation of fresh-cut fruits [130]. The most common bioactive compounds with proven efficacy are plant derivatives such as essential oils and phenolic compounds [131]. Antimicrobials have different concentrations and effects for inhibition or inactivation of microbial growth. The specific antimicrobial targets include cell walls, membranes, metabolic enzymes, protein synthesis, and genetic systems. However, the exact antimicrobial mechanisms have not been fully elucidated. They also depend on the type, genus, species, and strain of the microorganism and environmental conditions (pH, water activity, temperature, and atmosphere, among others) [132].

Essential oils from spices and medicinal plants are sources of many bioactive compounds of great interest in the food industry with antimicrobial attributes [133–135]. Essential oils are oily liquids formed from complex mixtures of volatile compounds with strong aromas that are responsible for the fragrances of flowers and other plants. Thei main compounds are of low molecular weight, such as hydrocarbons, terpenes, terpenoids, and their derivatives [136]. Essential oils are produced by many plants as secondary metabolites and extracted from leaves, husks, bark, flowers, buds, and seeds through steam distillation, hydrodistillation, or solvent extraction [10]. Essential oils are used as flavorings and aromatized in perfumery, cosmetics, and the food industry because they are generally recognized as safe.

Numerous studies have shown the potential of essential oils against Gram-negative and Gram-positive bacteria and fungi. The antibacterial mechanisms of essential oils are related to their hydrophobicity and the structures of their components. The lipophilic nature of essential oils allows them to pass through the cell wall and damage the cytoplasmic membrane while disrupting various layers of polysaccharides, fatty acids, and phospholipids, eventually rendering them permeable. They can also bind to proteins to prevent them from carrying out their normal functions of transporting molecules and ions. Additionally, the hydrophobic components present in the essential oil could change the permeability of the microbial cell membrane for cations such as $H+$ and $K+$, which modify the flow of protons, changing the cellular pH and affecting the chemical compositions of the cells and their activity. Loss of differential permeability results in an imbalance in intracellular osmotic pressure, which then disrupts intracellular organelles, leading to the release of cytoplasmic contents, lower proton motive force, a depleted ATP pool, and the denaturing of various enzymes and proteins, ultimately causing cell death [137,138].

For decades, essential oils have been identified as food additives to preserve the microbiological quality of fresh-cut fruits. Emerging technologies are currently being studied to improve solubility and reduce essential oils' strong aromas and flavors, such as edible coatings and films added with natural antimicrobials. Essential oils can help decrease the permeability of water vapor in hydrophilic films due to their lipidic properties, and provide other helpful properties, such as structural, optical, tensile, and antimicrobial ones [139]. For example, hydroxypropyl methylcellulose coatings added with Thai essential oil were applied to mango to decrease the losses in weight and firmness, and color change. Additionally, this treatment showed antifungal activity against *Colletotrichum gloeosporioides* and good sensory acceptance [140].

A nanoemulsion of orange peel essential oil was added to pectin-based edible coatings at 0.5 and 1.0% to evaluate the quality of orange slices stored at 4 °C for 17 days. The slices coated with 1.0% essential oil showed fewer mesophilic aerobic microorganisms and yeasts and molds loads compared to samples without essential oil. The findings revealed that orange peel essential oil nanoemulsion-based edible coatings could improve the shelf life of orange slices without compromising sensory qualities [141]. In a similar approach, gellan coatings with added geraniol (1.2 and 2.4 μL) were used to preserve the quality of fresh-cut strawberries for 7 days at 5 °C. The coatings at both doses reduced mesophilic bacterium, yeast, mold, and psychrophilic counts compared to untreated samples. However, the treatments did not improve the texture and sensory quality; the fruit showed higher firmness loss than the control fruit [142].

Phenolic compounds are a large and varied group of molecules found in plants which have different structures and functions. These compounds are synthesized by the plant's secondary metabolism during the plant's normal development and once the plant tissue undergoes diverse types of stress [143]. According to their structural characteristics, about 10,000 phenolic compounds are divided into several groups. The molecular structure of any phenolic compound includes an aromatic ring attached to at least one hydroxyl group. Phenolic acids and flavonoids are the most investigated food additives among the several families of phenolic compounds [144]. Antimicrobial activity has been associated with hydroxyl groups' presence, number, and positions in their structures. The hydroxyl groups can bind to essential enzymes of microbial metabolism and increase the affinity for cytoplasmatic membranes, which can cause microbial death, inhibition of virulence factors, and biofilm formation [145]. It has been demonstrated that adding a single hydroxyl group and particular lipophilicity to a molecule increases its antibacterial capabilities [144].

Byproducts from fruits, such as peel and seed, have many phenolic compounds with antimicrobial potential. For example, pomegranate peel extract rich in phenolic compounds, such as ellagitannins, showed in vitro bactericidal and bacteriostatic activity against five strains of *L. monocytogenes*, and at 12 g/L demonstrated vigorous antibacterial activity against bacterial load on fresh-cut apple, melon, and pear throughout a 7-day storage period [128]. However, this study did not evaluate the sensory acceptance of the product, which is an essential part of the quality of fresh-cut fruits. It is essential to consider that the results of the in vitro activity are always better than those in vivo, since the presence of organic matter increases the survival of the bacteria and decreases the contact of the extracts with the bacteria. In addition, it should also be considered that the extracts with the highest content of phenolic compounds generally present the greatest effects and the highest pH, which influences the chemical structures and functions of polyphenols.

Similarly, mango seed extract was used at 6.25 g/L, the highest concentration with sensory acceptability, to preserve fresh-cut mango quality and reduce inoculated bacteria [112]. In the same approach, aqueous extracts of nut byproducts, cashew nut shell and coconut shell, were applied to fresh-cut papaya, thereby reducing the population of *E. coli*, *L. monocytogenes*, and *S. enterica*. However, cashew nut shell extract affected the sensory characteristics—darkening the fresh-cut papaya tissues. Therefore, the authors proposed coconut shell, rich in luteolin, as an excellent antimicrobial additive in food applications [146]. Grape byproducts, stem and leaf extracts, were micro-encapsulated and added as biopreservatives in a fresh-cut fruit salad of grape, apple, and sweet corn. The extract capsules formed a film on the fruits, reduced the inoculated loads of *Aspergillus ochraceus* and *Alternaria alternata*, and reduced ochratoxin, a dangerous mycotoxin. When applied to fresh-cut fruit salad, the stem capsules were more effective than leaf capsules [147].

Using natural compounds such as essential oils and plant extracts is quite promising for preserving fresh-cut fruits; however, their applications still face some challenges [148]. The application of natural antimicrobials presents some limitations due to their potent aromas, high reactivity, hydrophobicity, decreased solubility, and potential interactions with food's carbohydrates and fatty acids that could modify its organoleptic qualities [149]. For example, essential oils are highly irritating, which can cause damage to fruit tissues, and have strong aromas and flavors that can be unpleasant for the consumer. Additionally, some plant extracts have strong colorations that can affect the color of food products [112]. In this sense, the optimal doses that are effective in inhibiting microbial growth and do not affect sensory quality should be evaluated. In addition, one must create a palatable combination with the aromas and flavor notes of essential oils or plant extracts and fresh-cut fruits. The appropriate combination might boost natural food preservation trends [10].

3.3. Physical Treatments

3.3.1. UV-C Radiation

Exploring more effective ways to maintain the quality of fresh-cut fruits has led us to look for other technologies, such as shortwave ultraviolet (UV-C, 190–280 nm) irradiation. This technique is approved by the Food and Drug Administration of the USA and is recognized as safe for food products because it is environmentally friendly and toxic-residue-free [150]. The uses that have been attributed to this technique in fresh-cut fruits are varied: inhibiting the browning process (reducing PPO), prolonging the shelf-life, maintaining optimal quality during storage, delaying ripening, reducing softening, and avoiding undesirable flavors [150–153]. Additionally, UV-C induces stress in fruit and consequently heightens the antioxidant defense system and some concentrations of secondary metabolites, such as phenolic compounds and pigments [154,155]. Additionally, UV-C radiation may extend the shelf life of fresh-cut products by its microbicidal effect due to the formation of pyrimidine dimers blocking the microbial cell replication due to DNA alteration [156].

A 4.54 kJ m^{-2} dose was used for the UV-C treatment of fresh-cut pomegranate arils storage for 14 days. The end of shelf life showed the lowest mesophile, yeast, and mold growth and the highest total antioxidant activity and phenolic content [157]. Recently, fresh-cut lotus (*Nelumbo nucifera* Gaertn.) root exposed to a UV-C lamp (75 W) for 10 min and then stored for 8 days exhibited a significantly low browning degree, soluble quinone content, and inactivation of enzymes activities (polyphenol oxidase, peroxidase, and phenylalanine ammonia lyase) [152]. UV-C (4.5 kJ m^{-2}) disinfection treatment and nano-coating lemon essential oil nanocapsules were used to preserve fresh-cut cucumber for 15 days. A good correlation was observed between increasing the fruit's shelf life and decreasing its enzymatic activity [158]. Similarly, UV-C radiation (4.0 kJ m^{-2}) for 5 min combined with calcium lactate was applied to fresh-cut kiwifruit slices for 7 days of storage. UV-C and calcium lactate treatment could synergistically maintain overall quality and improve the antioxidant capacity of kiwifruit slices [153]. These studies demonstrate the usefulness of the maintenance and quality of fresh-cut fruits when applying UV-C radiation. However, the application of UV-C radiation does not have the same impact on all fresh cut fruits, which is a limiting factor for the application of this technology. For example, the application of UV-C in fresh-cut pomegranate arils decreased the amounts of yeast and mold, whereas in kiwi slices this decrease was not observed [153]. The exposure time is also an important factor, since being exposed for several minutes could cause damage to the tissues, causing nutritional and physicochemical changes. For this reason, an alternative use is to apply UV-C radiation with other technologies [153,158].

3.3.2. High Hydrostatic Pressure

High hydrostatic pressure (HHP) generates safe and stable food products and does not damage the sensory or nutritional properties of the product [159,160]. This technology requires high initial capital investment. However, the costs of the HHP-processed products have improved during the last few years [161]. In addition, HHP applications are increasing the health benefits to consumers by adding value products [162,163]. In general, HHP uses isostatic pressures between 100 and 1000 MPa and does not require heat [164]. HHP is a non-thermal process, and the temperature of water used does not exceed 50 °C [165]. HHP processing may present positive effects such as eliminating pathogenic and deteriorative microorganisms to ensure food safety and extend the product's shelf-life [166]. However,

HPP technology negatively affects food matrices by altering the structures of enzymes responsible for desired compounds due to the breakage of weak interactions (hydrophobic and electrostatic interactions) [167].

HHP has shown some potential applications in maintaining and enhancing nutritional value and bioactive compounds through microbial and enzyme inactivation, extending the shelf lives of different fruits [161]. However, various quality changes have been reported after using HHP, such as darkening of avocado slices and persimmon [168,169]. Denoya et al. [170] suggested that increased pressure in the HHP conditions decreases the activity of PPO in fresh-cut fruits such as peach, and found evidence of the close relationship between the PPO residual activity and the degree of browning for the processed peach pieces. However, it has been reported that increasing pressure in fresh-cut Hachiya persimmon causes changes in tissue structure and physicochemical properties [171].

On the other hand, HHP with moderate pressure (up to 600 MPa) for short holding times on colorimetric and textural parameters on fresh-cut pumpkins and peach had no effect, and the esterification degree of pectin decreased after HHP [170,172,173]. However, moderate pressures produce lysis of the cell membranes, causing the food matrix's mechanical disruption. To avoid this scenario, it is recommended to use 200 MPa to maintain the best physicochemical properties and preserve the integrity of the bioactive compounds [170,171]. These alterations could be because mechanisms of enzymatic inactivation by HHP are very complex and depend on the product and conditions [174].

HHP technology is cold pasteurization that inactivates microorganisms, significantly extending the products' shelf-life and guaranteeing food safety [161]. For example, HHP processing can ensure the microbiological safety of fresh-cut pumpkins. HHP immediately reduced the total plate counts (TPC) to 1.61 and 1.52 log10 CFU/g at 450 and 550 MPa, respectively, and to 2.57 and 1.69 log10 CFU/g, respectively, after 45 days [175]. This increase in the TPC could be because the HHP treatment does not entirely inactive microorganisms. In general, the tendency in pressure-treated products at a higher pressure and holding time treatments is to significantly increase mesophilic aerobic microorganisms' lag phase times [175]. This may be due to the greater severity of membrane structure damage, an increase in pH, and the recovery and growth of injured cells [176].

3.3.3. Ozone

Ozone has been widely used in the food and agriculture industry and is "generally recognized as safe" (GRAS) by the FDA (US Food and Drug Administration, 2001). The ozone treatments are mainly applied as aqueous ozone and gaseous ozone. Aqueous ozone, a sanitizing agent, is a highly effective disinfectant at low concentrations and short contact times, capable of rapidly decomposing into oxygen, and therefore, does not leave a residue on treated products [177,178]. Therefore, washing with ozone is a highly efficient and safe disinfection method to preserve fresh-cut fruits and vegetables and can be a good substitute for washing with chlorine. Recent studies have indicated that ozone can delay texture deterioration of fresh-cut fruits and vegetables such as apples, green bell peppers, and onions [179–181]. Gaseous ozone treatment is developed as a sanitization method to avoid cross-contamination when used with large volumes of produce that could not be sanitized by further washing with disinfectants [182].

Some studies evaluated the effectiveness of ozone in various fruits and vegetables. The effects of ozone are associated with multiple reactions, including the inactivation of enzymes, alterations in nucleic acid concentrations, and oxidation of membrane lipids [183]. Ozone treatment could reduce weight loss, minimize tissue destruction, maintain cell integrity, and thus reduce moisture transpiration [184]. Ozone has been praised for its efficacy in maintaining the firmness of products. However, the firmness of products inevitably decays over storage time. In general, ozone has made a minor impact on this reduction and does not affect the texture of fresh-cut products during the storage period [178,184].

The degradation of polyphenols during ozonation produces several chemical reactions [184]. This occurs due to the penetration of ozone into the cells with an oxygen ion that leads to direct responses with the compounds and the hydroxyl radicals produced through the oxygen ion catalyzed by the hydroxide ion [178,184,185]. Several authors have reported the main quality parameters of fresh-cut fruits treated with aqueous or gaseous ozone (Table 4). Aqueous and gaseous ozone treatments do not induce stress in the respiration rate in fresh-cut products and cause the inactivation of biosynthetic enzymes, which will be responsible for various metabolic activities, inclusive of ethylene biosynthesis for a brief time, and enhancing the antioxidant capacity and total phenolic content without compromising on its sensorial quality during storage [184,186].

Table 4. Quality parameters and microbiological control of fresh-cut fruits treated with aqueous or gaseous ozone.

Fresh-Cut Fruit	Treatment	Results	References
Apple	Aqueous ozone 1.4 mg/L At 5 and 10 min	• Respiratory rate is maintained • Ethylene production maintained at low levels • PPO activity was decreased • Causes the loss of antioxidant compounds • Antioxidant capacity was increased	[184]
Apple	Aqueous ozone 1.4 mg/L at 5 min	• Softening during the storage is reduced • Water-soluble pectin was increased • Protopectin content, 4% KOH-soluble fraction, and cellulose content were decreased • 24% KOH-soluble fraction was maintained • Pectin methylesterase activity was increased • β-galactosidase and α-arabinofuranosidase activities were inhibited • Polygalacturonase activity was maintained	[179]
Papaya	Gaseous ozone 9.2 μL/L at 10, 20 and 30 min	• Total phenolic content was increased • Ascorbic acid content was decreased	[187]
Durian	Gaseous ozone 900 mg/L at 3 and 5 min	• The appearance of the flesh and funiculus was maintained • The respiration rate was reduced • Ethylene production was reduced • Total phenolic content was increased • Antioxidant capacity was increased	[186]
Durian	Gaseous ozone 900 mg/L for 14 days at 4 °C of storage	• Total bacteria count was 2.10 log CFU/g • The coliform population was 1.93 log CFU/g	[186]
Apple	Aqueous ozone 1.4 μL/L at 5 min for 12 days and 4 °C of storage	• Total bacteria count was 3.5 log CFU/g • Total molds count was 1.42 log CFU/g • Total yeasts count was 3.33 log CFU/g	[184]
Papaya	Gaseous ozone 9.2 μL/L for 10 and 30 min directly	• Total mesophilic bacteria value was 0.22 log CFU/g at 10 min • Coliform count was 1.12 log CFU/g at 30 min	[187]
Bell pepper	Gaseous ozone 9 ppm for 6 h	• *E. coli* O157 count was 2.89 log CFU/g • *S. Typhimurium* count was 2.56 log CFU/g • *L. monocytogenes* count was 3.06 log CFU/g	[182]

Ozone is a strong oxidant and an antimicrobial agent (Liu et al., 2016; Liu et al., 2021). The effectiveness of ozone in microorganisms increases proportionally to the increase in concentration and exposure time [184,186]. Oxidative damage by ozone produces irregular cell structure in bacteria, inducing ROS destruction of lipid and protein molecules embedded in the bacterial cell membrane [182,188]. The lipid peroxidation by ozone forms lipid hydroperoxides (LOOH) that produce lipid degradation, leading to cell wall rupture, cellular leakage, excessive nutrient loss, and cell death [182]. Studies showed that aqueous and gaseous ozone could be employed to achieve desired microbial safety (Table 4). Ozone is an alternative sanitizer used for several fresh-cut fruits, achieving microbial reductions and increasing shelf life. However, ozone is a highly reactive compound. It may cause physiological injury to the product, and it is necessary to keep ozone concentrations as low as possible [182]. Some fresh-cut fruits and vegetables, such as apples, green bell peppers, and onions, are susceptible to ozone, causing texture deterioration [180,181,184]. The effectiveness of ozone depends on product type, O_3 concentration, and treatment time.

4. Futures Trends

Emerging technologies, such as natural antioxidant and antimicrobial compounds, modified atmosphere, ozone, UV-C, and high hydrostatic pressure, are still not fully implemented in the food industry because of their high costs, or they do not guarantee the total safety of the product. Therefore, future research should combine various preservation methods, known as hurdle technologies, to generate a synergistic effect due to the different mechanisms of action to increase food products' nutritional, sensory, and microbiological quality. They should also carry out studies optimizing these technologies for industrial applications to determine doses and secondary effects on the nutritional and sensory quality of the products. This will allow us to offer methods that produce fresher, safer, more nutritious, more natural food with fewer synthetic additives.

According to the information collected in this review, there are still many areas of opportunity in terms of developing and applying technologies such as pulsed light and smart packaging to study the increase in the quality of fresh-cut fruits. Some advantages and disadvantages of the different emerging technologies that can be applied to fresh-cut fruits were already seen in this review; however, one of these technologies that was not mentioned in the previous points is pulsed light treatments. Pulsed light treatment (intense broad spectrum ranging from UV to near-infrared) is a decontamination method for foods (inactivates microorganisms) consisting of repeatedly providing light at certain intervals during storage to further extend the shelf lives of fresh-cut fruits [189]. However, only a few studies have been conducted using this technology on fresh-cut fruits; for example, raspberries and blueberries [190], mangoes [191,192], apples [193], strawberries [194], and cantaloupes [189]. Therefore, this technology could be used on other fresh-cut fruits to increase their shelf lives while preserving their nutritional quality and sensory aspects. Other technologies that were not seen in this review were treatments with cold plasma, heat shock, and hydrogen sulfide, among others, which are very interesting technologies, but the information on their effects on fresh-cut fruits is very limited. For this reason, it was difficult to cover these issues in the current review.

Another significant area of opportunity is being able to complement emerging technologies using so-called smart packaging. Smart packaging is a broad concept that encompasses several functions related to packaged goods, such as foods and beverages, to maintain integrity and prevent food spoilage (shelf life) through visual or other indicators. In addition, they respond to changes in the environmental conditions of the product or the packaging. Finally, they must communicate the product's condition and report on the opening and integrity of the seal. Some smart packages are in use today, and many others are under development. However, most studies focus on whole fruits, and there are almost no studies on fresh-cut fruits [195,196], so this could be an excellent area of opportunity.

5. Conclusions

Fresh-cut fruits are in great demand for their convenience and nutritional quality. However, they are highly perishable and susceptible to contamination and microbiological deterioration, which increases the risk of disease and compromises the nutritional and sensory quality of the product. The challenge for the food industries and the scientific community is the search for technologies to extend fresh-cut fruits' shelf lives. Emerging technologies are of recent creation and are alternatives to conventional technologies. They aim to eliminate microorganisms and maintain maximum nutritional and sensory quality. The objective, the nature of the fruit, its physiological characteristics, storage conditions, and efficacious doses, among other factors, must be considered to select the most suitable technology. This is because each fresh-cut fruit has distinctive characteristics, especially regarding bioactive compounds, since some products may contain them in greater or lesser amounts and even lack some of them. Within these technologies, active packaging, antioxidant and antimicrobial natural compounds, high hydrostatic pressure, UV-C radiation, and ozone are good candidates to extend the shelf lives of fresh-cut fruits and provide high microbiological and sensory quality during the preservation period.

Author Contributions: Writing-original draft preparation, A.T.B.-M., C.L.D.-T.-S. and R.D.I.-G.; conceptualization, O.M.-C., F.J.W.-C. and J.B.-F.; writing-review and editing, F.J.C.-M. and Y.I.C.-R.; supervision, S.R.-C. All authors have read and agreed to the published version of the manuscript.

Funding: This research received no external funding.

Institutional Review Board Statement: Not applicable.

Informed Consent Statement: Not applicable.

Conflicts of Interest: The authors declare no conflict of interest.

References

1. Giannakourou, M.C.; Tsironi, T.N. Application of processing and packaging hurdles for fresh-cut fruits and vegetables preservation. *Foods* **2021**, *10*, 830. [CrossRef]
2. Botondi, R.; Barone, M.; Grasso, C. A review into the effectiveness of ozone technology for improving the safety and preserving the quality of fresh-cut fruits and vegetables. *Foods* **2021**, *10*, 748. [CrossRef] [PubMed]
3. FDA. Draft Guidance for Industry: Guide to Minimize Food Safety Hazards of Fresh-Cut Produce. Available online: https://www.fda.gov/regulatory-information/search-fda-guidance-documents/draft-guidance-industry-guide-minimize-food-safety-hazards-fresh-cut-produce#:~:text=In%20this%20guidance%2C%20%E2%80%9Cfresh%2D,without%20additional%20processing%20(such%20as (accessed on 12 May 2022).
4. Yousuf, B.; Qadri, O.S.; Srivastava, A.K. Recent developments in shelf-life extension of fresh-cut fruits and vegetables by application of different edible coatings: A review. *LWT* **2018**, *89*, 198–209. [CrossRef]
5. Putnik, P.; Bursać Kovačević, D.; Herceg, K.; Levaj, B. Influence of cultivar, anti-browning solutions, packaging gasses, and advanced technology on browning in fresh-cut apples during storage. *J. Food Process Eng.* **2017**, *40*, e12400. [CrossRef]
6. Siddiq, M.; Harte, J.; Beaudry, R.; Dolan, K.; Singh, S.; Saha, K. Physicochemical properties of whole fruit and sensory quality of fresh-cut apples pre-treated with 1-Methylcyclopropene (1-MCP). *Int. J. Food Prop.* **2014**, *17*, 1081–1092. [CrossRef]
7. Raybaudi-Massilia, R.M.; Mosqueda-Melgar, J.; Soliva-Fortuny, R.; Martín-Belloso, O. Control of pathogenic and spoilage microorganisms in fresh-cut fruits and fruit juices by traditional and alternative natural antimicrobials. *Compr. Rev. Food Sci. Food Saf.* **2009**, *8*, 157–180. [CrossRef]
8. Putnik, P.; Roohinejad, S.; Greiner, R.; Granato, D.; Bekhit, A.E.-D.A.; Kovačević, D.B. Prediction and modeling of microbial growth in minimally processed fresh-cut apples packaged in a modified atmosphere: A review. *Food Control* **2017**, *80*, 411–419. [CrossRef]
9. Weisman, R.J.; Heinrich, A.; Letkiewicz, F.; Messner, M.; Studer, K.; Wang, L.; Regli, S. Estimating national exposures and potential bladder cancer cases associated with chlorination DBPs in US drinking water. *Environ. Health Perspect.* **2022**, *130*, 087002. [CrossRef] [PubMed]
10. Yousuf, B.; Wu, S.; Siddiqui, M.W. Incorporating essential oils or compounds derived thereof into edible coatings: Effect on quality and shelf life of fresh/fresh-cut produce. *Trends Food Sci. Technol.* **2021**, *108*, 245–257. [CrossRef]
11. Zhang, X.; Meng, W.; Chen, Y.; Peng, Y. Browning inhibition of plant extracts on fresh-cut fruits and vegetables—A review. *J. Food Process. Preserv.* **2022**, *46*, e16532. [CrossRef]
12. Kocira, A.; Kozłowicz, K.; Panasiewicz, K.; Staniak, M.; Szpunar-Krok, E.; Hortyńska, P. Polysaccharides as edible films and coatings: Characteristics and influence on fruit and vegetable quality—A review. *Agronomy* **2021**, *11*, 813. [CrossRef]
13. Farcuh, M.; Copes, B.; Le-Navenec, G.; Marroquin, J.; Cantu, D.; Bradford, K.J.; Guinard, J.-X.; Van Deynze, A. Sensory, physicochemical and volatile compound analysis of short and long shelf-life melon (*Cucumis melo* L.) genotypes at harvest and after postharvest storage. *Food Chem. X* **2020**, *8*, 100107. [CrossRef] [PubMed]

14. Wen, Y.T.; Liang, Y.Q.; Chai, W.M.; Wei, Q.M.; Yu, Z.Y.; Wang, L.J. Effect of ascorbic acid on tyrosinase and its anti-browning activity in fresh-cut Fuji apple. *J. Food Biochem.* **2021**, *45*, e13995. [CrossRef] [PubMed]
15. Lee, J.H.; Kasote, D.M.; Jayaprakasha, G.K.; Avila, C.A.; Crosby, K.M.; Patil, B.S. Effect of production system and inhibitory potential of aroma volatiles on polyphenol oxidase and peroxidase activity in tomatoes. *J. Sci. Food Agric.* **2021**, *101*, 307–314. [CrossRef] [PubMed]
16. Chen, X.; Ren, L.; Li, M.; Qian, J.; Fan, J.; Du, B. Effects of clove essential oil and eugenol on quality and browning control of fresh-cut lettuce. *Food Chem.* **2017**, *214*, 432–439. [CrossRef]
17. Zhou, X.; Xiao, Y.; Meng, X.; Liu, B. Full inhibition of Whangkeumbae pear polyphenol oxidase enzymatic browning reaction by L-cysteine. *Food Chem.* **2018**, *266*, 1–8. [CrossRef]
18. De Barcelos Costa, H.C.; Siguemoto, É.S.; Cavalcante, T.A.B.B.; de Oliveira Silva, D.; Vieira, L.G.M.; Gut, J.A.W. Effect of microwave-assisted processing on polyphenol oxidase and peroxidase inactivation kinetics of açai-berry (*Euterpe oleracea*) pulp. *Food Chem.* **2021**, *341*, 128287. [CrossRef] [PubMed]
19. Ji, D.; Wang, Q.; Lu, T.; Ma, H.; Chen, X. The effects of ultrasonication on the phytochemicals, antioxidant, and polyphenol oxidase and peroxidase activities in coffee leaves. *Food Chem.* **2022**, *373*, 131480. [CrossRef] [PubMed]
20. De Jesus, A.L.T.; Leite, T.S.; Cristianini, M. High isostatic pressure and thermal processing of açaí fruit (*Euterpe oleracea* Martius): Effect on pulp color and inactivation of peroxidase and polyphenol oxidase. *Food Res. Int.* **2018**, *105*, 853–862. [CrossRef] [PubMed]
21. Engmann, F.N.; Ma, Y.; Zhang, H.; Yu, L.; Deng, N. The application of response surface methodology in studying the effect of heat and high hydrostatic pressure on anthocyanins, polyphenol oxidase, and peroxidase of mulberry (*Morus nigra*) juice. *J. Sci. Food Agric.* **2014**, *94*, 2345–2356. [CrossRef] [PubMed]
22. Istiqamah, A.; Lioe, H.N.; Adawiyah, D.R. Umami compounds present in low molecular umami fractions of asam sunti–A fermented fruit of *Averrhoa bilimbi* L. *Food Chem.* **2019**, *270*, 338–343. [CrossRef] [PubMed]
23. Meng, K.; Hou, Y.; Han, Y.; Ban, Q.; He, Y.; Suo, J.; Rao, J. Exploring the functions of 9-lipoxygenase (DkLOX3) in ultrastructural changes and hormonal stress response during persimmon fruit storage. *Int. J. Mol. Sci.* **2017**, *18*, 589. [CrossRef] [PubMed]
24. Defilippi, B.G.; Ejsmentewicz, T.; Covarrubias, M.P.; Gudenschwager, O.; Campos-Vargas, R. Changes in cell wall pectins and their relation to postharvest mesocarp softening of "Hass" avocados (*Persea americana* Mill.). *Plant Physiol. Biochem.* **2018**, *128*, 142–151. [CrossRef] [PubMed]
25. Salinas-Hernández, R.M.; González-Aguilar, G.A.; Tiznado-Hernández, M.E. Utilization of physicochemical variables developed from changes in sensory attributes and consumer acceptability to predict the shelf life of fresh-cut mango fruit. *J. Food Sci. Technol.* **2015**, *52*, 63–77. [CrossRef]
26. Jayathunge, K.; Gunawardhana, D.; Illeperuma, D.; Chandrajith, U.; Thilakarathne, B.; Fernando, M.; Palipane, K. Physico-chemical and sensory quality of fresh cut papaya (*Carica papaya*) packaged in micro-perforated polyvinyl chloride containers. *J. Food Sci. Technol.* **2014**, *51*, 3918–3925. [CrossRef]
27. Sanchís, E.; Mateos, M.; Pérez-Gago, M.B. Physicochemical, sensory, and nutritional quality of fresh-cut "Rojo Brillante" persimmon affected by maturity stage and antibrowning agents. *Food Sci. Technol. Int.* **2016**, *22*, 574–586. [CrossRef] [PubMed]
28. Kitzberger, C.S.G.; da Silva, C.M.; dos Santos Scholz, M.B.; Ferreira, M.I.F.; Bauchrowitz, I.M.; Eilert, J.B.; dos Santos Neto, J. Physicochemical and sensory characteristics of plums accesses (*Prunus salicina*). *AIMS AgriFood* **2017**, *2*, 101–112.
29. Leguizamon-Delgado, M.A.; Duque-Cifuentes, A.L.; Quintero-Castaño, V.D. Physico-chemical and sensory evaluation of a mango-based fruit bar. *Dyna* **2019**, *86*, 276–283. [CrossRef]
30. Rana, S.S.; Pradhan, R.C.; Mishra, S. Image analysis to quantify the browning in fresh cut tender jackfruit slices. *Food Chem.* **2019**, *278*, 185–189. [CrossRef] [PubMed]
31. Zhao, Q.; Tang, S.; Fang, X.; Wang, Z.; Jiang, Y.; Guo, X.; Zhu, J.; Zhang, Y. The Effect of *Lactiplantibacillus plantarum* BX62 alone or in combination with chitosan on the qualitative characteristics of fresh-cut apples during cold storage. *Microorganisms* **2021**, *9*, 2404. [CrossRef] [PubMed]
32. Glicerina, V.; Siroli, L.; Betoret, E.; Canali, G.; Dalla Rosa, M.; Lanciotti, R.; Romani, S. Characterization and evaluation of the influence of an alginate, cocoa and a bilayer alginate–cocoa coating on the quality of fresh-cut oranges during storage. *J. Sci. Food Agric.* **2022**, *102*, 4454–4461. [CrossRef] [PubMed]
33. Yousuf, B.; Deshi, V.; Ozturk, B.; Siddiqui, M.W. Fresh-cut fruits and vegetables: Quality issues and safety concerns. In *Fresh-Cut Fruits and Vegetables*; Elsevier: Amsterdam, The Netherlands, 2020; pp. 1–15.
34. Zhao, P.; Ndayambaje, J.P.; Liu, X.; Xia, X. Microbial spoilage of fruits: A review on causes and prevention methods. *Food Rev. Int.* **2020**, 1–22. [CrossRef]
35. Zhang, H.; Yamamoto, E.; Murphy, J.; Locas, A. Microbiological safety of ready-to-eat fresh-cut fruits and vegetables sold on the Canadian retail market. *Int. J. Food Microbiol.* **2020**, *335*, 108855. [CrossRef] [PubMed]
36. Ortiz-Solà, J.; Valero, A.; Abadias, M.; Nicolau-Lapeña, I.; Viñas, I. Evaluation of water-assisted UV-C light and its additive effect with peracetic acid for the inactivation of *Listeria monocytogenes*, *Salmonella enterica* and murine norovirus on whole and fresh-cut strawberries during shelf-life. *J. Sci. Food Agric.* **2022**, 1–10. [CrossRef] [PubMed]
37. Carstens, C.K.; Salazar, J.K.; Darkoh, C. Multistate outbreaks of foodborne illness in the United States associated with fresh produce from 2010 to 2017. *Front. Microbiol.* **2019**, *10*, 2667. [CrossRef]

38. Hasan, S.K.; Ferrentino, G.; Scampicchio, M. Nanoemulsion as advanced edible coatings to preserve the quality of fresh-cut fruits and vegetables: A review. *Int. J. Food Sci. Technol.* **2020**, *55*, 1–10. [CrossRef]
39. Hodges, D.M.; Toivonen, P.M. Quality of fresh-cut fruits and vegetables as affected by exposure to abiotic stress. *Postharvest Biol. Technol.* **2008**, *48*, 155–162. [CrossRef]
40. Dea, S.; Brecht, J.K.; Nunes, M.C.N.; Baldwin, E.A. Occurrence of chilling injury in fresh-cut 'Kent'mangoes. *Postharvest Biol. Technol.* **2010**, *57*, 61–71. [CrossRef]
41. Marrero, A.; Kader, A.A. Optimal temperature and modified atmosphere for keeping quality of fresh-cut pineapples. *Postharvest Biol. Technol.* **2006**, *39*, 163–168. [CrossRef]
42. Cheng, Y.; Sun, C.; Zhai, X.; Zhang, R.; Zhang, S.; Sun, C.; Wang, W.; Hou, H. Effect of lipids with different physical state on the physicochemical properties of starch/gelatin edible films prepared by extrusion blowing. *Int. J. Biol. Macromol.* **2021**, *185*, 1005–1014. [CrossRef]
43. Galus, S.; Mikus, M.; Ciurzyńska, A.; Domian, E.; Kowalska, J.; Marzec, A.; Kowalska, H. The effect of whey protein-based edible coatings incorporated with lemon and lemongrass essential oils on the quality attributes of fresh-cut pears during storage. *Coatings* **2021**, *11*, 745. [CrossRef]
44. Ma, Y.; Zhao, Y.; Xie, J.; Sameen, D.E.; Ahmed, S.; Dai, J.; Qin, W.; Li, S.; Liu, Y. Optimization, characterization and evaluation of papaya polysaccharide-corn starch film for fresh cut apples. *Int. J. Biol. Macromol.* **2021**, *166*, 1057–1071. [CrossRef] [PubMed]
45. Flores-Martínez, N.; Valdez-Fragoso, A.; Jiménez-Islas, H.; Pérez-Pérez, M. Physical, barrier, mechanical and microstructural properties of Aloe vera-gelatin-glycerol edible films incorporated with *Pimenta dioica* L. Merrill essential oil. *Rev. Mex. Ing. Quim.* **2017**, *16*, 109–119. [CrossRef]
46. Díaz-Montes, E.; Castro-Muñoz, R. Edible films and coatings as food-quality preservers: An overview. *Foods* **2021**, *10*, 249. [CrossRef] [PubMed]
47. Pashova, S.; Radev, R.; Dimitrov, G.; Ivanov, J. Edible coatings in food industry related to circular economy. *Calitatea* **2018**, *19*, 111–117.
48. Valdés, A.; Ramos, M.; Beltrán, A.; Jiménez, A.; Garrigós, M.C. State of the art of antimicrobial edible coatings for food packaging applications. *Coatings* **2017**, *7*, 56. [CrossRef]
49. Chawla, R.; Sivakumar, S.; Kaur, H. Antimicrobial edible films in food packaging: Current scenario and recent nanotechnological advancements-a review. *Carbohydr. Polym. Technol. Appl.* **2021**, *2*, 100024. [CrossRef]
50. Radev, R.; Pashova, S. Application of edible films and coatings for fresh fruit and vegetables. *Qual. Access Success* **2020**, *21*, 108–112.
51. Hashemi, S.M.B.; Jafarpour, D. Bioactive edible film based on Konjac glucomannan and probiotic Lactobacillus plantarum strains: Physicochemical properties and shelf life of fresh-cut kiwis. *J. Food Sci.* **2021**, *86*, 513–522. [CrossRef]
52. Ganiari, S.; Choulitoudi, E.; Oreopoulou, V. Edible and active films and coatings as carriers of natural antioxidants for lipid food. *Trends Food Sci. Technol.* **2017**, *68*, 70–82. [CrossRef]
53. Restrepo, A.E.; Rojas, J.D.; García, O.R.; Sánchez, L.T.; Pinzón, M.I.; Villa, C.C. Mechanical, barrier, and color properties of banana starch edible films incorporated with nanoemulsions of lemongrass (*Cymbopogon citratus*) and rosemary (*Rosmarinus officinalis*) essential oils. *Food Sci. Technol. Int.* **2018**, *24*, 705–712. [CrossRef] [PubMed]
54. Ruan, C.; Zhang, Y.; Wang, J.; Sun, Y.; Gao, X.; Xiong, G.; Liang, J. Preparation and antioxidant activity of sodium alginate and carboxymethyl cellulose edible films with epigallocatechin gallate. *Int. J. Biol. Macromol.* **2019**, *134*, 1038–1044. [CrossRef] [PubMed]
55. Lazaridou, A.; Biliaderis, C.G. Edible films and coatings with pectin. *Pectin Technol. Physiol. Prop.* **2020**, 99–123. [CrossRef]
56. Basiak, E.; Lenart, A.; Debeaufort, F. Effect of starch type on the physico-chemical properties of edible films. *Int. J. Biol. Macromol.* **2017**, *98*, 348–356. [CrossRef] [PubMed]
57. Saha, A.; Tyagi, S.; Gupta, R.K.; Tyagi, Y.K. Natural gums of plant origin as edible coatings for food industry applications. *Crit. Rev. Biotechnol.* **2017**, *37*, 959–973. [CrossRef]
58. Viana, R.M.; Sá, N.M.; Barros, M.O.; de Fátima Borges, M.; Azeredo, H.M. Nanofibrillated bacterial cellulose and pectin edible films added with fruit purees. *Carbohydr. Polym.* **2018**, *196*, 27–32. [CrossRef]
59. Hassan, B.; Chatha, S.A.S.; Hussain, A.I.; Zia, K.M.; Akhtar, N. Recent advances on polysaccharides, lipids and protein based edible films and coatings: A review. *Int. J. Biol. Macromol.* **2018**, *109*, 1095–1107. [CrossRef]
60. Baldwin, E.A. Surface treatments and edible coatings in food preservation. In *Handbook of Food Preservation*; CRC Press: Boca Raton, FL, USA, 2020; pp. 507–528.
61. Feng, Z.; Wu, G.; Liu, C.; Li, D.; Jiang, B.; Zhang, X. Edible coating based on whey protein isolate nanofibrils for antioxidation and inhibition of product browning. *Food Hydrocoll.* **2018**, *79*, 179–188. [CrossRef]
62. Shendurse, A.; Gopikrishna, G.; Patel, A.; Pandya, A. Milk protein based edible films and coatings–preparation, properties and food applications. *J. Nutr. Health Food Eng.* **2018**, *8*, 219–226. [CrossRef]
63. Sharma, P.; Shehin, V.; Kaur, N.; Vyas, P. Application of edible coatings on fresh and minimally processed vegetables: A review. *Int. J. Veg. Sci.* **2019**, *25*, 295–314. [CrossRef]
64. Ju, J.; Xie, Y.; Guo, Y.; Cheng, Y.; Qian, H.; Yao, W. Application of edible coating with essential oil in food preservation. *Crit. Rev. Food Sci. Nutr.* **2019**, *59*, 2467–2480. [CrossRef] [PubMed]

65. Saleem, M.S.; Anjum, M.A.; Naz, S.; Ali, S.; Hussain, S.; Azam, M.; Sardar, H.; Khaliq, G.; Canan, İ.; Ejaz, S. Incorporation of ascorbic acid in chitosan-based edible coating improves postharvest quality and storability of strawberry fruits. *Int. J. Biol. Macromol.* **2021**, *189*, 160–169. [CrossRef] [PubMed]
66. Afifah, N.; Ratnawati, L.; Darmajana, D. Evaluation of Plasticizer Addition in Composite Edible Coating on Quality of Fresh-Cut Mangoes during Storage. In *IOP Conference Series: Earth and Environmental Science, Proceedings of the 2nd International Conference on Natural Products and Bioresource Sciences—2018, Tangerang, Indonesia, 1–2 November 2018*; IOP Publishing: Bristol, UK, 2019; p. 012029.
67. Chen, J.; Wu, A.; Yang, M.; Ge, Y.; Pristijono, P.; Li, J.; Xu, B.; Mi, H. Characterization of sodium alginate-based films incorporated with thymol for fresh-cut apple packaging. *Food Control* **2021**, *126*, 108063. [CrossRef]
68. Hammad, K.; Elsayed, N.; Elkashef, H. Development of a whey protein concentrate/apple pomace extract edible coating for shelf life extension of fresh-cut apple. *Int. Food Res. J.* **2021**, *28*, 377–385. [CrossRef]
69. Thivya, P.; Bhosale, Y.; Anandakumar, S.; Hema, V.; Sinija, V. Development of active packaging film from sodium alginate/carboxymethyl cellulose containing shallot waste extracts for anti-browning of fresh-cut produce. *Int. J. Biol. Macromol.* **2021**, *188*, 790–799. [CrossRef]
70. Özdemir, K.S.; Gökmen, V. Effect of chitosan-ascorbic acid coatings on the refrigerated storage stability of fresh-cut apples. *Coatings* **2019**, *9*, 503. [CrossRef]
71. Sumonsiri, N.; Danpongprasert, W.; Thaidech, K. Comparison of sweet orange (*Citrus sinensis*) and lemon (*Citrus limonum*) essential oils on qualities of fresh-cut apples during storage. *Sci. Study Res. Chem. Chem. Eng. Biotechnol. Food Ind.* **2020**, *21*, 47–57.
72. Zhang, C.; Yang, Z.; Shi, J.; Zou, X.; Zhai, X.; Huang, X.; Li, Z.; Holmes, M.; Daglia, M.; Xiao, J. Physical properties and bioactivities of chitosan/gelatin-based films loaded with tannic acid and its application on the preservation of fresh-cut apples. *LWT* **2021**, *144*, 111223. [CrossRef]
73. Passafiume, R.; Gaglio, R.; Sortino, G.; Farina, V. Effect of three different aloe vera gel-based edible coatings on the quality of fresh-cut "Hayward" kiwifruits. *Foods* **2020**, *9*, 939. [CrossRef]
74. Suriati, L.; Utama, I.; Harsojuwono, B.A.; Gunam, I.B.W. Ecogel incorporated with nano-additives to increase shelf-life of fresh-cut mango. *J. Appl. Hortic.* **2020**, *22*, 189–195.
75. Arnon-Rips, H.; Porat, R.; Poverenov, E. Enhancement of agricultural produce quality and storability using citral-based edible coatings; the valuable effect of nano-emulsification in a solid-state delivery on fresh-cut melons model. *Food Chem.* **2019**, *277*, 205–212. [CrossRef]
76. Tabassum, N.; Khan, M.A. Modified atmosphere packaging of fresh-cut papaya using alginate based edible coating: Quality evaluation and shelf life study. *Sci. Hortic.* **2020**, *259*, 108853. [CrossRef]
77. Kathiresan, S.; Lasekan, O. Effects of glycerol and stearic acid on the performance of chickpea starch-based coatings applied to fresh-cut papaya. *CYTA-J. Food.* **2019**, *17*, 365–374. [CrossRef]
78. Prakash, A.; Baskaran, R.; Vadivel, V. Citral nanoemulsion incorporated edible coating to extend the shelf life of fresh cut pineapples. *LWT* **2020**, *118*, 108851. [CrossRef]
79. Alharaty, G.; Ramaswamy, H.S. The effect of sodium alginate-calcium chloride coating on the quality parameters and shelf life of strawberry cut fruits. *J. Compos. Sci.* **2020**, *4*, 123. [CrossRef]
80. Farina, V.; Passafiume, R.; Tinebra, I.; Palazzolo, E.; Sortino, G. Use of aloe vera gel-based edible coating with natural anti-browning and anti-oxidant additives to improve post-harvest quality of fresh-cut 'fuji'apple. *Agronomy* **2020**, *10*, 515. [CrossRef]
81. Tapia-Blácido, D.R.; da Silva Ferreira, M.E.; Aguilar, G.J.; Costa, D.J.L. Biodegradable packaging antimicrobial activity. In *Processing and Development of Polysaccharide-Based Biopolymers for Packaging Applications*; Elsevier: Amsterdam, The Netherlands, 2020; pp. 207–238.
82. Zhang, S.; Wei, F.; Han, X. An edible film of sodium alginate/pullulan incorporated with capsaicin. *New J. Chem.* **2018**, *42*, 17756–17761. [CrossRef]
83. Marquez, G.R.; Di Pierro, P.; Mariniello, L.; Esposito, M.; Giosafatto, C.V.; Porta, R. Fresh-cut fruit and vegetable coatings by transglutaminase-crosslinked whey protein/pectin edible films. *LWT* **2017**, *75*, 124–130. [CrossRef]
84. Dhall, R. Advances in edible coatings for fresh fruits and vegetables: A review. *Crit. Rev. Food Sci. Nutr.* **2013**, *53*, 435–450. [CrossRef]
85. Aayush, K.; McClements, D.J.; Sharma, S.; Sharma, R.; Singh, G.P.; Sharma, K.; Oberoi, K. Innovations in the development and application of edible coatings for fresh and minimally processed Apple. *Food Control* **2022**, *141*, 109188. [CrossRef]
86. Nayak, S.L.; Sethi, S.; Sharma, R.; Prajapati, U. Active edible coatings for fresh fruits and vegetables. In *Polymers for Agri-Food Applications*; Springer: Berlin/Heidelberg, Germany, 2019; pp. 417–432.
87. Cozzolino, R.; Cefola, M.; Pace, B.; Malorni, L.; Martignetti, A.; Montemurro, N.; Pellicano, M.P. Quality, sensory and volatile profiles of fresh-cut big top nectarines cold stored in air or modified atmosphere packaging. *Int. J. Food Sci. Technol.* **2018**, *53*, 1736–1743. [CrossRef]
88. Iglesias, M.; Abadias, M.; Anguera, M.; Viñas, I. Efficacy of *Pseudomonas graminis* CPA-7 against *Salmonella* spp. and *Listeria monocytogenes* on fresh-cut pear and setting up of the conditions for its commercial application. *Food Microbiol.* **2018**, *70*, 103–112. [CrossRef]
89. United Nations. Sustainable Development Goals. Available online: https://www.un.org/sustainabledevelopment/sustainable-consumption-production/ (accessed on 23 March 2022).

90. Matar, C.; Guillard, V.; Gauche, K.; Costa, S.; Gontard, N.; Guilbert, S.; Gaucel, S. Consumer behaviour in the prediction of postharvest losses reduction for fresh strawberries packed in modified atmosphere packaging. *Postharvest Biol. Technol.* **2020**, *163*, 111119. [CrossRef]
91. Rux, G.; Bohne, K.; Huyskens-Keil, S.; Ulrichs, C.; Hassenberg, K.; Herppich, W. Effects of modified atmosphere and sugar immersion on physiology and quality of fresh-cut 'Braeburn' apples. *Food Packag. Shelf Life* **2021**, *29*, 100726. [CrossRef]
92. Iturralde-García, R.D.; García-Regueiro, J.A.; Castañé, C.; Riudavets, J. Sorption of carbon dioxide by chickpeas packaged in modified atmospheres. *J. Stored Prod. Res.* **2019**, *83*, 54–60. [CrossRef]
93. Dorostkar, M.; Moradinezhad, F.; Ansarifar, E. Influence of active modified atmosphere packaging pre-treatment on shelf life and quality attributes of cold stored apricot fruit. *Int. J. Fruit Sci.* **2022**, *22*, 402–413. [CrossRef]
94. Gorny, J.R. A summary of CA and MA requirements and recommendations for fresh-cut (minimally processed) fruits and vegetables. *Acta Hortic.* **2003**, *600*, 609–614. [CrossRef]
95. Ballard, C.R.; Junior, M.R.M. Health benefits of flavonoids. In *Bioactive Compounds*; Elsevier: Amsterdam, The Netherlands, 2019; pp. 185–201.
96. Wang, J.; Li, P.; Gong, B.; Li, S.; Ma, H. Phenol metabolism and preservation of fresh in-hull walnut stored in modified atmosphere packaging. *J. Sci. Food Agric.* **2017**, *97*, 5335–5342. [CrossRef] [PubMed]
97. Agar, I.; Streif, J.; Bangerth, F. Effect of high CO_2 and controlled atmosphere (CA) on the ascorbic and dehydroascorbic acid content of some berry fruits. *Postharvest Biol. Technol.* **1997**, *11*, 47–55. [CrossRef]
98. Sun, T.; Tadmor, Y.; Li, L. Pathways for carotenoid biosynthesis, degradation, and storage. In *Plant and Food Carotenoids*; Springer: Berlin/Heidelberg, Germany, 2020; pp. 3–23.
99. Mapelli-Brahm, P.; Barba, F.J.; Remize, F.; Garcia, C.; Fessard, A.; Khaneghah, A.M.; Sant'Ana, A.S.; Lorenzo, J.M.; Montesano, D.; Meléndez-Martínez, A.J. The impact of fermentation processes on the production, retention and bioavailability of carotenoids: An overview. *Trends Food Sci. Technol.* **2020**, *99*, 389–401. [CrossRef]
100. Soliva-Fortuny, R.C.; Elez-Martínez, P.; Martín-Belloso, O. Microbiological and biochemical stability of fresh-cut apples preserved by modified atmosphere packaging. *Innov. Food Sci. Emerg. Technol.* **2004**, *5*, 215–224. [CrossRef]
101. Odriozola-Serrano, I.; Soliva-Fortuny, R.; Martín-Belloso, O. Changes in bioactive composition of fresh-cut strawberries stored under superatmospheric oxygen, low-oxygen or passive atmospheres. *J. Food Compos. Anal.* **2010**, *23*, 37–43. [CrossRef]
102. Smith, A.; Waldron, K.; Maness, N.; Perkins-Veazie, P. Vegetable texture: Measurement and structural implications. In *Postharvest Phyusiology and Pathology of Vegetables*, 2nd ed.; Marcel Dekker, Inc.: New York, NY, USA, 2003; pp. 297–330.
103. Cliff, M.A.; Toivonen, P.M.; Forney, C.F.; Lu, C. Quality of fresh-cut apple slices stored in solid and micro-perforated film packages having contrasting O2 headspace atmospheres. *Postharvest Biol. Technol.* **2010**, *58*, 254–261. [CrossRef]
104. Rux, G.; Efe, E.; Ulrichs, C.; Huyskens-Keil, S.; Hassenberg, K.; Herppich, W.B. Effects of pre-processing short-term hot-water treatments on quality and shelf life of fresh-cut apple slices. *Foods* **2019**, *8*, 653. [CrossRef]
105. Siddiq, R.; Auras, R.; Siddiq, M.; Dolan, K.D.; Harte, B. Effect of modified atmosphere packaging (MAP) and NatureSeal®treatment on the physico-chemical, microbiological, and sensory quality of fresh-cut d'Anjou pears. *Food Packag. Shelf Life* **2020**, *23*, 100454. [CrossRef]
106. Feliziani, E.; Romanazzi, G. Postharvest decay of strawberry fruit: Etiology, epidemiology, and disease management. *J. Berry Res.* **2016**, *6*, 47–63. [CrossRef]
107. Mitcham, E.J. Strawberry. In *The Commercial Storage of Fruits, Vegetables, and Florist and Nursery Stocks*; Agriculture Handbook; Gross, K.C., Wang, C.Y., Saltveit, M., Eds.; United States Department of Agriculture, Agricultural Research Service: Davis, CA, USA, 2016; pp. 559–561.
108. Nakata, Y.; Izumi, H. Microbiological and quality responses of strawberry fruit to high CO_2 controlled atmosphere and modified atmosphere storage. *HortScience* **2020**, *55*, 386–391. [CrossRef]
109. Das, K.; Roychoudhury, A. Reactive oxygen species (ROS) and response of antioxidants as ROS-scavengers during environmental stress in plants. *Front. Environ. Sci.* **2014**, *2*, 53. [CrossRef]
110. Dias, C.; Fonseca, A.; Amaro, A.L.; Vilas-Boas, A.A.; Oliveira, A.; Santos, S.A.; Silvestre, A.J.; Rocha, S.M.; Isidoro, N.; Pintado, M. Natural-based antioxidant extracts as potential mitigators of fruit browning. *Antioxidants* **2020**, *9*, 715. [CrossRef]
111. Vijayalaxmi, S.; Jayalakshmi, S.; Sreeramulu, K. Polyphenols from different agricultural residues: Extraction, identification and their antioxidant properties. *J. Food Sci. Technol.* **2015**, *52*, 2761–2769. [CrossRef] [PubMed]
112. Bernal-Mercado, A.T.; Ayala-Zavala, J.F.; Cruz-Valenzuela, M.R.; Gonzalez-Aguilar, G.A.; Nazzaro, F.; Fratianni, F.; Miranda, M.R.A.d.; Silva-Espinoza, B.A. Using sensory evaluation to determine the highest acceptable concentration of mango seed extract as antibacterial and antioxidant agent in fresh-cut mango. *Foods* **2018**, *7*, 120. [CrossRef] [PubMed]
113. Rodríguez-Arzuaga, M.; Salsi, M.S.; Piagentini, A.M. Storage quality of fresh-cut apples treated with yerba mate (*Ilex paraguariensis*). *J. Food Sci. Technol.* **2021**, *58*, 186–196. [CrossRef] [PubMed]
114. Rodríguez-Arzuaga, M.; Piagentini, A.M. New antioxidant treatment with yerba mate (*Ilex paraguariensis*) infusion for fresh-cut apples: Modeling, optimization, and acceptability. *Food Sci. Technol. Int.* **2018**, *24*, 223–231. [CrossRef] [PubMed]
115. Akbari, E.; Gholami, M.; Ghobadi, C. Shelf-life and quality attributes in fresh-cut pear cv. Shahmive treated with different kinds of antioxidants. *J. Food Sci. Technol.* **2019**, *56*, 3998–4008. [CrossRef] [PubMed]
116. Hakkim, F.L.; Essa, M.M.; Arivazhagan, G.; Guizani, N.; Hyuk, S. Evaluation of food protective property of five natural products using fresh-cut apple slice model. *Pak. J. Biol. Sci.* **2012**, *15*, 10–18. [CrossRef] [PubMed]

117. Falagán, N.; Artés, F.; Aguayo, E. Natural additives to preserve quality and improve nutritional value of fresh-cut nectarine. *Food Sci. Technol. Int.* **2016**, *22*, 429–439. [CrossRef]
118. Zhu, L.; Hu, W.; Murtaza, A.; Iqbal, A.; Li, J.; Zhang, J.; Li, J.; Kong, M.; Xu, X.; Pan, S. Eugenol treatment delays the flesh browning of fresh-cut water chestnut (*Eleocharis tuberosa*) through regulating the metabolisms of phenolics and reactive oxygen species. *Food Chem. X* **2022**, *14*, 100307. [CrossRef] [PubMed]
119. Fan, P.; Huber, D.J.; Su, Z.; Hu, M.; Gao, Z.; Li, M.; Shi, X.; Zhang, Z. Effect of postharvest spray of apple polyphenols on the quality of fresh-cut red pitaya fruit during shelf life. *Food Chem.* **2018**, *243*, 19–25. [CrossRef] [PubMed]
120. Zheng, H.; Liu, W.; Liu, S.; Liu, C.; Zheng, L. Effects of melatonin treatment on the enzymatic browning and nutritional quality of fresh-cut pear fruit. *Food Chem.* **2019**, *299*, 125116. [CrossRef] [PubMed]
121. Supapvanich, S.; Yimpong, A.; Srisuwanwichan, J. Browning inhibition on fresh-cut apple by the immersion of liquid endosperm from mature coconuts. *J. Food Sci. Technol.* **2020**, *57*, 4424–4431. [CrossRef]
122. Xiao, Y.; Xie, J.; Wu, C.; He, J.; Wang, B. Effects of melatonin treatment on browning alleviation of fresh-cut foods. *J. Food Biochem.* **2021**, *45*, e13798. [CrossRef] [PubMed]
123. Adam, A.M.; Jeganathan, B.; Vasanthan, T.; Roopesh, M. Dipping fresh-cut apples in citric acid before plasma-integrated low-pressure cooling improves Salmonella and polyphenol oxidase inactivation. *J. Sci. Food Agric.* **2022**, *102*, 3425–3434. [CrossRef] [PubMed]
124. Yang, C.; Chen, T.; Shen, B.; Sun, S.; Song, H.; Chen, D.; Xi, W. Citric acid treatment reduces decay and maintains the postharvest quality of peach (*Prunus persica* L.) fruit. *Food Sci. Nutr.* **2019**, *7*, 3635–3643. [CrossRef] [PubMed]
125. Dan, Y.; Zhang, S.; Zhong, H.; Yi, H.; Sainz, M.B. Novel compounds that enhance Agrobacterium-mediated plant transformation by mitigating oxidative stress. *Plant Cell Rep.* **2015**, *34*, 291–309. [CrossRef]
126. Sun, Q.; Zhang, N.; Wang, J.; Zhang, H.; Li, D.; Shi, J.; Li, R.; Weeda, S.; Zhao, B.; Ren, S. Melatonin promotes ripening and improves quality of tomato fruit during postharvest life. *J. Exp. Bot.* **2015**, *66*, 657–668. [CrossRef]
127. Aghdam, M.S.; Fard, J.R. Melatonin treatment attenuates postharvest decay and maintains nutritional quality of strawberry fruits (*Fragaria* × *anannasa* cv. Selva) by enhancing GABA shunt activity. *Food Chem.* **2017**, *221*, 1650–1657. [CrossRef]
128. Belgacem, I.; Schena, L.; Teixidó, N.; Romeo, F.; Ballistreri, G.; Abadias, M. Effectiveness of a pomegranate peel extract (PGE) in reducing *Listeria monocytogenes* in vitro and on fresh-cut pear, apple and melon. *Eur. Food Res. Technol.* **2020**, *246*, 1765–1772. [CrossRef]
129. Ding, P.; Lee, Y. Use of essential oils for prolonging postharvest life of fresh fruits and vegetables. *Int. Food Res. J.* **2019**, *26*, 363–366.
130. Song, Z.; Li, F.; Guan, H.; Xu, Y.; Fu, Q.; Li, D. Combination of nisin and ε-polylysine with chitosan coating inhibits the white blush of fresh-cut carrots. *Food Control* **2017**, *74*, 34–44. [CrossRef]
131. Giacometti, J.; Kovačević, D.B.; Putnik, P.; Gabrić, D.; Bilušić, T.; Krešić, G.; Stulić, V.; Barba, F.J.; Chemat, F.; Barbosa-Cánovas, G. Extraction of bioactive compounds and essential oils from mediterranean herbs by conventional and green innovative techniques: A review. *Food Res. Int.* **2018**, *113*, 245–262. [CrossRef] [PubMed]
132. Rao, J.; Chen, B.; McClements, D.J. Improving the efficacy of essential oils as antimicrobials in foods: Mechanisms of action. *Annu. Rev. Food Sci. Technol.* **2019**, *10*, 365–387. [CrossRef]
133. Ahmadi, S.; Fazilati, M.; Nazem, H.; Mousavi, S.M. Green synthesis of magnetic nanoparticles using *Satureja hortensis* essential oil toward superior antibacterial/fungal and anticancer performance. *Biomed Res. Int.* **2021**, *2021*, 8822645. [CrossRef] [PubMed]
134. Ragno, R.; Papa, R.; Patsilinakos, A.; Vrenna, G.; Garzoli, S.; Tuccio, V.; Fiscarelli, E.; Selan, L.; Artini, M. Essential oils against bacterial isolates from cystic fibrosis patients by means of antimicrobial and unsupervised machine learning approaches. *Sci. Rep.* **2020**, *10*, 2653. [CrossRef] [PubMed]
135. Simirgiotis, M.J.; Burton, D.; Parra, F.; López, J.; Muñoz, P.; Escobar, H.; Parra, C. Antioxidant and antibacterial capacities of *Origanum vulgare* l. Essential oil from the arid andean region of chile and its chemical characterization by GC-MS. *Metabolites* **2020**, *10*, 414. [CrossRef]
136. Baptista-Silva, S.; Borges, S.; Ramos, O.L.; Pintado, M.; Sarmento, B. The progress of essential oils as potential therapeutic agents: A review. *J. Essent. Oil Res.* **2020**, *32*, 279–295. [CrossRef]
137. Marchese, A.; Arciola, C.R.; Coppo, E.; Barbieri, R.; Barreca, D.; Chebaibi, S.; Sobarzo-Sánchez, E.; Nabavi, S.F.; Nabavi, S.M.; Daglia, M. The natural plant compound carvacrol as an antimicrobial and anti-biofilm agent: Mechanisms, synergies and bio-inspired anti-infective materials. *Biofouling* **2018**, *34*, 630–656. [CrossRef]
138. Cui, H.; Zhang, C.; Li, C.; Lin, L. Antibacterial mechanism of oregano essential oil. *Ind. Crops Prod.* **2019**, *139*, 111498. [CrossRef]
139. Al-Tayyar, N.A.; Youssef, A.M.; Al-Hindi, R.R. Edible coatings and antimicrobial nanoemulsions for enhancing shelf life and reducing foodborne pathogens of fruits and vegetables: A review. *Sustain. Mater. Technol.* **2020**, *26*, e00215. [CrossRef]
140. Klangmuang, P.; Sothornvit, R. Active coating from hydroxypropyl methylcellulose-based nanocomposite incorporated with Thai essential oils on mango (cv. Namdokmai Sithong). *Food Biosci.* **2018**, *23*, 9–15. [CrossRef]
141. Radi, M.; Akhavan-Darabi, S.; Akhavan, H.R.; Amiri, S. The use of orange peel essential oil microemulsion and nanoemulsion in pectin-based coating to extend the shelf life of fresh-cut orange. *J. Food Process. Preserv.* **2018**, *42*, e13441. [CrossRef]
142. Tomadoni, B.; Moreira, M.d.R.; Pereda, M.; Ponce, A.G. Gellan-based coatings incorporated with natural antimicrobials in fresh-cut strawberries: Microbiological and sensory evaluation through refrigerated storage. *LWT* **2018**, *97*, 384–389. [CrossRef]

143. Lima, M.d.C.; de Sousa, C.P.; Fernandez-Prada, C.; Harel, J.; Dubreuil, J.; De Souza, E. A review of the current evidence of fruit phenolic compounds as potential antimicrobials against pathogenic bacteria. *Microb. Pathog.* **2019**, *130*, 259–270. [CrossRef] [PubMed]
144. Albuquerque, B.R.; Heleno, S.A.; Oliveira, M.B.P.; Barros, L.; Ferreira, I.C. Phenolic compounds: Current industrial applications, limitations and future challenges. *Food Funct.* **2021**, *12*, 14–29. [CrossRef]
145. Miklasińska-Majdanik, M.; Kępa, M.; Wojtyczka, R.D.; Idzik, D.; Wąsik, T.J. Phenolic compounds diminish antibiotic resistance of *Staphylococcus aureus* clinical strains. *Int. J. Environ. Res. Public Health* **2018**, *15*, 2321. [CrossRef] [PubMed]
146. Prakash, A.; Nithyanand, P.; Vadivel, V. In vitro antibacterial activity of nut by-products against foodborne pathogens and their application in fresh-cut fruit model. *J. Food Sci. Technol.* **2018**, *55*, 4304–4310. [CrossRef]
147. Badr, A.N.; Gromadzka, K.; Shehata, M.G.; Stuper-Szablewska, K.; Drzewiecka, K.; Abdel-Razek, A.G.; Youssef, M.M. Encapsulated Bioactive Ingredients of grape by-products applicate in fresh-cut fruit and juices diminished the ochratoxins. *J. Food Process. Preserv.* **2021**, *45*, e15112. [CrossRef]
148. Falleh, H.; Jemaa, M.B.; Saada, M.; Ksouri, R. Essential oils: A promising eco-friendly food preservative. *Food Chem.* **2020**, *330*, 127268. [CrossRef]
149. Maurya, A.; Prasad, J.; Das, S.; Dwivedy, A.K. Essential oils and their application in food safety. *Front. Sustain. Food Syst.* **2021**, *5*, 653420. [CrossRef]
150. Han, C.; Zhen, W.; Chen, Q.; Fu, M. UV-C irradiation inhibits surface discoloration and delays quality degradation of fresh-cut stem lettuce. *LWT* **2021**, *147*, 111533. [CrossRef]
151. Li, M.; Li, X.; Han, C.; Ji, N.; Jin, P.; Zheng, Y. UV-C treatment maintains quality and enhances antioxidant capacity of fresh-cut strawberries. *Postharvest Biol. Technol.* **2019**, *156*, 110945. [CrossRef]
152. Wang, D.; Chen, L.; Ma, Y.; Zhang, M.; Zhao, Y.; Zhao, X. Effect of UV-C treatment on the quality of fresh-cut lotus (*Nelumbo nucifera* Gaertn.) root. *Food Chem.* **2019**, *278*, 659–664. [CrossRef] [PubMed]
153. Hu, X.; Chen, Y.; Wu, X.; Liu, W.; Jing, X.; Liu, Y.; Yan, J.; Liu, S.; Qin, W. Combination of calcium lactate impregnation with UV-C irradiation maintains quality and improves antioxidant capacity of fresh-cut kiwifruit slices. *Food Chem. X* **2022**, 100329. [CrossRef] [PubMed]
154. González-Aguilar, G.A.; Villegas-Ochoa, M.A.; Martínez-Téllez, M.; Gardea, A.; Ayala-Zavala, J.F. Improving antioxidant capacity of fresh-cut mangoes treated with UV-C. *J. Food Sci.* **2007**, *72*, S197–S202. [CrossRef]
155. Collado, E.; Venzke Klug, T.; Martínez-Hernández, G.B.; Artés-Hernández, F.; Martínez-Sánchez, A.; Aguayo, E.; Artés, F.; Fernández, J.A.; Gómez, P.A. UV-C pretreatment of fresh-cut faba beans (*Vicia faba*) for shelf life extension: Effects of domestic microwaving for consumption. *Food Sci. Technol. Int.* **2020**, *26*, 140–150. [CrossRef]
156. Nakajima, S.; Lan, L.; Kanno, S.-i.; Takao, M.; Yamamoto, K.; Eker, A.P.; Yasui, A. UV light-induced DNA damage and tolerance for the survival of nucleotide excision repair-deficient human cells. *J. Biol. Chem.* **2004**, *279*, 46674–46677. [CrossRef]
157. Maghoumi, M.; Gomez, P.A.; Artés-Hernández, F.; Mostofi, Y.; Zamani, Z.; Artés, F. Hot water, UV-C and superatmospheric oxygen packaging as hurdle techniques for maintaining overall quality of fresh-cut pomegranate arils. *J. Sci. Food Agric.* **2013**, *93*, 1162–1168. [CrossRef]
158. Zambrano-Zaragoza, M.L.; Quintanar-Guerrero, D.; González-Reza, R.M.; Cornejo-Villegas, M.A.; Leyva-Gómez, G.; Urbán-Morlán, Z. Effects of UV-C and edible nano-coating as a combined strategy to preserve fresh-cut cucumber. *Polymers* **2021**, *13*, 3705. [CrossRef]
159. Martín-Vertedor, D.; Rodrigues, N.; Marx, Í.M.; Veloso, A.C.; Peres, A.M.; Pereira, J.A. Impact of thermal sterilization on the physicochemical-sensory characteristics of Californian-style black olives and its assessment using an electronic tongue. *Food Control* **2020**, *117*, 107369. [CrossRef]
160. Nissim, Y.; Shloberg, M.; Biton, I.; Many, Y.; Doron-Faigenboim, A.; Zemach, H.; Hovav, R.; Kerem, Z.; Avidan, B.; Ben-Ari, G. High temperature environment reduces olive oil yield and quality. *PLoS ONE* **2020**, *15*, e0231956. [CrossRef]
161. De la Peña-Armada, R.; Villanueva-Suárez, M.J.; Molina-García, A.D.; Rupérez, P.; Mateos-Aparicio, I. Novel rich-in-soluble dietary fiber apple ingredient obtained from the synergistic effect of high hydrostatic pressure aided by Celluclast®. *LWT* **2021**, *146*, 111421. [CrossRef]
162. Arshadi, M.; Attard, T.M.; Lukasik, R.M.; Brncic, M.; da Costa Lopes, A.M.; Finell, M.; Geladi, P.; Gerschenson, L.N.; Gogus, F.; Herrero, M. Pre-treatment and extraction techniques for recovery of added value compounds from wastes throughout the agri-food chain. *Green Chem.* **2016**, *18*, 6160–6204. [CrossRef]
163. Huang, H.-W.; Hsu, C.-P.; Wang, C.-Y. Healthy expectations of high hydrostatic pressure treatment in food processing industry. *J. Food Drug Anal.* **2020**, *28*, 1–13. [CrossRef]
164. Naderi, N.; House, J.D.; Pouliot, Y.; Doyen, A. Effects of high hydrostatic pressure processing on hen egg compounds and egg products. *Compr. Rev. Food Sci. Food Saf.* **2017**, *16*, 707–720. [CrossRef] [PubMed]
165. Yamamoto, K. *Food Processing by High Hydrostatic Pressure*; Oxford University Press: Oxford, UK, 2017.
166. Serment-Moreno, V.; Jacobo-Velázquez, D.A.; Torres, J.A.; Welti-Chanes, J. Microstructural and physiological changes in plant cell induced by pressure: Their role on the availability and pressure-temperature stability of phytochemicals. *Food Eng. Rev.* **2017**, *9*, 314–334. [CrossRef]

167. Escobedo-Avellaneda, Z.; Pérez-Simón, I.; Lavilla-Martín, M.; Baranda-González, A.; Welti-Chanes, J. Enzymatic and phytochemical stabilization of orange–strawberry–banana beverages by high hydrostatic pressure and mild heat. *Food Sci. Technol. Int.* **2017**, *23*, 185–193. [CrossRef]
168. Vázquez-Gutiérrez, J.L.; Hernando, I.; Quiles, A. Changes in tannin solubility and microstructure of high hydrostatic pressure–treated persimmon cubes during storage at 4 C. *Eur. Food Res. Technol.* **2013**, *237*, 9–17. [CrossRef]
169. Woolf, A.B.; Wibisono, R.; Farr, J.; Hallett, I.; Richter, L.; Oey, I.; Wohlers, M.; Zhou, J.; Fletcher, G.C.; Requejo-Jackman, C. Effect of high pressure processing on avocado slices. *Innov. Food Sci. Emerg. Technol.* **2013**, *18*, 65–73. [CrossRef]
170. Denoya, G.I.; Polenta, G.A.; Apóstolo, N.M.; Budde, C.O.; Sancho, A.M.; Vaudagna, S.R. Optimization of high hydrostatic pressure processing for the preservation of minimally processed peach pieces. *Innov. Food Sci. Emerg. Technol.* **2016**, *33*, 84–93. [CrossRef]
171. Vázquez-Gutiérrez, J.L.; Quiles, A.; Vonasek, E.; Jernstedt, J.A.; Hernando, I.; Nitin, N.; Barrett, D.M. High hydrostatic pressure as a method to preserve fresh-cut Hachiya persimmons: A structural approach. *Food Sci. Technol. Int.* **2016**, *22*, 688–698. [CrossRef] [PubMed]
172. Miguel-Pintado, C.; Nogales, S.; Fernández-León, A.M.; Delgado-Adámez, J.; Hernández, T.; Lozano, M.; Cañada-Cañada, F.; Ramírez, R. Effect of hydrostatic high pressure processing on nectarine halves pretreated with ascorbic acid and calcium during refrigerated storage. *LWT-Food Sci. Technol.* **2013**, *54*, 278–284. [CrossRef]
173. Hu, X.; Ma, T.; Ao, L.; Kang, H.; Hu, X.; Song, Y.; Liao, X. Effect of high hydrostatic pressure processing on textural properties and microstructural characterization of fresh-cut pumpkin (*Cucurbita pepo*). *J. Food Process Eng.* **2020**, *43*, e13379. [CrossRef]
174. Chakraborty, S.; Kaushik, N.; Rao, P.S.; Mishra, H. High-pressure inactivation of enzymes: A review on its recent applications on fruit purees and juices. *Compr. Rev. Food Sci. Food Saf.* **2014**, *13*, 578–596. [CrossRef] [PubMed]
175. Zhou, C.-L.; Liu, W.; Zhao, J.; Yuan, C.; Song, Y.; Chen, D.; Ni, Y.-Y.; Li, Q.-H. The effect of high hydrostatic pressure on the microbiological quality and physical–chemical characteristics of Pumpkin (*Cucurbita maxima* Duch.) during refrigerated storage. *Innov. Food Sci. Emerg. Technol.* **2014**, *21*, 24–34. [CrossRef]
176. Dong, P.; Kong, M.; Yao, J.; Zhang, Y.; Liao, X.; Hu, X.; Zhang, Y. The effect of high hydrostatic pressure on the microbiological quality and physicochemical properties of lotus root during refrigerated storage. *Innov. Food Sci. Emerg. Technol.* **2013**, *19*, 79–84. [CrossRef]
177. Botondi, R.; Moscetti, R.; Massantini, R. A comparative study on the effectiveness of ozonated water and peracetic acid in the storability of packaged fresh-cut melon. *J. Food Sci. Technol.* **2016**, *53*, 2352–2360. [CrossRef] [PubMed]
178. Miller, F.A.; Silva, C.L.; Brandao, T.R. A review on ozone-based treatments for fruit and vegetables preservation. *Food Eng. Rev.* **2013**, *5*, 77–106. [CrossRef]
179. Liu, C.; Chen, C.; Zhang, Y.; Jiang, A.; Hu, W. Aqueous ozone treatment inhibited degradation of cellwall polysaccharides in fresh-cut apple during cold storage. *Innov. Food Sci. Emerg. Technol.* **2021**, *67*, 102550. [CrossRef]
180. Ummat, V.; Singh, A.; Sidhu, G. Effect of aqueous ozone on quality and shelf life of shredded green bell pepper (*Capsicum annuum*). *J. Food Process. Preserv.* **2018**, *42*, 1–14. [CrossRef]
181. Chen, C.; Liu, C.; Jiang, A.; Zhao, Q.; Liu, S.; Hu, W. Effects of ozonated water on microbial growth, quality retention and pesticide residue removal of fresh-cut onions. *Ozone Sci. Eng.* **2020**, *42*, 399–407. [CrossRef]
182. Alwi, N.A.; Ali, A. Reduction of *Escherichia coli* O157, *Listeria monocytogenes* and *Salmonella enterica* sv. Typhimurium populations on fresh-cut bell pepper using gaseous ozone. *Food Control* **2014**, *46*, 304–311. [CrossRef]
183. Kim, J.-G.; Yousef, A.E.; Khadre, M.A. *Ozone and Its Current and Future Application in the Food Industry*; Elsevier: Amsterdam, The Netherlands, 2003.
184. Liu, C.; Ma, T.; Hu, W.; Tian, M.; Sun, L. Effects of aqueous ozone treatments on microbial load reduction and shelf life extension of fresh-cut apple. *Int. J. Food Sci. Technol.* **2016**, *51*, 1099–1109. [CrossRef]
185. Chitravathi, K.; Chauhan, O.; Raju, P.; Madhukar, N. Efficacy of aqueous ozone and chlorine in combination with passive modified atmosphere packaging on the postharvest shelf-life extension of green chillies (*Capsicum annuum* L.). *Food Bioproc Technol.* **2015**, *8*, 1386–1392. [CrossRef]
186. Sripong, K.; Uthairatanakij, A.; Jitareerat, P. Impact of gaseous ozone on microbial contamination and quality of fresh-cut durian. *Sci. Hortic.* **2022**, *294*, 110799. [CrossRef]
187. Yeoh, W.K.; Ali, A.; Forney, C.F. Effects of ozone on major antioxidants and microbial populations of fresh-cut papaya. *Postharvest Biol. Technol.* **2014**, *89*, 56–58. [CrossRef]
188. Keutgen, A.J.; Pawelzik, E. Influence of pre-harvest ozone exposure on quality of strawberry fruit under simulated retail conditions. *Postharvest Biol. Technol.* **2008**, *49*, 10–18. [CrossRef]
189. Koh, P.C.; Noranizan, M.A.; Karim, R.; Nur Hanani, Z.A. Sensory quality and flavour of alginate coated and repetitive pulsed light treated fresh-cut cantaloupes (*Cucumis melo* L. Var. Reticulatus Cv. Glamour) during storage. *J. Food Sci. Technol.* **2019**, *56*, 2563–2575. [CrossRef] [PubMed]
190. Huang, Y.; Sido, R.; Huang, R.; Chen, H. Application of water-assisted pulsed light treatment to decontaminate raspberries and blueberries from *Salmonella*. *Int. J. Food Microbiol.* **2015**, *208*, 43–50. [CrossRef] [PubMed]
191. De Almeida Lopes, M.M.; Silva, E.O.; Laurent, S.; Charles, F.; Urban, L.; de Miranda, M.R.A. The influence of pulsed light exposure mode on quality and bioactive compounds of fresh-cut mangoes. *J. Food Sci. Technol.* **2017**, *54*, 2332–2340. [CrossRef] [PubMed]

192. De Sousa, A.E.D.; Fonseca, K.S.; da Silva Gomes, W.K.; Monteiro da Silva, A.P.; de Oliveira Silva, E.; Puschmann, R. Control of browning of minimally processed mangoes subjected to ultraviolet radiation pulses. *J. Food Sci. Technol.* **2017**, *54*, 253–259. [CrossRef] [PubMed]
193. Moreira, M.R.; Álvarez, M.V.; Martín-Belloso, O.; Soliva-Fortuny, R. Effects of pulsed light treatments and pectin edible coatings on the quality of fresh-cut apples: A hurdle technology approach. *J. Sci. Food Agric.* **2017**, *97*, 261–268. [CrossRef] [PubMed]
194. Avalos-Llano, K.R.; Martín-Belloso, O.; Soliva-Fortuny, R. Effect of pulsed light treatments on quality and antioxidant properties of fresh-cut strawberries. *Food Chem.* **2018**, *264*, 393–400. [CrossRef] [PubMed]
195. Alam, A.U.; Rathi, P.; Beshai, H.; Sarabha, G.K.; Deen, M.J. Fruit quality monitoring with smart packaging. *Sensors* **2021**, *21*, 1509. [CrossRef]
196. Chaudhary, V.; Punia Bangar, S.; Thakur, N.; Trif, M. Recent advancements in smart biogenic packaging: Reshaping the future of the food packaging industry. *Polymers* **2022**, *14*, 829. [CrossRef]

horticulturae

MDPI

Article

Quality Grading Algorithm of *Oudemansiella raphanipes* Based on Transfer Learning and MobileNetV2

Tongkai Li [1,2], Huamao Huang [2], Yangyang Peng [1], Hui Zhou [1], Haiying Hu [3,*] and Ming Liu [1,*]

1 Guangdong Key Laboratory for New Technology Research of Vegetables, Vegetable Research Institute, Guangdong Academy of Agricultural Sciences, Guangzhou 510640, China
2 School of Physics and Optoelectronics, South China University of Technology, Guangzhou 510640, China
3 School of Civil Engineering and Transportation, South China University of Technology, Guangzhou 510640, China
* Correspondence: cthyhu@scut.edu.cn (H.H.); liuming@gdaas.cn (M.L.)

Abstract: As a traditional edible and medicinal fungus in China, *Oudemansiella raphanipes* has high economic benefits. In order to achieve the automatic classification of *Oudemansiella raphanipes* into four quality levels using their image dataset, a quality grading algorithm based on neural network models was proposed. At first, the transfer learning strategy and six typical convolution neural network models, e.g., VGG16, ResNet50, InceptionV3, NasNet-Mobile, EfficientNet, and MobileNetV2, were used to train the datasets. Experiments show that MobileNetV2 has good performance considering both testing accuracy and detection time. MobileNetV2 only needs 37.5 ms to classify an image, which is shorter by 11.76%, 28.57%, 46.42%, 59.45%, and 79.73%, respectively, compared with the classification times of InceptionV3, EfficientNetB0, ResNet50, NasNet-Mobile, and VGG16. Based on the original MobileNetV2 model, four optimization methods, including data augmentation, hyperparameter selecting, an overfitting control strategy, and a dynamic learning rate strategy, were adopted to improve the accuracy. The final classification accuracy can reach as high as 98.75%, while the detection time for one image is only 22.5 ms and the model size is only 16.48 MB. This quality grading algorithm based on an improved MobileNetV2 model is feasible and effective for *Oudemansiella raphanipes*, satisfying the needs in the production line.

Keywords: *Oudemansiella raphanipes*; quality grading; convolutional neural network; transfer learning

Citation: Li, T.; Huang, H.; Peng, Y.; Zhou, H.; Hu, H.; Liu, M. Quality Grading Algorithm of *Oudemansiella raphanipes* Based on Transfer Learning and MobileNetV2. *Horticulturae* **2022**, *8*, 1119. https://doi.org/10.3390/horticulturae8121119

Academic Editors: Paola Sánchez-Bravo and Luis Noguera-Artiaga

Received: 8 November 2022
Accepted: 25 November 2022
Published: 29 November 2022

1. Introduction

Oudemansiella raphanipes is one of the traditional Chinese edible and medicinal mushrooms and has high development value [1]. It is not only rich in many nutrients, such as vitamins, amino acids, and rich trace elements, but also has some medical effects, such as anti-inflammatory and analgesic effects. Owing to the significantly different prices in the market for different qualities of *Oudemansiella raphanipes*, it is very important for quality grading to distinguish the high-quality products from the poor-quality ones. However, the traditional manual quality grading process requires a high labor force and is low in efficiency, which seriously restricts the economic benefits. Therefore, it is urgent to develop an automated and intelligent quality grading algorithm for *Oudemansiella raphanipes*.

The quality grading of *Oudemansiella raphanipes* belongs to the field of image classification. In recent years, with the development of deep learning, convolutional neural networks (CNNs) can fully extract the high-dimensional features of objects. The classification algorithm based on deep learning has been widely used for various crops, such as grains [2–4], fruits [5–8], and vegetables [9–11]. For instance, Sujatha R et al. compared the performance of machine learning (support vector machine, random forest, stochastic gradient descent) and deep learning (InceptionV3, VGG16, VGG19) in terms of citrus plant disease detection, showing that VGG16 had the best classification accuracy of 89.5% [12]. Esgario J. G. M. et al. estimated the stress severity caused by biotic agents on coffee leaves using ResNet50, showing an accuracy of 95.24% for the biotic stress classification and 86.51% for severity estimation [13]. Arwatchananukul S et al. recognized 15 species of Paphiopedilum orchid using the InceptionV3 feature extractor, showing a recognition rate of up to 98.6% [14]. Ji M. et al. proposed a binary relevance multi-label learning algorithm combined with NasNet to simultaneously recognize 7 crop species, classify 10 crop diseases, and estimate the severity of 3 kinds of crop disease, showing a test accuracy of 85.28% [15]. Suharjito et al. classified the ripeness levels of a fresh fruit bunch of oil palm using EfficientNetB0 with a data augmentation method named "9-angle crop", showing a good test accuracy of 89.8% on Keras datasets [16].

In terms of the classification of mushrooms, Preechasuk J. et al. detected the 45 species of edible and poisonous mushrooms by a 6-layer CNN with the average accuracy of 74% on the dataset of 8556 images [17]. Zahan N. et al. used InceptionV3, VGG16, and ResNet50 to identify the 45 species of edible, inedible, and poisonous mushrooms, and InceptionV3 achieved the highest accuracy of 88.40% on the dataset of 8190 images [18]. Zhao M. et al. proposed a four-type classification method based on a compound knowledge distillation algorithm, using ResNet50 as the teacher model and ResNet18 as the student model to perform weight training and compound distillation on the dataset of 3000 images, in which the testing accuracy was 96.89% and the detection time was 32 ms [19]. Ketwongsa W. et al. classified five species of edible and poisonous mushrooms based on AlexNet, ResNet50, and GoogleNet, and a proposed modified AlexNet model using a dataset of 623 images; the proposed model had the fastest training time while maintaining an accuracy of 98.5% [20].

In a production line, the quality grading of mushrooms requires not only a high accuracy, but also a small detection time and few computing resources. However, the deep learning CNN often demands a large detection time and a high level of computing resources. MobileNet is the first generation of lightweight neural networks [21] and has demonstrated its high accuracy as well as small detection time and low requirement of computing resources on various datasets [22–26]. Compared with other neural networks [27–33], the depth separable convolution in MobileNet leads to a smaller number of parameters and, thus, a smaller amount of calculation. As a result, the high accuracy MobileNet can also achieve a fast speed on a CPU and a small storage space, which is suitable for practical industrial computers.

Moreover, the collection of labelled datasets is time-consuming and labor-intensive work; thus, sometimes only a small dataset can be obtained. However, a deep CNN trained from scratch with small datasets has difficulty obtaining high accuracy. As a result, transfer learning was used to obtain knowledge from a big dataset, and then to complete new tasks using small datasets [9].

In this paper, we provide an algorithm for the quality grading of *Oudemansiella raphanipes*, which is a good case study when developing a new image classification application. At first, six typical CNNs, including VGG16, ResNet50, InceptionV3, NasNet-Mobile, EfficientNet, and MobileNetV2, were pretrained on the dataset of ImageNet and transferred to our dataset of *Oudemansiella raphanipes*. Balancing the three criterions of high accuracy, small detection time, and small mode size, MobileNetV2 was selected and improved for the quality grading of *Oudemansiella raphanipes*. The final classification accuracy can reach as high as 98.75%, while the detection time for one image is only 22.5 ms and the model size is only 16.48 MB. This algorithm is efficient, fast, and resource-saving, which meets the demand of the price-sensitive market and is suitable for deployment on the production line.

2. Materials and Methods

2.1. Datasets

Figure 1 shows our experimental setup for image acquisition. Each *Oudemansiella raphanipes* was put into a shielding box, in which a light source (four 45-watt fluorescent lamps with a color temperature of 5000 K) was illuminated from the top and a color camera

(Canon 760D with EOSUtility software) took the images with the vertical shooting height of 45 cm. The resolution of each image is 3984 × 2656.

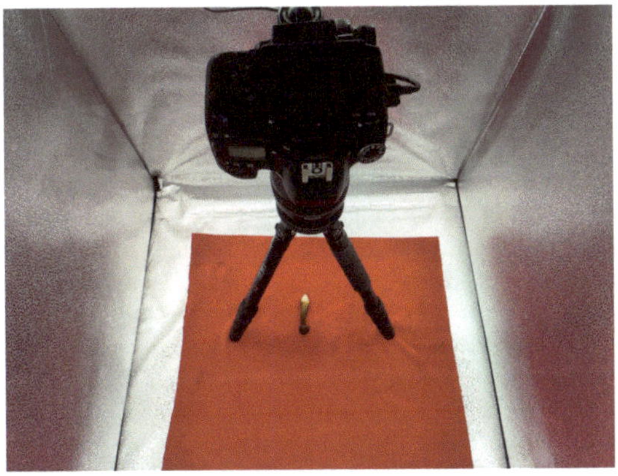

Figure 1. Experimental setup for image acquisition.

All these images were preliminarily labeled by three experts and divided into four quality levels: Class 1, Class 2, Class 3, and Class 4, according to the size, shape, color, texture, viewpoint. After preliminarily labelling, the three experts inspected each category and corrected the wrongly labelled images after discussion. Finally, each category contained 1000 images, the quality level of which was agreed and confirmed by all three experts. Figure 2 gives a typical example of each category.

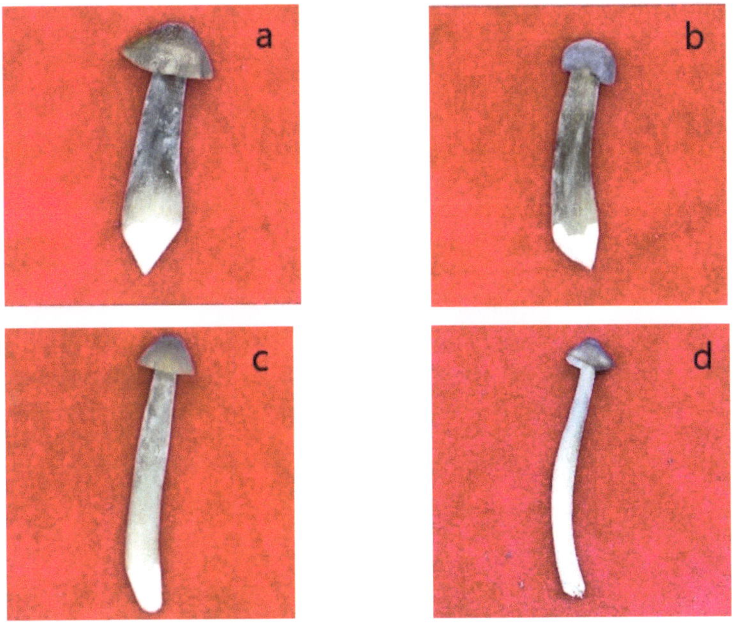

Figure 2. Four quality levels of *Oudemansiella raphanipes*. (**a**) Class 1; (**b**) Class 2; (**c**) Class 3; (**d**) Class 4.

Before training, the input image needed to be preprocessed. Firstly, the image was downscaled in equal proportion so that it could be used as the input of a neural network. For example, in the MobileNetV2 network [21], the image size needs to be adjusted to be 224 × 224 × 3 pixels. Secondly, the images of each category were randomly grouped into a training dataset, a validation dataset, and a testing dataset at a ratio of 7:2:1. After grouping, there were 2800 images in the training set (700 images for each category), 800 images in the validation set (200 images for each category), and 400 images in the testing set (100 images for each category).

2.2. Transfer Leaning

Figure 3 gives the flowchart of the transfer learning method. At first, the selected base model, e.g., MobileNetV2, was trained on a big dataset, e.g., ImageNet [34], and its network weights were determined. Following the pretrained model, we added a head model and a classifier layer. The head model had a global average pooling layer, a flatten layer, and a fully connected layer. The global average pooling layer was used to downsample the output of the pretrained model, the flattening layer was used to flatten the multi-dimensional vector into one dimension, and the fully connected layer adopted the ReLU as the activation function to integrate and map the image features into values. The classifier layer adopted the softmax activation function to classify the values' output from the fully connected layer. Subsequently, the whole model composed of the pretrained model, the head model, and the classifier layer was trained on the small dataset acquired in Section 2.1. Note that the network weights of the pretrained model are kept frozen in the small dataset training, while those of the head model and the classifier layer should be trained.

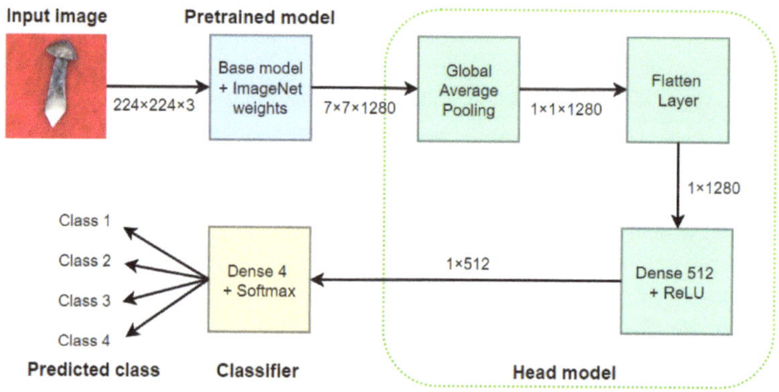

Figure 3. Transfer leaning method for MobileNetV2.

2.3. Computing Resources

One benefit of transfer leaning is that the required computing resources are less than the traditional CNN training. The computing resources used in this paper are shown in Table 1.

Table 1. Computing resources.

Equipment	Environment
CPU	Intel(R) Core(TM) i5-8250U, 1.80 GHz
Memory	8 GB
Operation system	Windows 10 (64bit)
Python	Version 3.7.0
Tensorflow	Version 2.3.0
Compiler environment	PyCharm 2020

3. Results

3.1. Comparison of MobileNetV2 with other Five Typical CNN Models

Six typical CNN models, including VGG16 [27], ResNet50 [28], InceptionV3 [29], NasNet-Mobile [30], EfficientNetB0 [31], and MobileNetV2 [21], were evaluated based on the transfer learning from a big dataset, that is, ImageNet [34], to a small dataset, that is, the dataset acquired in Section 2.1.

In the training process, the Adam optimizer was adopted, the learning rate was selected as 0.01, the batch size was 16, and the number of epochs was 20. Figure 4 gives the validation accuracies and losses. VGG16 can reach the highest validation accuracy of 98.88% and the lowest validation loss. ResNet50 fluctuates greatly in the early epochs and its validation accuracy is only 93.68%. EfficientNetB0 also has a relatively lower accuracy of 94.37%. The convergence characteristics of InceptionV3, NasNet-Mobile, and MobileNetV2 are similar, and their validation accuracies are 95.96%, 95.42%, and 96.04%, respectively.

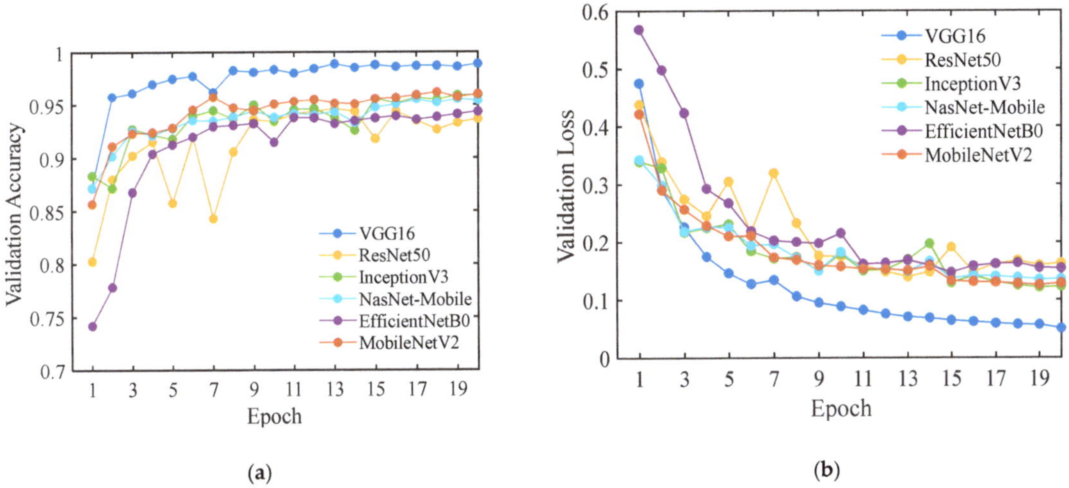

Figure 4. The validation (**a**) accuracies and (**b**) losses of VGG16, ResNet50, InceptionV3, NasNet-Mobile, EfficientNetB0, and MobileNetV2.

The trained models were applied on the testing dataset and their confusion matrixes are presented in Figure 5. Each of the six CNN models have good recognition accuracies for the four classes, except that there are many errors in the identification of Class 1 from Class 2. It also seems that the classification errors between Class 1 and Class 2 are higher than those between Class 3 and 4. However, the classification errors between the relatively high-quality *Oudemansiella raphanipes* (Class 1 and 2) and the relatively low-quality ones (Class 3 and 4) is very low, indicating that the characteristics of these two levels of products are obvious and easy to distinguish.

The testing accuracy on the total testing dataset, the average detection time for a single image, and the model size are given in Table 2. In terms of testing accuracy, VGG16 still performs the best and can reach an accuracy of 97.75%. InceptionV3, NasNet-Mobile, and MobileNetV2 also perform well with accuracies of 94.25%, 94.25%, and 94.75%, respectively, while EfficientNetB0 has a relatively low testing accuracy of 93%. Owing to the errors in the identification of Class 1 (see Figure 5b), ResNet50 has an overall accuracy of only 90%. In terms of detection time, MobileNetV2 only needs 37.5 ms to classify an image, which is lower than that of the other five models. This detection time is shortened by 11.76%, 28.57%, 46.42%, 59.46%, and 79.73%, respectively, compared with those of InceptionV3, EfficientNetB0, ResNet50, NasNet-Mobile, and VGG16. In terms of the size of the model, MobileNetV2, representing a typical lightweight CNN, takes up the least memory space because the number of parameters is much smaller than that of ordinary CNN models. It only needs 16.44 MB of storage space, which is smaller than the 23.35 MB of EfficientNetB0, 23.70 MB of NasNet-Mobile, 56.2 MB of VGG16, 90.38 MB of ResNet50, and 83.74 MB of Inception V3.

		Predicted Class			
		1	**2**	**3**	**4**
Actual Class	**1**	**96%**	4%	0	0
	2	1%	**99%**	0	0
	3	0	0	**98%**	1%
	4	0	0	1%	**99%**

(a)

		Predicted Class			
		1	**2**	**3**	**4**
Actual Class	**1**	**64%**	34%	2%	0
	2	2%	**98%**	0	0
	3	0	0	**98%**	2%
	4	0	0	0	**100%**

(b)

		Predicted Class			
		1	**2**	**3**	**4**
Actual Class	**1**	**89%**	10%	0	1%
	2	2%	**98%**	0	0
	3	0	3%	**91%**	6%
	4	0	0	1%	**99%**

(c)

		Predicted Class			
		1	**2**	**3**	**4**
Actual Class	**1**	**93%**	7%	0	0
	2	9%	**90%**	1%	0
	3	0	0	**95%**	5%
	4	0	0	1%	**99%**

(d)

		Predicted Class			
		1	**2**	**3**	**4**
Actual Class	**1**	**90%**	9%	1%	0
	2	9%	**91%**	0	0
	3	1%	0	**94%**	5%
	4	0	0	3%	**97%**

(e)

		Predicted Class			
		1	**2**	**3**	**4**
Actual Class	**1**	**93%**	7%	0	0
	2	8%	**92%**	0	0
	3	0	1%	**98%**	1%
	4	0	0	4%	**96%**

(f)

Figure 5. The confusion matrixes of (**a**) VGG16, (**b**) ResNet50, (**c**) InceptionV3, (**d**) NasNet-Mobile, (**e**) EfficientNetB0, and (**f**) MobileNetV2 applied on testing dataset.

Table 2. The testing accuracy, average detection time, and model size of VGG16, ResNet50, InceptionV3, NasNet-Mobile, EfficientNetB0, and MobileNetV2.

CNN Models	Test Accuracy (%)	Average Detection Time (ms)	Model Size (MB)
VGG16	97.75	185	56.20
ResNet50	90.00	70	90.38
InceptionV3	94.25	42.5	83.74
NasNet-Mobile	94.25	92.5	23.70
EfficientNetB0	93.00	52.5	23.35
MobileNetV2	94.75	37.5	16.44

In the production line of *Oudemansiella raphanipes*, the appropriate quality grading algorithm should not only consider the accuracy but also the matching degree of hardware equipment. Although VGG16 performs excellently when testing accuracy, it is not suitable for our grading owing to its long detection time and high requirement of computing resources. ResNet50 performs poorly on the dataset and cannot meet the accuracy requirements. Both InceptionV3 and NasNet-Mobile have a moderate accuracy, but the former requires a large model size, while the latter requires a long detection time. EfficientNetB0 balances the testing accuracy, the detection time, and the model size, but the performance is still not good enough. Compared with the above five models, MobileNetV2 can greatly shorten the detection time and can be deployed in productional computers, or even mobile phones, with poor computing resources, while its accuracy is only 3% lower than that of VGG16. These advantages are in line with the production practice. As a result, we selected MobileNetV2 as the basic model for the following improvements to increase the testing accuracy.

3.2. *Optimizations of MobileNetV2*

Four optimization methods, including data augmentation, hyperparameter selecting, an overfitting control strategy, and a dynamic learning rate strategy, were applied on the MobileNetV2 model to further improve its classification accuracy while giving full play to the advantages of portability and rapidity. In the experiments, each model was trained three times and the average values of validation accuracies and losses were taken to prevent random errors.

3.2.1. Optimization 1: Data Augmentation

A small amount of training data have difficulty providing enough information for the model to learn. Therefore, data augmentation is often used to expand the amount of data so as to make the model fully learn the image features and to improve the generalization ability of the model.

For our dataset of *Oudemansiella raphanipes*, three methods, including 90° rotation, horizontal flip, and vertical flip, were randomly used to enrich the dataset. The dataset was expanded to 8000 images. Therefore, we had 5600 images in the training set (1400 images for each category), 1600 images in the verification set (400 images for each category), and 800 images in the testing set (200 images for each category).

3.2.2. Optimization 2: Hyperparameter Selecting

In the training process of a neural network model, the appropriate hyperparameters can effectively accelerate the convergence of the model and improve its accuracy. Based on the basic model after data augmentation, four important hyperparameters, including the optimizer, learning rate, batch size, and number of epochs, were investigated.

- Optimizer

The Adam, SGD, and Adagrad optimizers were investigated. The Adam optimizer performs the best on the validation accuracy and loss.

- Learning Rate

The learning rates of 0.1, 0.01, 0.001, and 0.0001 were evaluated. If the learning rate is 0.001, both the validation accuracy and loss have relatively stable convergence trends, which can not only avoid the large fluctuations when the learning rate is larger, but also avoid the slow convergence speed when the learning rate is smaller.

- Batch Size

The batch sizes of 8, 16, 32, and 64 were examined. A small batch size, such as 8 or 16, leads to a short training time but a significant fluctuation in the function curves of validation accuracy and loss. A large batch size, such as 32 or 64, leads to a stable and fast convergence speed but a relatively long training time. Considering the validation accuracy and training time, the batch size of 32 was adopted.

- Number of Epochs

The number of epochs was set as large as 40 for performance evaluation. The validation accuracy and loss curves decrease rapidly in the first 20 epochs. After that, the curves are convergent, indicating that the performance of the model is no longer improved. Balancing the training time and validation accuracy, the number of epochs was set to 20 in the following experiments.

3.2.3. Optimization 3: Overfitting Control Strategy

Overfitting is a condition where the model fits well with the training dataset but does not have good generalization ability, thus it provides low accuracy on new datasets. The appropriate overfitting control strategy can improve the validation and testing accuracies. On the basis of the optimized model after hyperparameter selection, two overfitting control strategies, including the batch normalization (BN) layer [35] and the dropout layer [36], were discussed.

We added the BN layer after the pretrained model to normalize its last convolution layer and the dropout layer after the fully connected layer in the head model. The probability in the dropout layer was set to 0.5, which means that the neuron had a 50% probability of being dropout. Figure 6 presents the validation accuracies and losses of different cases with or without the two strategies. It is shown that adding both the BN and dropout layers can effectively restrain the curve fluctuation and improve the validation accuracy as well as reduce the validation loss. It is better to apply the two strategies simultaneously than to apply only one alone. Compared with the dropout layer, the BN layer plays a more important role in accelerating the convergence speed. In addition, the training accuracy after applying both strategies is 99.02%, while the validation accuracy is 98.50%, with a difference of 0.52%. This value is lower than the difference of 2.18% in the model with optimization 2, proving the effectiveness of the overfitting control strategies.

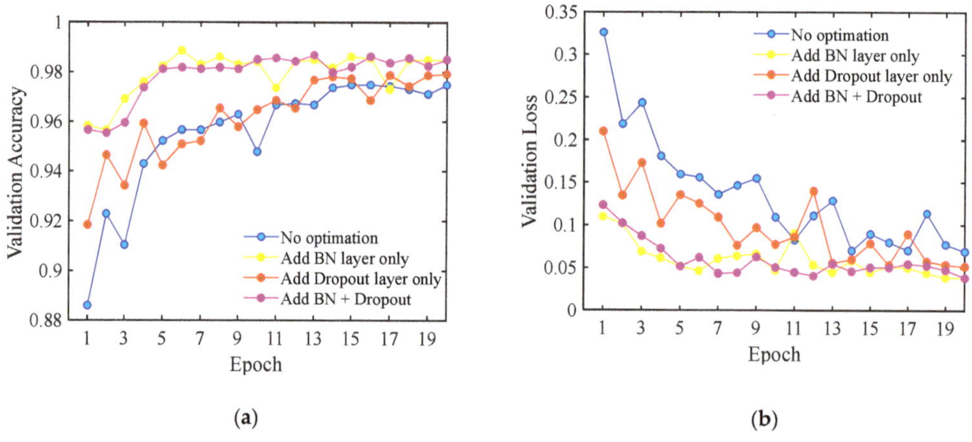

Figure 6. The validation (**a**) accuracies and (**b**) losses of the cases with different overfitting control strategies.

3.2.4. Optimization 4: Dynamic Learning Rate Strategy

Whether the learning rate is good or not depends on two aspects: (a) the convergence speed should be as fast as possible and (b) the optimal solution can be found accurately. From the results in Section 3.2.2 (not shown here), a large learning rate makes the validation loss curve decline rapidly in the early epochs of training, but it also causes fluctuation in the later epochs, which makes it difficult for the model to converge to the optimal solution; a small learning rate makes the validation loss curve more stable in the later epochs, but there is a problem of slow convergence. If both a large learning rate and a small learning rate are combined in the training process to provide the full effect of their respective advantages, the training performance can be further improved. This is the adjustment strategy of the dynamic learning rate. Generally, the dynamic learning rate means that a large learning rate is set in the early epochs of training and is attenuated to be a relatively small value in the later epochs.

Three adjustment strategies, including exponential decay, cosine annealing decay, and adaptive decay, were evaluated.

- Exponential Decay

Exponential decay obeys $\alpha_t = \alpha_0 \beta^t$, where α_0 is the initial learning rate, β is the decay coefficient, and t is the number of iterations. The characteristic of exponential decay is that the variation rate of the learning rate decreases with the increase in iterations.

- Cosine Annealing Decay

Cosine annealing decay uses the cosine function to reduce the learning rate. The learning rate decreases slowly in the early epochs, rapidly in the middle, and slowly again in the later epochs. Compared with the exponential decay, cosine annealing decay can maintain a large learning rate at the beginning of training, thus accelerating the convergence of the model.

- Adaptive Decay

Adaptive decay is a dynamic learning rate strategy, which usually takes the value of the validation loss as the monitoring index for dynamic adjustment. If the validation loss does not decrease in several epochs, the learning rate would decay to a fraction of the original learning rate. This strategy can be dynamically adjusted according to the validation results and has more flexibility.

Figure 7 presents the validation accuracies and losses for the following four cases: (a) a fixed learning rate of 0.001, (b) the exponential decay with α_0 is 0.001 and β is 0.9, (c) the cosine annealing decay with the initial learning rate of 0.001 and the end value of 0, (d) the adaptive decay in which the learning rate would decay to 1/10 of the previous value if the validation loss does not decrease for two consecutive epochs. It is shown that the adaptive decay strategy has a low validation loss and a high validation accuracy, which are, respectively, 0.0310 and 99.12% after 20 epochs. In the training process, the adaptive adjustment triggered two times and the learning rate attenuated to 1e-4 and 1e-5 in the 7th and 18th epoch, respectively.

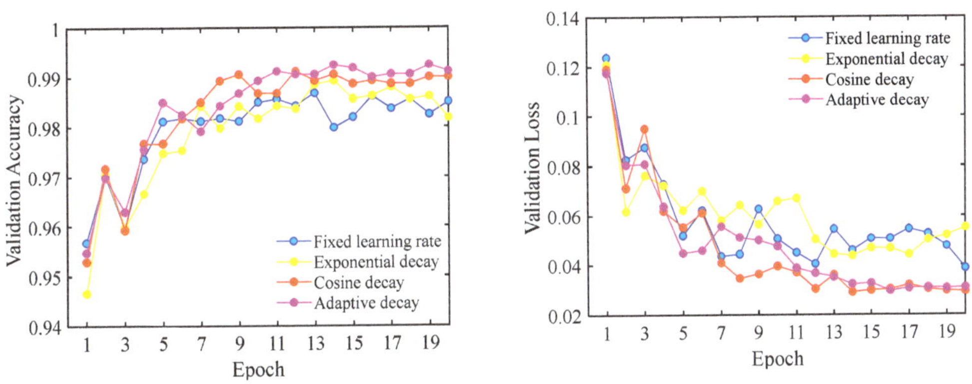

(a) (b)

Figure 7. The validation (**a**) accuracies and (**b**) losses of the cases with different learning rates.

3.2.5. Performance Comparison after the Four Optimizations

To improve the quality grading performance of *Oudemansiella raphanipes*, the original MobileNetV2 model in Section 3.1 (Model 0), was continuously optimized by the data augmentation in Section 3.2.1 (Model 1), the hyperparameter selection in Section 3.2.2 (Model 2), the overfitting control strategy in Section 3.2.3 (Model 3), and the dynamic learning rate strategy in Section 3.2.4 (Model 4). In this section, the performances of these models are compared and discussed.

Figure 8 presents the validation accuracy and loss curves of the five models. With the increase in the optimization methods, the validation accuracy increases and the loss decreases. Moreover, their convergence speed is significantly accelerated. Taking Model 4 at the 20th epoch as an example, its validation accuracy and loss are 99.12% and 0.0295, respectively, showing a 3.35% increment and 0.1091 decrement, respectively, compared with those of Model 0 as shown in Figure 4.

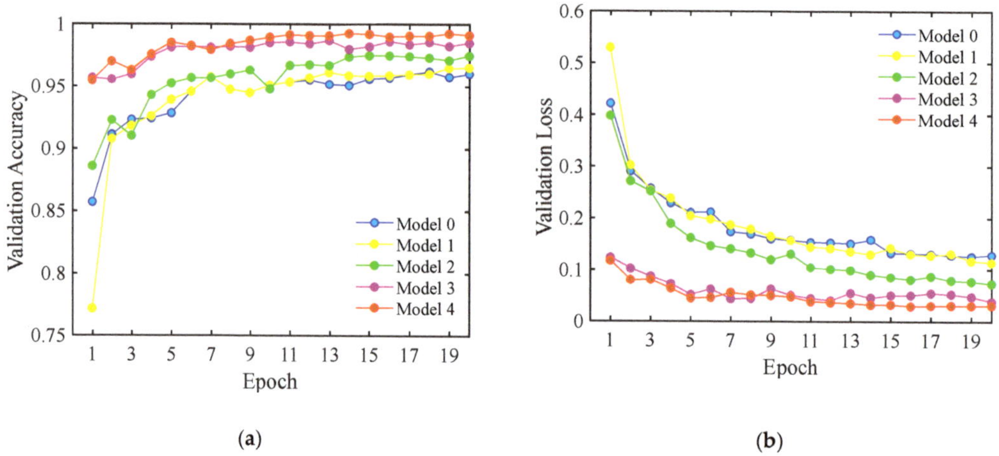

Figure 8. The validation (**a**) accuracies and (**b**) losses of different models.

Figure 9 gives the confusion matrixes of the four optimized models. Compared with the results of Model 0 shown in Figure 5d, the testing accuracies of all the four classes are improved. In particular, the problem of a low accuracy for distinguishing Class 1 from Class 2 in the original Model 0 is solved in the updated Model 4, in which the testing accuracies for Classes 1, 2, 3, and 4 are 98%, 99%, 98%, and 100%, respectively. Table 3 lists the testing accuracies, detection time for a single image, model size, F1 scores, and Kappa coefficients for the testing dataset. The F1 score, which takes into account both the precision and recall, can be calculated by Equation (1) [17,18,20].

$$\text{F1 Score} = 2 \times \frac{\text{precision} \times \text{recall}}{\text{precision} + \text{recall}}, \tag{1}$$

where precision = TP/(TP + FP), recall = TP/(TP + FN), TP represents the quantity that the real value is 1 and the predicted value is 1, FP represents the quantity that the real value is 0 and the predicted value is 1, and FN represents the quantity that the real value is 1 and the predicted value is 0. The Kappa coefficient, which is a statistical indicator used for a consistency test; that is, whether the predicted results are consistent with the actual classification results, can be calculated by Equation (2) [19].

$$\text{Kappa coefficient} = \frac{P_o - P_e}{1 - P_e}, \tag{2}$$

where Po is the testing accuracy and Pe represents the sum of the actual number and predicted number of respective categories divided by the square of the total number of samples. It is shown that, with the increase in the optimization methods, all of the testing accuracies, F1 scores, and Kappa coefficients increase. Taking Model 4 as an example, its testing accuracy, F1 score, and Kappa coefficient are 98.75%, 98.75%, and 0.983, respectively, showing 4.00%, 3.75%, and 0.053 increments, respectively, compared with those of Model 0. The large F1 score and Kappa coefficient prove that the predicted results are almost perfectly consistent with the actual classification results. In addition, the detection time for a single image in Model 4 is 22.5 ms, which is 40% less than the 37.5 ms in Model 0, while the model size stays almost the same.

		Predicted Class			
		1	2	3	4
Actual Class	1	**98%**	2%	0	0
	2	8.5%	**91.5%**	0	0
	3	0	0	**97%**	3%
	4	0	0	1.5%	**98.5%**

(a)

		Predicted Class			
		1	2	3	4
Actual Class	1	**100%**	0	0	0
	2	7.5%	**92.5%**	0	0
	3	0	0	**98%**	2%
	4	0	0	1%	**99%**

(b)

		Predicted Class			
		1	2	3	4
Actual Class	1	**96%**	3%	1%	0
	2	1.5%	**98.5%**	0	0
	3	0	0	**98.5%**	1.5%
	4	0	0	1%	**99%**

(c)

		Predicted Class			
		1	2	3	4
Actual Class	1	**98%**	2%	0	0
	2	1%	**99%**	0	0
	3	0	0	**98%**	2%
	4	0	0	0	**100%**

(d)

Figure 9. The confusion matrixes of (**a**) Model 1, (**b**) Model 2, (**c**) Model 3, and (**d**) Model 4.

Table 3. The validation and testing performance of different models.

Model	Testing Accuracy (%)	Average Detection Time (ms)	Model Size (MB)	F1 Score (%)	Kappa Coefficient
Model 0	94.75	37.5	16.44	95.00	0.930
Model 1	96.25	45	16.44	96.30	0.948
Model 2	97.38	52.5	16.44	96.95	0.965
Model 3	98.00	23.75	16.48	98.00	0.973
Model 4	98.75	22.5	16.48	98.75	0.983

4. Discussion

The motivation of this work is for the quality grading of *Oudemansiella raphanipes* in a production line. The *Oudemansiella raphanipes* are placed on a carrier plate one by one and transported to an image-acquisition box, in which a color camera takes images of the *Oudemansiella raphanipes* and send the images to an industrial computer. The quality grading algorithm implements the classification and, then, the *Oudemansiella raphanipes* of different grades are grouped into different collection bins. As a result, the application of a quality grading algorithm requires not only a high accuracy but also a small detection time. Moreover, if cheaper computer resources are expected, a smaller model size is also required. On the premise of a relatively good accuracy, the basic MobileNetV2 model has the advantages of fast and efficient training, a small detection time, and small storage space compared with VGG16, ResNet50, InceptionV3, NasNet-Mobile, and EfficientNet. As a result, we selected MobileNetV2 as the basic model. Subsequently, the basic model was optimized by four strategies, including data augmentation, hyperparameter selecting, the overfitting control strategy, and the dynamic learning rate strategy; thereby, the convergence speed was accelerated and the classification accuracy was improved. The performance of the optimized MobileNetV2, of which the testing accuracy is 98.75%, the detection time for a single image is 22.5 ms, and the model size is 16.48 MB, satisfies the needs in the production line. Moreover, compared with the published papers on the classification of mushrooms using CNN [17–20], shown in Table 4, our improved MobileNetV2 still has the highest testing accuracy, the lowest detection time, and the best consistency of the predicted results with the actual classification results.

Table 4. Performance comparison between published papers and this paper.

Reference	Model	Testing Accuracy (%)	Detection Time (ms)	F1 Score (%)	Kappa Coefficient
[17]	6-layer CNN	74	/	74	/
[18]	InceptionV3	88.40	/	88	/
[19]	ResNet18	96.89	32	/	0.959
[20]	Modified AlexNet	98.5	/	99.09	/
This paper	Improved MobileNetV2	98.75	22.5	98.75	0.983

Although the final classification accuracy can reach as high as 98.75%, the performance of the quality grading algorithm was only evaluated on the dataset of images with a simple background (see Figure 2). This simple background is applicable in a production line as described above. However, if the images are taken with a complex background, e.g., multiple *Oudemansiella raphanipes* in the soil, the classification accuracy would significantly decrease. The quality grading algorithm for *Oudemansiella raphanipes* in the soil is in progress and will be published in the near future.

5. Conclusions

To obtain a quality grading algorithm of Oudemansiella raphanipes by a deep CNN model, transfer leaning was adopted. Six typical CNN models, including VGG16, ResNet50, InceptionV3, NasNet-Mobile, EfficientNet, and MobileNetV2, were pretrained on the dataset of ImageNet and transferred to our dataset of Oudemansiella raphanipes. Considering both the testing accuracy and detection time, the MobileNetV2 was selected for further optimization. Four optimization methods, including data augmentation, hyperparameter selecting, the overfitting control strategy, and the dynamic learning rate strategy, were applied. As a result, the improved MobileNetV2 has advantages of a high efficiency, high speed, and small storage space. The testing accuracy is 98.75%, the detection time for a single image is 22.5 ms, and the model size is 16.48 MB. These characteristics are in line with the production practice.

Author Contributions: Conceptualization, H.H. (Huamao Huang) and M.L.; methodology, T.L. and H.H. (Huamao Huang); investigation, T.L.; data curation, T.L., Y.P. and H.Z.; writing—original draft preparation, T.L.; writing—review and editing, H.H. (Haiying Hu), H.H. (Huamao Huang), and M.L.; supervision, H.H. (Huamao Huang); project administration, H.H. (Haiying Hu) and M.L.; funding acquisition, M.L. All authors have read and agreed to the published version of the manuscript.

Funding: This research was funded by the Project of Collaborative Innovation Center of Guangdong Academy of Agricultural Sciences (XTXM202212), the Heyuan Science and Technology Bureau (HeKe2021038), and the Open Research Fund of Guangdong Key Laboratory for New Technology Research of Vegetables (2022zz-07).

Data Availability Statement: Not applicable.

Conflicts of Interest: The authors declare no conflict of interest.

References

1. Zhao, Y.; Wang, Y.; Li, K.; Mazurenko, I. Effect of *Oudemansiella Raphanipes* Powder on Physicochemical and Textural Properties, Water Distribution and Protein Conformation of Lower-Fat Pork Meat Batter. *Foods* **2022**, *11*, 2623. [CrossRef] [PubMed]
2. Azadnia, R.; Kheiralipour, K. Recognition of Leaves of Different Medicinal Plant Species Using a Robust Image Processing Algorithm and Artificial Neural Networks Classifier. *J. Appl. Res. Med. Aromat. Plants* **2021**, *25*, 100327. [CrossRef]
3. Seydi, S.T.; Amani, M.; Ghorbanian, A. A Dual Attention Convolutional Neural Network for Crop Classification Using Time-series Sentinel-2 Imagery. *Remote Sens.* **2022**, *14*, 498. [CrossRef]
4. Moreno-Revelo, M.Y.; Guachi-Guachi, L.; Gómez-Mendoza, J.B.; Revelo-Fuelagán, J.; Peluffo-Ordóñez, D.H. Enhanced Convolutional-Neural-Network Architecture for Crop Classification. *Appl. Sci.* **2021**, *11*, 4292. [CrossRef]

5. Yu, Q.; Yang, H.; Gao, Y.; Ma, X.; Chen, G.; Wang, X. LFPNet: Lightweight Network on Real Point Sets for Fruit Classification and Segmentation. *Comput. Electron. Agric.* **2022**, *194*, 106691. [CrossRef]
6. Ghazal, S.; Qureshi, W.S.; Khan, U.S.; Iqbal, J.; Rashid, N.; Tiwana, M.I. Analysis of Visual Features and Classifiers for Fruit Classification Problem. *Comput. Electron. Agric.* **2021**, *187*, 106267. [CrossRef]
7. Hossain, M.S.; Al-Hammad, M.; Muhammad, G. Automatic Fruit Classification Using Deep Learning for Industrial Applications. *IEEE Trans. Industr. Inform.* **2018**, *15*, 1027–1034. [CrossRef]
8. Altaheri, H.; Alsulaiman, M.; Muhammad, G. Date Fruit Classification for Robotic Harvesting in a Natural Environment Using Deep Learning. *IEEE Access* **2019**, *7*, 117115–117133. [CrossRef]
9. Ahmed, M.I.; Mamun, S.M.; Asif, A.U.Z. DCNN-Based Vegetable Image Classification Using Transfer Learning: A Comparative Study. In Proceedings of the 5th International Conference on Computer, Communication and Signal Processing (ICCCSP), Chennai, India, 24–25 May 2021.
10. Ulloa, C.C.; Krus, A.; Barrientos, A.; Del Cerro, J.; Valero, C. Robotic Fertilization in Strip Cropping Using a CNN Vegetables Detection-Characterization Method. *Comput. Electron. Agric.* **2022**, *193*, 106684. [CrossRef]
11. Hameed, K.; Chai, D.; Rassau, A. A Comprehensive Review of Fruit and Vegetable Classification Techniques. *Image Vis. Comput.* **2018**, *80*, 24–44. [CrossRef]
12. Sujatha, R.; Chatterjee, J.M.; Jhanjhi, N.Z.; Brohi, S.N. Performance of Deep Learning vs Machine Learning in Plant Leaf Disease Detection. *Microprocess. Microsyst.* **2021**, *80*, 103615. [CrossRef]
13. Esgario, J.G.M.; Krohling, R.A.; Ventura, J.A. Deep Learning for Classification and Severity Estimation of Coffee Leaf Biotic Stress. *Comput. Electron. Agric.* **2020**, *169*, 105162. [CrossRef]
14. Arwatchananukul, S.; Kirimasthong, K.; Aunsri, N. A New Paphiopedilum Orchid Database and Its Recognition Using Convolutional Neural Network. *Wireless. Pers. Commun.* **2020**, *115*, 3275–3289. [CrossRef]
15. Ji, M.; Zhang, K.; Wu, Q.; Deng, Z. Multi-label Learning for Crop Leaf Diseases Recognition and Severity Estimation Based on Convolutional Neural Networks. *Soft Comput.* **2020**, *24*, 15327–15340. [CrossRef]
16. Suharjito Elwirehardja, G.N.; Prayoga, J.S. Oil Palm Fresh Fruit Bunch Ripeness Classification on Mobile Devices Using Deep Learning Approaches. *Comput. Electron. Agr.* **2021**, *188*, 106359. [CrossRef]
17. Preechasuk, J.; Chaowalit, O.; Pensiri, F.; Visutsak, P. Image Analysis of Mushroom Types Classification by Convolution Neural Networks. In Proceedings of the 2nd Artificial Intelligence and Cloud Computing Conference (AICCC), Kobe, Japan, 21–23 December 2019.
18. Zahan, N.; Hasan, M.Z.; Malek, M.A.; Reya, S.S. A Deep Learning-Based Approach for Edible, Inedible and Poisonous Mushroom Classification. In Proceedings of the International Conference on Information and Communication Technology for Sustainable Development (ICICT4SD), Dhaka, Bangladesh, 27–28 February 2021.
19. Zhao, M.; Li, Y.; Xu, P.; Song, T.; Li, H. Image Classification Method for *Oudemansiella Raphanipes* Using Compound Knowledge Distillation Algorithm. *Trans. Chin. Soc. Agric. Eng.* **2021**, *37*, 303–309.
20. Ketwongsa, W.; Boonlue, S.; Kokaew, U. A New Deep Learning Model for the Classification of Poisonous and Edible Mushrooms Based on Improved AlexNet Convolutional Neural Network. *Appl. Sci.* **2022**, *12*, 3409. [CrossRef]
21. Sandler, M.; Howard, A.; Zhu, M.; Zhmoginov, A.; Chen, L.C. Mobilenetv2: Inverted Residuals and Linear Bottlenecks. In Proceedings of the 2018 IEEE Conference on Computer Vision and Pattern Recognition (CVPR), Salt Lake City, UT, USA, 18–23 June 2018.
22. Srinivasu, P.N.; SivaSai, J.G.; Ijaz, M.F.; Bhoi, A.K.; Kim, W.; Kang, J.J. Classification of Skin Disease Using Deep Learning Neural Networks with MobileNetV2 and LSTM. *Sensors* **2021**, *21*, 2852. [CrossRef]
23. Lin, Z.; Guo, W. Cotton Stand Counting from Unmanned Aerial System Imagery Using MobileNet and CenterNet Deep Learning Models. *Remote Sens.* **2021**, *13*, 2822. [CrossRef]
24. Kim, M.; Kwon, Y.; Kim, J.; Kim, Y. Image Classification of Parcel Boxes Under the Underground Logistics System Using CNN MobileNet. *Appl. Sci.* **2022**, *12*, 3337. [CrossRef]
25. Sun, H.; Zhang, S.; Ren, R.; Su, L. Maturity Classification of "Hupingzao" Jujubes with an Imbalanced Dataset Based on Improved MobileNetV2. *Agriculture* **2022**, *12*, 1305. [CrossRef]
26. Li, Y.; Xue, J.; Wang, K.; Zhang, M.; Li, Z. Surface Defect Detection of Fresh-Cut Cauliflowers Based on Convolutional Neural Network with Transfer Learning. *Foods* **2022**, *11*, 2915. [CrossRef] [PubMed]
27. Simonyan, K.; Zisserman, A. Very Deep Convolutional Networks for Large-Scale Image Recognition. In Proceedings of the 2015 International Conference on Learning Representations (ICLR), San Diego, CA, USA, 7–9 May 2015.
28. He, K.; Zhang, X.; Ren, S.; Sun, J. Deep Residual Learning for Image Recognition. In Proceedings of the 2016 IEEE Conference on Computer Vision and Pattern Recognition (CVPR), Las Vegas, NV, USA, 27–30 June 2016.
29. Szegedy, C.; Vanhoucke, V.; Ioffe, S.; Shlens, J.; Wojna, Z. Rethinking the Inception Architecture for Computer Vision. In Proceedings of the 2016 IEEE Conference on Computer Vision and Pattern Recognition (CVPR), Las Vegas, NV, USA, 27–30 June 2016.
30. Zoph, B.; Vasudevan, V.; Shlens, J.; Le, Q.V. Learning Transferable Architectures for Scalable Image Recognition. In Proceedings of the 2018 IEEE Conference on Computer Vision and Pattern Recognition (CVPR), Salt Lake City, UT, USA, 18–23 June 2018.
31. Tan, M.; Le, Q. Efficientnet: Rethinking Model Scaling for Convolutional Neural Networks. In Proceedings of the 2019 International Conference on Machine Learning (ICML), Long Beach, CA, USA, 9–15 June 2019.

32. Mehta, S.; Rastegari, M. MobileViT: Light-Weight, General-Purpose, and Mobile-Friendly Vision Transformer. *arXiv* **2021**, arXiv:2110.02178.
33. Kogilavani, S.V.; Prabhu, J.; Sandhiya, R.; Kumar, M.S.; Subramaniam, U.; Karthick, A.; Imam, S.B.S. COVID-19 Detection Based on Lung CT Scan Using Deep Learning Techniques. *Comput. Math. Methods Med.* **2022**, *2022*, 7672196. [CrossRef]
34. Deng, J.; Dong, W.; Socher, R.; Li, L.J.; Li, K.; Fei-Fei, L. Imagenet: A Large-Scale Hierarchical Image Database. In Proceedings of the 2009 IEEE Conference on Computer Vision and Pattern Recognition (CVPR), Miami, FL, USA, 20–25 June 2009.
35. Ioffe, S.; Szegedy, C. Batch normalization: Accelerating Deep Network Training by Reducing Internal Covariate Shift. In Proceedings of the 32nd International Conference on Machine Learning (ICML 2015), Lille, France, 6–11 July 2015.
36. Srivastava, N.; Hinton, G.; Krizhevsky, A.; Sutskever, I.; Salakhutdinov, R. Dropout: A Simple Way to Prevent Neural Networks from Overfitting. *J. Mach. Learn. Res.* **2014**, *15*, 1929–1958.

MDPI AG
Grosspeteranlage 5
4052 Basel
Switzerland
Tel.: +41 61 683 77 34

Horticulturae Editorial Office
E-mail: horticulturae@mdpi.com
www.mdpi.com/journal/horticulturae

www.ingramcontent.com/pod-product-compliance
Lightning Source LLC
Chambersburg PA
CBHW041405210926
51850CB00013B/3182
* 9 7 8 3 7 2 5 8 2 9 9 3 4 *